Beyond Earth

A CHRONICLE OF DEEP SPACE EXPLORATION, 1958–2016

by
Asif A. Siddiqi

NATIONAL AERONAUTICS AND SPACE ADMINISTRATION

Office of Communications
NASA History Division
Washington, DC 20546

NASA SP-2018-4041

Library of Congress Cataloging-in-Publication Data

Names: Siddiqi, Asif A., 1966– author. | United States. NASA History Division, issuing body. | United States. NASA History Program Office, publisher.

Title: Beyond Earth : a chronicle of deep space exploration, 1958–2016 / by Asif A. Siddiqi.

Other titles: Deep space chronicle

Description: Second edition. | Washington, DC : National Aeronautics and Space Administration, Office of Communications, NASA History Division, [2018] | Series: NASA SP ; 2018-4041 | Series: The NASA history series | Includes bibliographical references and index.

Identifiers: LCCN 2017058675 (print) | LCCN 2017059404 (ebook) | ISBN 9781626830424 | ISBN 9781626830431 | ISBN 9781626830431?q(paperback)

Subjects: LCSH: Space flight—History. | Planets—Exploration—History.

Classification: LCC TL790 (ebook) | LCC TL790 .S53 2018 (print) | DDC 629.43/509—dc23 | SUDOC NAS 1.21:2018-4041

LC record available at *https://lccn.loc.gov/2017058675*

Original Cover Artwork provided by Ariel Waldman

The artwork titled Spaceprob.es is a companion piece to the Web site that catalogs the active human-made machines that freckle our solar system. Each space probe's silhouette has been paired with its distance from Earth via the Deep Space Network or its last known coordinates.

This publication is available as a free download at *http://www.nasa.gov/ebooks*.

ISBN 978-1-62683-043-1

9 781626 830431

For
my beloved
father

Dr. Hafiz G. A. Siddiqi
Whose achievements I can only hope to emulate

Contents

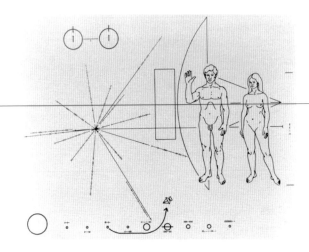

Preface

January 31, 1958 marked a significant beginning for space exploration. More than the historic and successful launch of Explorer 1, the first U.S. satellite, it was the beginning of an unprecedented era of exploration and understanding of our own planet and the distant worlds beyond. The more we uncover about the mysteries and beauty of space, the more we are inspired to go farther. Yet, with all we have learned, we still cannot even imagine what future generations will find.

Spacecraft from NASA and others have shown us the intricacies within clouds and terrain of distant planets that were only a dot in an astronomer's telescope a few decades ago. We have seen the birth of stars, black holes, and found exoplanets orbiting stars in systems remarkably similar to ours. Future missions will take us forward in history, as we seek to uncover the very origins of our universe.

We may not know precisely what—or who—we will find out there, but we can be sure that space exploration will continue to surprise and inspire us, as it did for those who came before and those who will follow. Along the way there will be missteps, some more devastating than others. That is the price of doing what's never been done before—a price that sometimes is tragically paid at the highest cost by the courageous. But like those early days of the space program, we are as motivated to succeed by the missions that do not make it as those that do. And we learn from them, coming back stronger and smarter.

In this book, the history of NASA's six decades of exploration beyond Earth and its Moon to other planets and their moons is laid out. The story follows spacecraft to the Sun, comets, minor and dwarf planets and, ultimately, beyond our solar system. As we marvel at the ingenuity of the early pioneers of the Space Age, we realize how much

they achieved with what, comparatively, was so little. Computers were human, and when the machines did take over calculations, they also took up entire rooms with processing capability less than smartphones in your pockets today.

Yet some of NASA's greatest achievements took place during this period: Mariner IV, which took the first pictures of the surface of Mars in 1965; the global view of Mars from Mariner 9 in 1971; and the Viking landers of the 1970s, which executed the first planetary soft-landings of American spacecraft. The crowning achievements of America's mid-century robotic space exploration were the Pioneer and Voyager missions which were sent to the far boundaries of our solar system using early 1970s technology. As this is being written, Voyager 1 and Voyager 2 continue to send us data from beyond the outer planets from the boundary region of the Sun's sphere of influence, the heliosphere. But space does not belong to the United States alone. We have evolved from the earliest days of the Space Race, when being "first" brought serious geopolitical consequences, to our current era of international partnerships that have taken us farther together than we could have gone alone.

In the modern era of exploration, which itself will look outdated in a generation, we have discovered the extraordinary rings and moons of Saturn with NASA's Cassini spacecraft and the Huygens lander built by the European Space Agency. We marvel at images of the swirling storms on Jupiter sent back to Earth by our Juno mission. And we constantly find new science from the Curiosity rover that's been trekking across the surface of Mars for more than five years.

Our robotic emissaries have made tremendous journeys over the past six decades. They carry the vision and inspiration of humankind beyond our physical ability to make the trip—yet. This book celebrates the extraordinary men and women who have looked up and wondered what's out there and then found the answer. In only 60 years, our technology has evolved from a simple, modified Geiger counter launched into Earth orbit to sublime technologies sending full-color, high-resolution images and data from the edge of the universe. The next 60 should be exponentially rewarding.

– Dr. Thomas H. Zurbuchen
NASA Associate Administrator
at Science Mission Directorate

Introduction

Humans abandoned their nomadic habits and moved into settlements about 40 to 50 thousand years ago. We have been using tools even longer. But our ability to send one of our tools into the heavens is of much more recent origin, spanning only the past 60 years. Yet, in that time, we have created new tools—we call them robotic spacecraft—and sent them into the cosmos, far beyond Earth. Of course, many never got very far. That's the cost of hubris and ambition. But most did. And many never came back to Earth and never will. In that sense, we as a species have already left a mark on the heavens; these small objects that dot the cosmos are a permanent legacy of our species, existing for millions of years, even if we as a planet were to disappear. This book that you hold in your hands (or are reading in digital form) is a chronicle of all these tools, both failed and successful, that humans have flung into the heavens beyond Earth.

The text in front of you is a completely updated and revised version of a monograph published in 2002 by the NASA History Office under the original title *Deep Space Chronicle: A Chronology of Deep Space and Planetary Probes, 1958–2000*. This new edition not only adds all events in robotic deep space exploration after 2000 and up to the end of 2016, but it also completely corrects and updates all accounts of missions from prior years. The information in the monograph is current as of mid-2017 when I completed writing.

What Does This Publication Include?

This monograph contains brief descriptions of all robotic deep space missions attempted by humans since the opening of the space age in 1957. The missions are listed chronologically in order of their launch dates (i.e., not their target body encounters).

Different people have different criteria for which kind of spacecraft to include in a list of "deep space probes." In the list that follows, I have included all spacecraft that satisfied the following guidelines:

1. Any probe that was launched to an "encounter" with a "target."
   a. An "encounter" includes the following events:
      i. flybys
      ii. orbiting
      iii. atmospheric entry and impacts
      iv. soft-landing
   b. "Targets" include the following:
      i. the planets of the Solar System (Mercury, Venus, Mars, Jupiter, Saturn, Uranus, and Neptune)
      ii. the Earth's Moon
      iii. minor planets or asteroids
      iv. natural satellites of the planets and asteroids
      v. comets
      vi. dwarf planets (such as Pluto)
2. Any probe that was deliberately sent into heliocentric (solar) orbit.
3. Any probe that was sent into a halo (Lissajous) orbit around any of the libration points involving Earth, the Moon, and the Sun.
4. Any probe that was launched as part of a science, lunar, or planetary program to at least lunar distance in order to simulate a deep space trajectory (such as, for example, Geotail, Zond 4, and a few early Surveyor Model mockups).

I have included probes whether they succeeded in their objectives or not. Thus, some probes never got a few meters beyond the launch pad while others are heading into the outer reaches of the solar system.

It should be noted that the criteria for inclusion in this volume does not always coincide with NASA's own programmatic distinctions about what constitutes a planetary science mission. For example, this volume includes missions such as Wind, ACE, MAP, and SIRTF, none of which was funded through NASA's solar system exploration line. The criteria for inclusion here is simply whether the mission was intended to operate *beyond* Earth orbit and satisfied the above four requirements, regardless of who funded it or what kind of science it generated.

Where Is the Information From?

For statistical data on U.S. probes (such as launch vehicle numbers, launch times, list of instruments, etc.), I have used original NASA sources such as Public Affairs releases, press kits, postflight mission reports, and official histories. These records are deposited in the NASA Historical Reference Collection at NASA Headquarters in Washington, DC, or are available online at various NASA or government databases. For missions after approximately 2000, there is a proliferation of official mission websites, hosted by the organization sponsoring the missions, including, for example, the Jet Propulsion Laboratory (JPL), the Applied Physics Laboratory (APL), and the S. A. Lavochkin Scientific-Production Association (or NPO imeni Lavochkina). I have used these as primary sources of information. For some of the initial Earth orbital parameters of many deep space probes, a very useful source has been the online newsletter "Jonathan's Space Report" prepared by Jonathan McDowell.

For Soviet/Russian sources, I have used only Russian-language sources, such as the journal *Kosmicheskaya issledovaniya* (Space Research), official organizational histories, reliable memoirs, or the semi-official journal *Novosti kosmonavtiki* (News of Cosmonautics).

In the bibliography at the end of the monograph, I list a few published secondary sources that have been useful in verifying or framing data. Every attempt has been made to present accurate information, but with a project of this size and scope, there will naturally be errors.

I have avoided as much as possible using unofficial amateur Web sites (such as Wikipedia) or secondary history books, such as Andrew Wilson's otherwise quite wonderful *Solar System Log*. These sources are good for a quick overview but they often reproduce errors (especially in timings, details, and distances) that have now been repeatedly and erroneously propagated in the Internet era. The one exception is Don Mitchell's excellent website on Soviet lunar and planetary exploration, found at *http://mentallandscape.com/V_Venus.htm*. I highly recommend it.

What Kind of Information Have I Included?

In terms of the mission descriptions, I have kept the focus on *dynamic* mission events (course corrections, orbital insertion, reentry, landing, equipment deployment, etc.) rather than mission planning or *scientific* results, although in many cases I have included brief summaries of both. But I do not make any claim to comprehensiveness in terms of the scientific outcomes of the missions. This monograph is more about *what happened* rather than what was discovered. In the interest of space, the mission descriptions have been kept relatively short, filled with detail, and to the point.

Conflicting Information

There are many, many areas where different sources have supplied different information, especially for some early Soviet probes launched between 1960 and 1965. The precise instrument complement of these probes (1M, 1V, 2MV, and 3MV series) is not known very well because in many cases, scientific instruments that were meant for the spacecraft were removed before launch. I have listed all the *originally* slated instruments meant for those vehicles even if they were later removed before launch. Undoubtedly, there are mistakes and inconsistencies in the lists presented here but I have made every effort to be as accurate as possible.

A Note About Terminology

Mission Designations

I have made every attempt to use the names of spacecraft and missions that were contemporaneous to the time of the mission and assigned by the agency or organization implementing the missions.

In the 1960s, NASA routinely used Roman numerals for missions (Mariner IV, Mariner V, etc.) in their official documentation, but these were discontinued in the 1970s. Readers will note that I have used this convention for all missions until and including 1970 but after that switched to Latin numbers (Mariner 9, Mariner 10, etc.). This division is somewhat arbitrary but was necessary not to confuse readers.

The practice of giving spacecraft "official" names is complicated by the fact that beginning with the launch of Sputnik in 1957 and until the late 1980s, the Soviet Union never announced or acknowledged a mission if it failed to reach Earth orbit. In addition, for those lunar and planetary probes that *did* reach Earth orbit but failed to leave it, the Soviets adopted two strategies:

- between 1960 and 1963, the Soviets simply never admitted their existence and made no announcement; and
- beginning November 1963, the Soviet media began to give these stranded-in-Earth-orbit spacecraft "Kosmos" numbers. So, the deep space vehicle launched on 11 November 1963 that reached Earth orbit but failed to leave for deep space was simply given the "next" Kosmos number, in this case "21." By giving it such a nondescript name ("Kosmos 21"), Soviet officials sought to draw attention away from such failures. This practice was followed well into the late 1980s.

For both of these three types of missions, I have used the following convention:

[Program, Spacecraft design designation, serial number]
OR
Kosmos number [Program]

I do *not* use terms such as "Marsnik 1" or "Mars 1960A" (listed in the National Space Science Data Center, for example, to denote the spacecraft launched on 10 October 1960). Since the Soviets never used such names, it would be entirely inaccurate to ascribe such designations. Such fictitious names (such as "Sputnik 27") unfortunately proliferate online but are Western inventions.

Launch Sites

For Soviet and Russian launch sites, the following conventions apply:

"Site A / B" implies that the probe was launched from Site A, Launch Unit B

Mission Goals

There are good reasons not to use terms such as "flyby" or "flight" since spacecraft do not fly. As one of the reviewers for this manuscript pointed out, these terms are remnants of the era of aeronautics. As such, more appropriate terms would be "swingby" (instead of "flyby") and "mission" (instead of "flight"). However, because terms such as "flyby" and "flight" are still widely used by many space agencies globally, this manuscript has retained their use, despite their imprecise nature.

Acknowledgments

I wish to thank all at the NASA History Division who were patient with me throughout this process, particularly Chief Historian Bill Barry and Steve Garber. A special note of gratitude to Roger Launius who conceived the original project in 1999.

For help with the manuscript itself, I need to acknowledge the comments and criticisms of Jason Callahan, Dwayne Day, Chris Gamble, Marc Rayman, and Randii Wessen. Their comments were immensely helpful to this project and made this a much better manuscript than I alone could have made it. I would also like to thank Don Mitchell, Sven Grahn, and Timothy Varfolomeyev for sharing images from their collection. Also, a note of gratitude to Jonathan McDowell for sharing his insights. Despite the help of all these individuals, any mistakes are, however, mine.

A very special note of thanks to Ariel Waldman for kindly providing the source image for the cover of this publication.

Thanks also go to the Communications Support Service Center (CSSC) team of talented professionals who brought this project from manuscript to finished publication. J. Andrew Cooke carefully copyedited the detailed text, Michele Ostovar did an expert job laying out the design and creating the e-book version, Kristin Harley performed the exacting job of creating the index, and printing specialist Tun Hla oversaw the production of the traditional hard copies. Supervisor Maxine Aldred helped by overseeing all of this CSSC production work.

Firsts

in the History of Deep Space Exploration

Absolute Firsts

First attempt to launch a probe into deep space:
USA / **Able 1 [Pioneer 0]** / 17 August 1958

First probe to reach escape velocity:
USSR / **Soviet Space Rocket [Luna 1]** /
2 January 1959

First spacecraft to impact on another celestial body:
USSR / **Second Soviet Space Rocket
[Luna 2]** / 14 September 1959 (Moon)

First successful planetary mission:
USA / **Mariner II** / 14 December 1962 (Venus)

First spacecraft to impact another planet:
USSR / **Venera 3** / 1 March 1966 (Venus)

First spacecraft to make a survivable landing on a
celestial body:
USSR / **Luna 9** / 3 February 1966 (Moon)

First spacecraft to orbit a celestial body other than
Earth or the Sun:
USSR / **Luna 10** / 2 April 1966 (Moon)

First successful planetary atmospheric entry probe:
USSR / **Venera 4** / 18 October 1967 (Venus)

First liftoff from another celestial body:
USA / **Surveyor VI** / 17 November 1967 (Moon)

First transmission from the surface of another planet:
USSR / **Venera 7** / 15 December 1970 (Venus)

First robotic spacecraft to recover and return samples
from another celestial body:
USSR / **Luna 16** / 12–21 September 1970 (Moon)

First wheeled vehicle on another celestial body:
USSR / **Lunokhod 1** / 17 November 1970 (Moon)

First spacecraft to fly through the asteroid belts:
USA / **Pioneer 10** / out in February 1973

First spacecraft to use gravity assist to change its
interplanetary trajectory:
USA / **Mariner 10** / 5 February 1974 (at Venus)

First spacecraft to fly past multiple planets:
USA / **Mariner 10** / 5 February 1974 (Venus)
and 29 March 1974 (Mercury)

First spacecraft to transmit photos from the surface of
another planet:
USSR / **Venera 10** / 22 October 1975 (Venus)

First spacecraft to orbit a libration point:
USA / **ISEE-3** / 20 November 1978 (Sun-Earth L1)

First spacecraft to fly past a comet:
USA / **ISEE-3** / 11 September 1985 (Comet Giacobini-Zinner)

First spacecraft to use Earth for a gravity assist:
ESA / **Giotto** / 2 July 1990

First spacecraft to use aerobraking to reduce velocity:
Japan / **Hiten** / 19 March 1991

First spacecraft to fly past an asteroid:
USA / **Galileo** / 26 October 1991 (951 Gaspra)

First wheeled vehicle on a planet:
USA / **Sojourner** / 5 July 1997 (Mars)

First spacecraft to use ion propulsion as primary propulsion in deep space:
USA / **Deep Space 1** / 24 November 1998

First spacecraft to orbit an asteroid:
USA / **NEAR Shoemaker** / 14 February 2000 (433 Eros)

First spacecraft to land on an asteroid:
USA / **NEAR Shoemaker** / 12 February 2001 (433 Eros)

First spacecraft to return extraterrestrial material from beyond lunar orbit:
USA / **Genesis** / Returned 8 September 2004

First spacecraft to use a solar sail as primary propulsion in deep space:
Japan / **IKAROS** / 9 July 2010

First spacecraft to orbit a body in the main asteroid belt:
USA / **Dawn** / 16 July 2011 (4 Vesta)

First spacecraft to orbit a dwarf planet:
USA / **Dawn** / 7 March 2015 (1 Ceres)

The Moon

First lunar probe attempt:
USA / **Able 1 [Pioneer 0]** / 17 August 1958

First spacecraft to impact the Moon:
USSR / **Second Soviet Space Rocket [Luna 2]** / 14 September 1959

First spacecraft to fly by the Moon:
USSR / **Automatic Interplanetary Station [Luna 3]** / 6 October 1959

First to photograph farside of the Moon:
USSR / **Automatic Interplanetary Station [Luna 3]** / 6 October 1959

First survivable landing on the Moon:
USSR / **Luna 9** / 3 February 1966

First soft-landing on the Moon:
USA / **Surveyor I** / 2 June 1966

First spacecraft to orbit the Moon:
USSR / **Luna 10** / 2 April 1966

First liftoff from the Moon:
USA / **Surveyor VI** / 17 November 1967

First successful circumlunar mission:
USSR / **Zond 6** / 14–21 September 1968

First robotic return of soil sample from the Moon:
USSR / **Luna 16** / 12–21 September 1970

First wheeled vehicle on Moon:
USSR / **Lunokhod 1** / 17 November 1970

The Sun

First probe into heliocentric orbit:
USSR / **Soviet Space Rocket** / 2 January 1959

First spacecraft to view the poles of the Sun:
ESA / **Ulysses** / September 2000 to January 2001

Mercury

First spacecraft to fly by Mercury:
USA / **Mariner 10** / 29 March 1974

First spacecraft to orbit Mercury:
USA / **MESSENGER** / 18 March 2011

Venus

First attempt to send a spacecraft to Venus:
USSR / **Heavy Satellite** / 4 February 1961

First spacecraft to successfully fly past Venus:
USA / **Mariner II** / 14 December 1962

First spacecraft to impact Venus:
USSR / **Venera 3** / 1 March 1966

First successful atmospheric entry into Venus:
USSR / **Venera 4** / 18 October 1967

First soft-landing and return of surface data from Venus:
USSR / **Venera 7** / 15 December 1970

First surface photos of Venus:
USSR / **Venera 10** / 22 October 1975

First spacecraft to orbit Venus:
USSR / **Venera 10** / 22 October 1975

First spacecraft to image the entire surface of Venus:
USA / **Magellan** / 1990–1994

Mars

First Mars probe attempt:
USSR / [**Mars, 1M no. 1**] / 10 October 1960

First successful mission to Mars:
USA / **Mariner IV** / Launched 15 July 1965

First spacecraft to orbit Mars:
USA / **Mariner 9** / 14 November 1971

First spacecraft to impact Mars:
USSR / **Mars 2** / 27 November 1971

First successful soft-landing on the Martian surface:
USA / **Viking 1** / 20 July 1976

First wheeled vehicle on Mars:
USA / **Sojourner** / 5 July 1997

Jupiter

First spacecraft to fly by Jupiter:
USA / **Pioneer 10** / 4 December 1973

First atmospheric entry into Jupiter:
USA / **Galileo Probe** / 7 December 1995

First spacecraft to orbit Jupiter:
USA / **Galileo Orbiter** / 8 December 1995

First spacecraft to carry out detailed investigations of Jupiter's interior:
USA / **Juno** / 2016–present

Saturn

First spacecraft to fly by Saturn:
USA / **Pioneer 11** / 1 September 1979

First spacecraft to orbit Saturn:
USA / **Cassini** / 1 July 2004

First spacecraft to soft-land on Titan:
ESA / **Huygens** / 14 January 2005

Uranus

First spacecraft to fly by Uranus:
USA / **Voyager 2** / 24 January 1986

Neptune

First spacecraft to fly by Neptune:
USA / **Voyager 2** / 25 August 1989

Pluto

First spacecraft to fly by Pluto:
USA / **New Horizons** / 14 July 2015

Comets

First spacecraft to fly past a comet:
USA / **ISEE-3** / 11 September 1985 (Comet 21P/
Giacobini-Zinner)

First spacecraft to enter the coma of a comet:
USA / **Deep Space 1** / 22 September 2001
(Comet 19P/Borrelly)

First spacecraft to impact a comet:
USA / **Deep Impact** / 4 July 2005 (Comet 9P/
Tempel)

First spacecraft to return material from a comet:
USA / **Stardust** / Returned 15 January 2006
(Comet 81P/Wild)

First spacecraft to orbit a cometary nucleus:
ESA / **Rosetta** / 10 September 2014 (Comet 67P/
Churyumov-Gerasimenko)

First spacecraft to land on a comet:
ESA / **Philae** / 12 November 2014 (Comet 67P/
Cburyumov-Gerasimenko)

Dwarf Planets

First spacecraft to orbit a dwarf planet:
USA / **Dawn** / March 7, 2015 (1 Ceres)

Asteroids

First spacecraft to fly past an asteroid:
USA / **Galileo** / 26 October 1991 (951 Gaspra)

First spacecraft to orbit an asteroid:
USA / **NEAR Shoemaker** / 14 February 2000
(433 Eros)

First spacecraft to land on an asteroid:
USA / **NEAR Shoemaker** / 13 February 2001
(433 Eros)

First spacecraft to return material from an asteroid:
Japan / **Hayabusa** / 13 June 2010 (25143 Hideo
Itokawa)

First spacecraft to orbit an asteroid in the main asteroid belt:
USA / **Dawn** / 16 July 2011 (4 Vesta)

Lagrange Points

First spacecraft to orbit a libration point (L1 [Sun–Earth]):
USA / **ISEE-3** / 20 November 1978

First spacecraft to orbit libration point L2 (Sun–Earth):
USA / **WMAP** / 1 October 2001

First spacecraft to orbit libration point L2 (Earth–Moon):
USA / **ARTEMIS P1** / 25 August 2010

First spacecraft to orbit libration point L1 (Earth–Moon):
USA / **ARTEMIS P2** / 22 October 2010

1958

1

[Pioneer 0]

Nation: USA (1)
Objective(s): lunar orbit
Spacecraft: Able 1
Spacecraft Mass: 38.5 kg
Mission Design and Management: ARPA / AFBMD
Launch Vehicle: Thor Able 1 (Thor no. 127)
Launch Date and Time: 17 August 1958 / 12:18 UT
Launch Site: Cape Canaveral / Launch Complex 17A
Scientific Instruments:

1. magnetometer
2. micrometeoroid detector
3. 2 temperature sensors
4. TV camera

Results: On 27 March 1958, the U.S. Department of Defense announced the launch of four to five lunar probes later in the year, all under the supervision of the Advanced Research Projects Agency (ARPA) as part of scientific investigations during the International Geophysical Year. Of these, one or two (later confirmed as two) would be carried out by the Army's Ballistic Missile Agency and the other three by the Air Force's Ballistic Missile Division. This launch was the first of three Air Force attempts, and the first attempted deep space launch by any country. The Able 1 spacecraft, a squat conical fiberglass structure built by Space Technology Laboratories (STL), carried a crude infrared TV scanner. The simple thermal radiation device carried a small parabolic mirror for focusing reflected light from the lunar surface onto a cell that would transmit voltage proportional to the light it received. Engineers painted a pattern of dark and light stripes on the spacecraft's outer surface to regulate internal temperature.

The spacecraft was also disinfected with ultraviolet light prior to launch. The launch vehicle was a three-stage variant of the Thor intermediate range ballistic missile (IRBM) with elements appropriated from the Vanguard rocket used on its second and third stages. The entire project involved 3,000 people from 52 scientific and industrial firms, all but 6 of which were located in southern California. According to the ideal mission profile, Able 1 was designed to reach the Moon's vicinity 2.6 days after launch following which the TX-8-6 solid propellant motor would fire to insert the payload into orbit around the Moon. Orbital altitude would have been 29,000 kilometers with an optimal lifetime of about two weeks. The actual mission, however, lasted only 73.6 seconds, the Thor first stage having exploded at an altitude of 15.2 kilometers altitude. Telemetry was received from the payload for at least 123 seconds after the explosion, probably until impact in the Atlantic. Investigators concluded that the accident had been caused by a turbopump gearbox failure. The mission was not named at the time but has been retroactively known as "Pioneer 0."

2

[Luna, Ye-1 no. 1]

Nation: USSR (1)
Objective(s): lunar impact
Spacecraft: Ye-1 (no. 1)
Spacecraft Mass: c. 360 kg (including the power
    sources installed on the upper stage)
Mission Design and Management: OKB-1
Launch Vehicle: 8K72 (no. B1-3)
Launch Date and Time: 23 September 1958 / 07:03:23 UT
Launch Site: NIIP-5 / Site 1/5

Scientific Instruments:

Ye-1:

1. flux-gate magnetometer
2. sodium-iodide scintillation counter
3. 2 gas discharge Geiger counters
4. 2 micrometeorite counters
5. Cherenkov detector
6. 4 ion traps

Blok Ye (upper stage):

1. sodium vapor experiment
2. scintillation counter

Results: The Soviet government approved a modest plan for initial exploration of the Moon in March 1958. Engineers conceived of four initial probes, the Ye-1 (for lunar impact), Ye-2 (to photograph the farside of the Moon), Ye-3 (to photograph the farside of the Moon with advanced imaging equipment), and Ye-4 (lunar impact with a nuclear explosion). The Ye-1 was a simple probe, a pressurized spherical object made from aluminum-magnesium alloy slightly bigger than the first Sputnik. The goals were to detect the magnetic field of the Moon, study the intensity and variation of cosmic rays, record photons in cosmic rays, detect lunar radiation, study the distribution of heavy nucleii in primary cosmic radiation, study the gas component of interplanetary matter, study corpuscular solar radiation, and record the incidence of meteoric particles. The Blok Ye upper stage (with the 8D714 engine) carried additional instrumentation, including radio transmitters and one kilogram of sodium to create an artificial comet on the outbound trajectory that could be photographed from Earth. During the first Ye-1 launch, at T+87 seconds, the launch vehicle's strapon boosters began to develop longitudinal resonant vibrations. The rocket eventually disintegrated at T+93 seconds, destroying its payload. The problem was traced to violent pressure oscillations in the combustion chamber of one of the strapon booster engines. This generated a resonant frequency vibration throughout the frame causing it to shake violently. A fix was proposed by reducing the thrust at T+85 seconds when the rocket reached maximum dynamic pressure.

3

Able 2 [Pioneer]

Nation: USA (2)
Objective(s): lunar orbit
Spacecraft: Able 2
Spacecraft Mass: 38.3 kg
Mission Design and Management: NASA / AFBMD
Launch Vehicle: Thor Able I (Thor Able I no. 1 / Thor no. 130/DM-1812-6)
Launch Date and Time: 11 October 1958 / 08:42:13 UT
Launch Site: Cape Canaveral / Launch Complex 17A
Scientific Instruments:

1. ion chamber
2. magnetometer
3. micrometeoroid detector
4. TV camera
5. 2 temperature sensors

Results: Although the USAF actually conducted the mission, this was the first U.S. space mission technically under the aegis of the recently formed National Aeronautics and Space Administration (NASA). The spacecraft was very similar in design to the Able 1 probe and like its predecessor, built by Space Technology Laboratories (STL). The probe was designed to record micrometeoroid impacts, take magnetic field and radiation measurements, and obtain a "facsimile image of the surface of the moon." During the launch, the Thor second stage shut down 10 seconds early due to incorrect information from an accelerometer measuring incremental velocity. The launch vehicle thus imparted insufficient velocity for the probe to escape Earth's gravity. An attempt to insert the spacecraft into high Earth orbit at 128,700 × 32,200 kilometers by using its Thiokol-built retromotor failed because internal temperatures had fallen too low for the batteries to provide adequate power. The probe did, however, reach an altitude of 114,750 kilometers (according to NASA information from February 1959) by 11:42 UT on launch day, verifying the existence of the Van Allen belts and returning

Thor-Able I with the Pioneer I spacecraft atop, prior to launch at Eastern Test Range at what is now Kennedy Space Center. Pioneer I, launched on 11 October 1958, was the first spacecraft launched by the 11-day-old National Aeronautics and Space Administration (NASA). Although the spacecraft failed to reach the Moon, it did transmit 43 hours of data. *Credit: NASA*

other useful data on the boundary of the geomagnetic cavity. It reentered 43 hours 17 minutes after launch. Investigators later concluded that an accelerometer had mistakenly cut off the Able stage because of an incorrect setting of a valve. In a press release on October 11 soon after the launch, the U.S. Department of Defense officially bestowed the name "Pioneer" to the probe, although it has often been retroactively known as "Pioneer 1." The name was apparently suggested not by any NASA official but by one Stephen A. Saliga, an official in charge of Air Force exhibits at Wright-Patterson Air Force Base in Dayton, Ohio, who designed a display to coincide with the launch.

4

[Luna, Ye-1 no. 2]

Nation: USSR (2)
Objective(s): lunar impact
Spacecraft: Ye-1 (no. 2)
Spacecraft Mass: c. 360 kg (including power sources installed on the upper stage)
Mission Design and Management: OKB-1
Launch Vehicle: 8K72 (no. B1-4)
Launch Date and Time: 11 October 1958 / 21:41:58 UT
Launch Site: NIIP-5 / Site 1/5
Scientific Instruments:

Ye-1:
1. flux-gate magnetometer
2. sodium-iodide scintillation counter
3. 2 gas discharge Geiger counters
4. 2 micrometeorite counters
5. Cherenkov detector
6. 4 ion traps

Blok Ye (upper stage):
1. sodium vapor experiment
2. scintillation counter

Results: The second attempt to send a Ye-1 probe to impact on the Moon also never left Earth's atmosphere. The 8K72 launch vehicle exploded at T+104 seconds, once again, due to longitudinal resonant vibrations in the strapon boosters.

5

Pioneer II

Nation: USA (3)
Objective(s): lunar orbit
Spacecraft: Able 3
Spacecraft Mass: 39.6 kg
Mission Design and Management: NASA / AFBMD
Launch Vehicle: Thor Able I (Thor Able I no. 2 / Thor no. 129/DM-1812-6)
Launch Date and Time: 8 November 1958 / 07:30:20 UT

Launch Site: Cape Canaveral / Launch Complex 17A

Scientific Instruments:

1. ionization chamber
2. magnetometer
3. temperature sensors
4. micrometeoroid sensor
5. proportional counter telescope
6. TV system

Results: This was the second official NASA deep space launch although operations on the ground were handled by the Air Force. For this third launch of an STL-built lunar orbiter, engineers introduced a number of changes to the Thor Able launcher. The probe also now included a new TV scanner and a new type of battery, as well as a new cosmic ray telescope to study the Cherenkov Effect. Pioneer II, like its predecessors, never reached its target. A signal from the ground shut down the Thor launch vehicle's second stage earlier than planned. Additionally, when the X-248 third stage engine separated, it failed to fire. As a result, the probe burned up in Earth's atmosphere only 42 minutes 10 seconds after launch at 28.6° E longitude. During its brief mission, it reached an altitude of 1,530 kilometers (as announced in December 1959) and sent back data that suggested that Earth's equatorial region had higher flux and energy levels than previously thought. The information also indicated that micrometeoroid density was higher near Earth than in space. Investigators concluded that the third stage engine had failed to fire because of a broken wire. A NASA press release from Administrator T. Keith Glennan (1905–1995) soon after the launch officially named the probe "Pioneer II."

6

[Luna, Ye-1 no. 3]

Nation: USSR (3)

Objective(s): lunar impact

Spacecraft: Ye-1 (no. 3)

Spacecraft Mass: c. 360 kg (including power sources installed on the upper stage)

Mission Design and Management: OKB-1

Launch Vehicle: 8K72 (no. B1-5)

Launch Date and Time: 4 December 1958 / 18:18:44 UT

Launch Site: NIIP-5 / Site 1/5

Scientific Instruments:

Ye-1:

1. flux-gate magnetometer
2. sodium-iodide scintillation counter
3. 2 gas discharge Geiger counters
4. 2 micrometeorite counters
5. Cherenkov detector
6. 4 ion traps

Blok Ye (upper stage):

1. sodium vapor experiment
2. scintillation counter

Results: This was the third failure in a row in Soviet attempts to send a Ye-1 lunar impact probe to the Moon. The thrust level of the core engine (8D75) of the R-7 booster dropped abruptly at T+245.4 seconds to about 70% of optimal levels, leading to premature engine cutoff. The payload never reached escape velocity. Later investigation showed that a pressurized seal in the hydrogen peroxide pump of the main engine had lost integrity in vacuum conditions. The malfunction caused the main turbine to cease working and thus led to engine failure.

7

Pioneer III

Nation: USA (4)

Objective(s): lunar flyby

Spacecraft: Pioneer III

Spacecraft Mass: 5.87 kg

Mission Design and Management: NASA / ABMA / JPL

Launch Vehicle: Juno II (no. AM-11)

Launch Date and Time: 6 December 1958 / 05:44:52 UT

Launch Site: Cape Canaveral / Launch Complex 5

Scientific Instruments:

1. photoelectric sensor trigger
2. two Geiger-Mueller counters

Pioneer III being assembled by technicians. *Credit: NASA*

Results: This was the first of two U.S. Army launches to the Moon, subsequent to three attempts by the Air Force. Pioneer III was a spin-stabilized probe (up to 400 rpm) whose primary goal was to fly by the Moon. Two special 6-gram weights were to be spun out on 1.5-meter wires to reduce spin to 6 rpm once the mission was under way. The spacecraft carried an optical sensor to test a future imaging system. If the sensor received a collimated beam of light from a source (such as the Moon) that was wide enough to pass through the lens and fall simultaneously on two photocells, then the sensor would send a signal to switch on the imaging system (which was actually not carried on this spacecraft). Unfortunately, the main booster engine shut down 4 seconds earlier than planned due to premature propellant depletion. Once put on its trajectory, it was determined that Pioneer III was about 1,030 km/hour short of escape velocity. It eventually reached a maximum altitude of 102,322 kilometers and subsequently plummeted and burned up over Africa 38 hours 6 minutes after launch. In addition, the de-spin mechanism failed to operate, preventing the test of the optical system. The radiation counters, however, returned important data. Dr. William H. Pickering (1910–2004), in a paper presented to an IGY Symposium on 29 December 1958, noted that "[w]hile the results of the launch were disappointing…the dividend of radiation measurements of the Van Allen belt gained as the payload returned to earth were of great value in defining this energy field." This data contributed to the major scientific discovery of dual bands of radiation around Earth.

During the Pioneer IV mission, NASA tested a new space communications system. One component of the system was this 26-meter diameter antenna at Goldstone, California, the first of several antennas that would later constitute nodes in NASA's Deep Space Network. *Credit: NASA/JPL-Caltech*

1959

8

Soviet Space Rocket [Luna 1]

Nation: USSR (4)

Objective(s): lunar impact

Spacecraft: Ye-1 (no. 4)

Spacecraft Mass: 361.3 kg (including power sources installed on the upper stage)

Mission Design and Management: OKB-1

Launch Vehicle: 8K72 (no. B1-6)

Launch Date and Time: 2 January 1959 / 16:41:21 UT

Launch Site: NIIP-5 / Site 1/5

Scientific Instruments:

Ye-1:

1. flux-gate magnetometer
2. sodium-iodide scintillation counter
3. 2 gas discharge Geiger counters
4. 2 micrometeorite counters
5. Cherenkov detector
6. 4 ion traps

Blok Ye (upper stage):

1. sodium vapor experiment
2. scintillation counter

Results: Due to a ground error in an antenna that transmitted guidance information to the launch vehicle, the Blok Ye upper stage of the launch vehicle fired longer than intended—731.2 seconds—and thus, imparted 175 meters/second extra velocity to its payload. Because of this error, the Ye-1 probe missed the Moon. Nevertheless, the probe became the first human-made object to attain Earth escape velocity. The spacecraft (which with its entire launch vehicle was referred as "Soviet Space Rocket" in the Soviet press) eventually passed by the Moon at a distance of 6,400 kilometers about 34 hours following launch at

02:59 UT on January 4. (Some sources say the range was as high as 7,500 kilometers). Before the lunar encounter, at 00:57 UT on 3 January 1959 the attached upper stage released one kilogram of sodium at a distance of 113,000 kilometers from Earth which was photographed by Soviet ground-based astronomers although the quality of the images was poor, partly due to poor weather conditions. Ground controllers lost contact with the Soviet Space Rocket (retroactively named "Luna 1" in 1963) approximately 62 hours after launch due to loss of battery power when the payload was 597,000 kilometers from Earth, after which the probe became the first spacecraft to enter orbit around the Sun.

9

Pioneer IV

Nation: USA (5)

Objective(s): lunar flyby

Spacecraft: Pioneer IV

Spacecraft Mass: 6.08 kg

Mission Design and Management: NASA / ABMA / JPL

Launch Vehicle: Juno II (no. AM-14)

Launch Date and Time: 3 March 1959 / 05:10:56 UT

Launch Site: Cape Canaveral / Launch Complex 5

Scientific Instruments:

1. photoelectric sensor trigger
2. two Geiger-Mueller counters

Results: This was the last of 5 American lunar probes launched as part of a series during the International Geophysical Year (although the year officially ended a few months prior). Its design was very similar to Pioneer III; a key difference was the addition of a "monitor" to measure the voltage of the main radio transmitter, which had failed for unknown reasons on Pioneer III. A de-spin mechanism was on board to slow the spin-stabilized vehicle from its initial spin of 480 rpm down to about 11 rpm about 11 hours after launch. Although it did not achieve its primary objective to photograph the Moon during a flyby, Pioneer IV became the first U.S. spacecraft to reach Earth escape velocity. During the launch, the Sergeants on the second stage did not cut off on time, causing the azimuths and elevation angles of the trajectory to change. The spacecraft thus passed by the Moon at a range of about 60,000 kilometers (instead of the planned 32,000 kilometers), i.e., not close enough for the imaging scanner to function. Closest approach was at 10:24 UT on 4 March 1959, about 41 hours after launch. Its tiny radio transmitted information for 82 hours before contact was lost at a distance of 655,000 kilometers from Earth—the greatest tracking distance for a human-made object to date. The probe eventually entered heliocentric orbit becoming the first American spacecraft to do so. Scientists received excellent data that suggested that the intensity of the upper belt of the Van Allen belts had changed since Pioneer III (probably attributable to a recent solar flare) and that there might be a third belt at a higher altitude to the others.

10

[Luna, Ye-1A no. 5]

Nation: USSR (5)
Objective(s): lunar impact
Spacecraft: Ye-1A (no. 5)
Spacecraft Mass: c. 390 kg (including power sources installed on the upper stage)
Mission Design and Management: OKB-1
Launch Vehicle: 8K72 (no. I1-7)
Launch Date and Time: 18 June 1959 / 08:08 UT
Launch Site: NIIP-5 / Site 1/5

Scientific Instruments:
Ye-1:
1. flux-gate magnetometer
2. sodium-iodide scintillation counter
3. 2 micrometeorite counters
4. Cherenkov detector
5. 4 ion traps
6. 6 Geiger counters
Blok Ye (upper stage):
1. sodium vapor experiment
2. Cherenkov detector
3. 2 scintillation counters

Results: The Soviet Ye-1A probe, like the Ye-1, was designed for lunar impact. Engineers had incorporated some minor modifications to the scientific instruments (a modified antenna housing for the magnetometer, six instead of four gas discharge counters, and an improved piezoelectric detector) as a result of information received from the first Soviet Space Rocket (retroactively known as "Luna 1") and the American Pioneer IV. The launch was originally scheduled for 16 June but had to be postponed for two days due to the negligence of a young lieutenant who inadvertently permitted fueling of the upper stage with the wrong propellant. During the actual launch, one of the gyroscopes of the inertial guidance system failed at T+153 seconds, and the wayward booster was subsequently destroyed by command from the ground.

11

Second Soviet Space Rocket [Luna 2]

Nation: USSR (6)
Objective(s): lunar impact
Spacecraft: Ye-1A (no. 7)
Spacecraft Mass: 390.2 kg (including power sources installed on the upper stage)
Mission Design and Management: OKB-1
Launch Vehicle: 8K72 (no. I1-7b)
Launch Date and Time: 12 September 1959 / 06:39:42 UT

Launch Site: NIIP-5 / Site 1/5

Scientific Instruments:

Ye-1:

1. flux-gate magnetometer
2. sodium-iodide scintillation counter
3. 2 micrometeorite counters
4. Cherenkov detector
5. 4 ion traps
6. 6 Geiger counters

Blok Ye (upper stage):

1. sodium vapor experiment
2. Cherenkov detector
3. scintillation counter

Results: After an aborted launch on 9 September, the Ye-1A probe successfully lifted off and reached Earth escape velocity. Officially named the "Second Soviet Space Rocket" by *Pravda* the day after launch, the spacecraft released its one kilogram of sodium at 18:42:42 UT on 12 September at a distance of 156,000 kilometers from Earth in a cloud that expanded out to 650 kilometers in diameter, clearly visible from the ground. This sixth attempt at lunar impact was much more accurate than its predecessors; the spacecraft successfully impacted the surface of the Moon at 21:02:23 UT on 14 September 1959, thus becoming the first object of human origin to make contact with another celestial body. The Blok Ye upper stage impacted about 30 minutes later, also at a velocity of just over 3 kilometers/second. The probe's impact point was approximately at 30° N / 0° longitude on the slope of the Autolycus crater, east of Mare Serenitatis. Luna 2 (as it was called after 1963) deposited Soviet emblems on the lunar surface carried in 9 × 15-centimeter sized metallic spheres. The spacecraft's magnetometer measured no significant lunar magnetic field as close as 55 kilometers to the lunar surface. The radiation detectors also found no hint of a radiation belt. These were the first measurements of physical fields for celestial bodies other than Earth. The ion traps on board Luna 2 made the first *in situ* measurements of the extended plasma envelope of Earth, suggesting the existence of what was later called the plasmapause.

12

Automatic Interplanetary Station [Luna 3]

Nation: USSR (7)

Objective(s): lunar flyby

Spacecraft: Ye-2A (no. 1)

Spacecraft Mass: 278.5 kg

Mission Design and Management: OKB-1

Launch Vehicle: 8K72 (no. I1-8)

Launch Date and Time: 4 October 1959 / 00:43:40 UT

Launch Site: NIIP-5 / Site 1/5

Scientific Instruments:

1. photographic-TV imaging system
2. 4 micrometeoroid counters
3. 4 ion traps
4. Cherenkov radiation detector
5. sodium iodide scintillation counter
6. 3 gas discharge counters

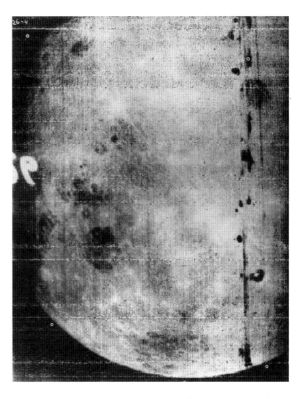

This is a processed version of one of the exposures from the Luna 3 mission. Credit: *Don Mitchell*

A model of the Ye-2A class spacecraft, later known as Luna-3, on display at the Memorial Museum of Cosmonautics in Moscow. *Credit: https://bit.ly/2wPp8AI*

Results: This spacecraft, of the Ye-2A class, was the first Soviet probe designed to take pictures of the farside of the Moon using the Yenisey unit (using the AFA-Ye1 camera), which consisted of two lenses of 200 mm (wide angle) and 500 mm (high resolution) focal lengths, and a capacity to read up to 40 images on a 35 mm film roll. The dual lenses exposed adjacent frames simultaneously. It was also the first spacecraft to be fully powered by solar panels. Strictly speaking, the probe was not meant to reach Earth escape velocity; instead, the launch vehicle inserted the spacecraft, called the Automatic Interplanetary Station (*Avtomaticheskaya mezhplanetnaya stantsiya*, AMS) in the Soviet press, into a highly-elliptical orbit around Earth at 48,280 × 468,300 kilometers, sufficient to reach lunar distance. During the coast to the Moon, the AMS suffered overheating problems and poor communications, but the vehicle eventually passed over the Moon's southern polar cap at a range of 7,900 kilometers at 14:16 UT on 6 October before climbing up (northward) over the Earth-Moon plane. At a distance of 65,200 kilometers from the Moon at 03:30 UT on 7 October, having been properly oriented by its Chayka attitude control system (the first successful 3-axis stabilization used on a spacecraft), the twin-lens 35 mm camera began taking the first of 29 pictures of the farside of the Moon. This session lasted a total of about 40 minutes, by the end of which the spacecraft was 68,400 kilometers from the lunar surface. The exposed film was then developed, fixed, and dried automatically, following which a special light-beam of up to 1,000 lines per image scanned the film for transmission to Earth. Images were received the next day (after a few aborted attempts) at two locations, a primary at Simeiz in Crimea known as IP-41Ye and a backup in the Soviet far east, at Yelizovo in Kamchatka, known as NIP-6. On the ground, there were two systems for recording the images from

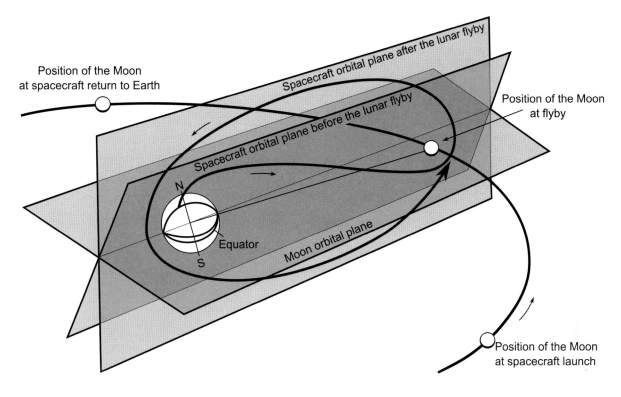

Diagram showing Luna 3's trajectory out to lunar distance. *Credit: https://bit.ly/2NkheZx*

The Yenisey "photo-television unit" on board Luna 3 was developed by the Leningrad-based VNII-380 institute. It included a dual-objective camera (AFA-Ye1) developed by the Krasnogorsk Mechanical Plant. The camera took two pictures simultaneously using a 200 mm lens and a 500 mm lens. *Credit: T. Varfolomeyev*

AMS, Yenisey I (fast mode, at 50 lines/second) and Yenisey II (at slow mode at 0.8 lines/second). Seventeen of the images were of usable quality, and for the first time, they showed parts of the Moon never before seen by human eyes. The trajectory of AMS was specifically designed so that images would show at least half of the Moon, one-third of which was on the near side, so as to provide a point of reference for evaluating formations on the farside. Controllers were unable to regain contact with the spacecraft after the spacecraft entered and then exited Earth's shadow. Luna 3 circled Earth at least 11 times and probably reentered sometime in early 1960. Post-Cold War revelations confirmed that the film type used on AMS, known as "ASh" by the Soviets, was actually unexposed film that was repurposed from a CIA reconnaissance balloon (from Project Genetrix) that had drifted into Soviet territory in the late 1950s. This film had been stored at the A. F. Mozhayskiy Military Academy in Leningrad when the space camera manufacturers at the VNII-380 institute had stumbled upon it. Further investigation showed that the temperature-resistant and radiation-hardened film would be perfect for the AFA-Ye1 camera, and the rest was history. The spacecraft, named Luna 3 after 1963, photographed about 70% of the farside and found fewer mare areas on the farside, prompting scientists to revise their theories of lunar evolution.

13

Able IVB [Pioneer]

Nation: USA (6)
Objective(s): lunar orbit
Spacecraft: P-3 / Able IVB
Spacecraft Mass: 168.7 kg
Mission Design and Management: NASA / AFBMD
Launch Vehicle: Atlas Able (Atlas Able no. 1 / Atlas D no. 20)
Launch Date and Time: 26 November 1959 / 07:26 UT
Launch Site: Cape Canaveral / Launch Complex 14
Scientific Instruments:

1. high-energy radiation counter
2. ionization chamber
3. Geiger-Mueller tube
4. low-energy radiation counter
5. a flux-gate magnetometer and a search-coil magnetometer
6. photo scanning device
7. micrometeoroid detector
8. aspect indicator (a photoelectric cell)
9. radio receiver to detect natural radio waves
10. transponder to measure electron densities

Results: This was the first of three spacecraft designed by Space Technology Laboratories for a lunar assault in 1959–1960; two of them had originally been slated for Venus orbit (in June 1959) but mission planners had redirected their missions after the success of the Soviet Automatic Interplanetary Station ("Luna 3") mission. All the scientific experiments and internal instrumentation were powered by Nickel-Cadmium batteries charged from 1,100 solar cells on four paddles which made the vehicle resemble the recently-launched Explorer VI. The imaging system, the same one used on Explorer VI, comprised a tiny 1.13-kilogram scanning device developed by STL that was "to be used in [an] attempt to get a crude outline of the moon's surface if the probe achieve[d] lunar orbit." Each probe also carried a hydrazine monopropellant tank with two thrust chambers (each 9 kgf), one of which was for lunar orbit insertion at a range of 8,000 kilometers from the Moon. Ideal lunar orbital parameters were planned as 6,400 × 4,800 kilometers. The mission also inaugurated the first use of the Atlas-with-an-upper-stage combination, affording increased payload weight. During this first launch, which took place on Thanksgiving Day 1959, the nose fairing began to break away just 45 seconds after liftoff, still during first stage operation. Aerodynamic forces then caused the third stage and payload to break away and explode. The ground lost contact with the tumbling booster at T+104 seconds. Investigation showed that the 3-meter fiberglass shroud failed because there had been no measures to account for pressure differentials as the rocket rapidly gained altitude after liftoff.

1960

14

Pioneer V

Nation: USA (7)
Objective(s): heliocentric orbit
Spacecraft: P-2 / Able 6
Spacecraft Mass: 43.2 kg
Mission Design and Management: NASA / AFBMD
Launch Vehicle: Thor Able IV (Thor Able IV no. 4 / Thor no. 219/DM-1812-6A)
Launch Date and Time: 11 March 1960 / 13:00:07 UT
Launch Site: Cape Canaveral / Launch Complex 17A
Scientific Instruments:

1. magnetometer
2. ionization chamber
3. Geiger-Mueller tube
4. micrometeoroid momentum spectrometer
5. omnidirectional proportional counter telescope

Results: Launched on a direct solar orbit trajectory, Pioneer V successfully reached heliocentric orbit between Earth and Venus to demonstrate deep space technologies and to provide the first map of the interplanetary magnetic field. The spacecraft had originally been intended for a Venus encounter but the mission was switched to a direct entry into solar orbit. Pioneer V carried Telebit, the first digital telemetry system operationally used on a U.S. spacecraft—it was first tested on Explorer VI. The system used a 5-watt or a 150-watt transmitter, with a 5-watt transmitter acting as driver. Information rates varied from 1 to 64 bits/second. Controllers maintained contact with Pioneer V until 11:31 UT on 26 June 1960 to a record distance of 36.4 million kilometers from Earth (later beaten by Mariner II). The probe, using its 18.1-kilogram suite of scientific instruments, confirmed the existence of a previously conjectured weak interplanetary magnetic field. Information from the magnetometer was unfortunately unusable due to the instrument's position within the spacecraft. Pioneer V remains a derelict spacecraft circling the Sun.

15

[Luna, Ye-3 no. 1]

Nation: USSR (8)
Objective(s): lunar farside photography
Spacecraft: Ye-3 (no. 1)
Spacecraft Mass: [unknown]
Mission Design and Management: OKB-1
Launch Vehicle: 8K72 (no. 1l-9)
Launch Date and Time: 15 April 1960 / 15:06:44 UT
Launch Site: NIIP-5 / Site 1/5
Scientific Instruments:

1. photographic-TV imaging system
2. micrometeoroid detector
3. cosmic ray detector

Results: This spacecraft was launched to return more detailed photos of the lunar farside, after the spectacular success of Luna 3. The Ye-3 class vehicle was essentially a Ye-2A probe with a modified radio-telemetry system, but with the original imaging system. (A more advanced Ye-3 type with a new imaging system had been abandoned earlier). During the launch, the probe received insufficient velocity (too low by 110 meters/second) after premature third stage engine cutoff (3 seconds short). The spacecraft reached an altitude of 200,000 kilometers and then fell back to Earth and burned up in Earth's atmosphere, much like some of the early American Pioneer probes. The most likely reentry point was over central Africa.

16

[Luna, Ye no. 2]

Nation: USSR (9)
Objective(s): farside lunar photography
Spacecraft: Ye-3 (no. 2)
Spacecraft Mass: [unknown]
Mission Design and Management: OKB-1
Launch Vehicle: 8K72 (no. Ll-9a)
Launch Date and Time: 19 April 1960 / 16:07:43 UT
Launch Site: NIIP-5 / Site 1/5
Scientific Instruments:

1. photographic-TV imaging system
2. micrometeoroid detector
3. cosmic ray detector

Results: This was the last of the "first generation" Soviet probes to the Moon. Like its immediate predecessor, it was designed to photograph the farside of the Moon. Unfortunately, the probe never left Earth's atmosphere. Instead, immediately after launch, at T+10 seconds, the launch vehicle began to fall apart (its Blok D strapon actually began separating 0.02 seconds after launch). As each strapon fell away, parts of the booster landed separately over a large area near the launch site, breaking up between 21.15 and 40.3 seconds of launch. Thundering explosions broke windows in many nearby buildings.

17

Able VA [Pioneer]

Nation: USA (8)
Objective(s): lunar orbit
Spacecraft: P-30 / Able VA
Spacecraft Mass: 175.5 kg
Mission Design and Management: NASA / AFBMD
Launch Vehicle: Atlas Able (Atlas Able no. 2 / Atlas D no. 80)
Launch Date and Time: 25 September 1960 / 15:13 UT

Launch Site: Cape Canaveral / Launch Complex 12
Scientific Instruments:

1. high-energy radiation counter
2. ionization chamber
3. Geiger-Mueller tube
4. low-energy radiation counter
5. two magnetometers
6. scintillation spectrometer
7. micrometeoroid detector
8. plasma probe
9. Sun scanner

Results: This probe, Able VA, had a slightly different instrument complement compared to its predecessor Able IVB (launched in November 1959), but had similar mission goals. Able VA was to enter lunar orbit about 62.5 hours after launch with parameters of 4,000 × 2,250 kilometers and a period of 10 hours. After launch, while the first stage performed without problems, the Able second stage ignited abnormally and shut down early because of an oxidizer system failure. Ground controllers were still able to fire the third stage engine, making this small STL-built engine the first rocket engine to successfully ignite and operate in space. Because of the second stage failure, the spacecraft failed to reach sufficient velocity and burned up in Earth's atmosphere 17 minutes after launch. Later, on 15 November 1960, NASA announced that two objects from the Able VA payload had been found in Transvaal, South Africa.

18

[Mars, 1M no. 1]

Nation: USSR (10)
Objective(s): Mars flyby
Spacecraft: 1M (no. 1)
Spacecraft Mass: 480 kg
Mission Design and Management: OKB-1
Launch Vehicle: Molniya + Blok L (8K78 no. L1-4M)
Launch Date and Time: 10 October 1960 / 14:27:49 UT
Launch Site: NIIP-5 / Site 1/5

The 4-stage 8K78 launch vehicle (and its various modifications) launched most Soviet lunar and planetary probes in the 1960s until the advent of the Proton booster in the late 1960s. *Credit: T. Varfolomeyev*

Scientific Instruments:

1. infrared spectrometer [removed before launch]
2. ultraviolet spectrometer [removed before launch]
3. micrometeorite detectors
4. ion traps
5. magnetometer
6. cosmic ray detectors
7. Yenisey imaging system [removed before launch]

Results: This was the first of two Soviet Mars spacecraft intended to fly past Mars. They were also the first attempt by humans to send spacecraft to

the vicinity of Mars. Although the spacecraft initially included a TV imaging system (similar to the one carried on Luna 3), a UV spectrometer, and a spectroreflectometer (to detect organic life on Mars), mass constraints forced engineers to delete these instruments a week before launch. A possibly apocryphal story has it that once removed from the spacecraft, the spectroreflectometer was tested not far from the Tyuratam launch site but failed to detect any life. The spacecraft itself was a cylinder, about a meter in diameter with all the basic systems required of interplanetary travel—a means to regulate temperatures, batteries charged by solar panels, a long-distance communication system, three-axis stabilization, and a mid-course correction engine (the S5.9). The mission profile called for the probe to first enter Earth orbit and then use a new fourth stage (called "Blok L") capable of firing in vacuum, to gain enough additional velocity for a Mars encounter. During the launch, violent vibrations caused a gyroscope to malfunction. As a result, the booster began to veer from its planned attitude. The guidance system failed at T+309.9 seconds and the third stage (Blok I) engine was shut down at T+324.2 seconds, after the trajectory deviated to greater than 7° (pitch). The payload eventually burned up in Earth's atmosphere over eastern Siberia without reaching Earth orbit. The Mars flyby had been planned for 13 May 1961.

19

[Mars, 1M no. 2]

Nation: USSR (11)
Objective(s): Mars flyby
Spacecraft: 1M (no. 2)
Spacecraft Mass: 480 kg
Mission Design and Management: OKB-1
Launch Vehicle: Molniya + Blok L (8K78 no. L1-5M)
Launch Date and Time: 14 October 1960 / 13:51:03 UT
Launch Site: NIIP-5 / Site 1/5

Scientific Instruments:

1. infrared spectrometer [removed before launch]
2. ultraviolet spectrometer [removed before launch]
3. micrometeorite detectors
4. ion traps
5. magnetometer
6. cosmic ray detectors
7. imaging system [removed before launch]

Results: Besides a slightly uprated S5.9A main engine, this vehicle was identical to its predecessor, launched four days before. And like its predecessor, it never reached Earth orbit. During the launch trajectory, there was a failure in the third stage (Blok I) engine at T+290 seconds due to frozen kerosene in the pipeline feeding its turbopump (which prevented a valve from opening). The third and fourth stages, along with the payload, burned up in Earth's upper atmosphere over eastern Siberia. The Mars flyby had been planned for 15 May 1961.

20

Able VB [Pioneer]

Nation: USA (9)
Objective(s): lunar orbit
Spacecraft: P-31 / Able VB
Spacecraft Mass: 176 kg
Mission Design and Management: NASA / AFBMD
Launch Vehicle: Atlas Able (Atlas Able no. 3 / Atlas D no. 91)
Launch Date and Time: 15 December 1960 / 09:11 UT
Launch Site: Cape Canaveral / Launch Complex 12

Scientific Instruments:

1. micrometeoroid detector
2. high-energy radiation counter
3. ionization chamber
4. Geiger-Mueller tube
5. low-energy radiation counter
6. a flux-gate magnetometer and a spin-search coil magnetometer
7. Sun scanner (or photoelectric cell)
8. plasma probe
9. scintillation spectrometer
10. solid state detector

Results: The mission of Able VB, like its two unsuccessful predecessors, was to enter lunar orbit. Scientific objectives included studying radiation near the Moon, recording the incidence of micrometeoroids, and detecting a lunar magnetic field. Planned lunar orbital parameters, to be achieved about 60 hours after launch, were 4,300 × 2,400 kilometers with a period of 9–10 hours. The spacecraft had a slightly different scientific instrument complement than its predecessors, including a plasma probe experiment designed by NASA's Ames Research Center that was to provide data on the energy and momentum distribution of streams of protons with energies above a few kilovolts per particle in the vicinity of the Moon. Unfortunately, the Atlas Able booster suffered a malfunction 66.68 seconds after launch and then exploded at T+74 seconds at an altitude of about 12.2 kilometers. Later investigation indicated that the Able upper stage prematurely ignited while the first stage was still firing. This was the third and last attempt by NASA to launch a probe to orbit the Moon in the 1959–1960 period.

1961

Heavy Satellite [Venera]

Nation: USSR (12)
Objective(s): Venus impact
Spacecraft: 1VA (no. 1)
Spacecraft Mass: c. 645 kg
Mission Design and Management: OKB-1
Launch Vehicle: Molniya + Blok L (8K78 no. L1-7V)
Launch Date and Time: 4 February 1961 / 01:18:04 UT
Launch Site: NIIP-5 / Site 1/5
Scientific Instruments:

1. infrared spectrometer
2. ultraviolet spectrometer
3. micrometeorite detectors
4. 2 ion traps
5. magnetometer
6. cosmic ray detectors

Results: This was the first attempt to send a spacecraft to Venus. Original intentions had been to send the 1V spacecraft to descend and take pictures of the Venusian surface, but this proved to be far too ambitious a goal. Engineers instead downgraded the mission and used the 1VA spacecraft for a simple Venus atmospheric entry and impact. The 1VA was essentially a modified 1M spacecraft used for Martian exploration (albeit with a different main engine, the S5.14 with a thrust of 200 kgf). The spacecraft contained a small globe containing various souvenirs and medals commemorating the mission. It was also the first Soviet mission to use an intermediate Earth orbit to launch a spacecraft into interplanetary space. Although the booster successfully placed the probe into Earth orbit, the fourth stage (the Blok L) never fired to send the spacecraft to Venus. A subsequent investigation showed that there had been a failure in the PT-200 DC transformer that ensured power supply to the Blok L guidance system. The part had evidently not been designed for work in vacuum. The spacecraft+upper stage stack reentered on 26 February 1961. The Soviets announced the total weight of the combination as 6,483 kilograms without specifying any difference between the payload and the upper stage. In the Soviet press, the satellite was usually referred to as *Tyazhelyy sputnik* or "Heavy Satellite."

Automatic Interplanetary Station [Venera 1]

Nation: USSR (13)
Objective(s): Venus impact
Spacecraft: 1VA (no. 2)
Spacecraft Mass: 643.5 kg
Mission Design and Management: OKB-1
Launch Vehicle: Molniya + Blok L (8K78 no. L1-6V)
Launch Date and Time: 12 February 1961 / 00:34:38 UT
Launch Site: NIIP-5 / Site 1/5
Scientific Instruments:

1. infrared spectrometer
2. ultraviolet spectrometer
3. micrometeorite detectors
4. ion traps
5. magnetometer
6. cosmic ray detectors

Results: This was the second of two Venus impact probes that the Soviets launched in 1961. This time, the probe—which many years later was retroactively named "Venera 1"—successfully exited Earth orbit and headed towards Venus. On the way to Venus, on 12 February, data indicated unstable operation of the system designed to keep the spacecraft permanently oriented to the Sun, needed to generate energy from its solar panels. The spacecraft was programmed so that if such a problem occurred, it would automatically orient itself toward the Sun using gyroscopes, and then shut down non-essential systems. Unfortunately, it automatically shut down its communications system for five days until the next planned communications system, because it detected higher than usual temperatures in the spacecraft. The extra heat was due to the failure of mechanical thermal shutters designed to regulate heat in the vehicle. Despite these problems, the spacecraft responded properly during a communications session on 17 February 1961 at a distance of 1.9 million kilometers when scientific data on interplanetary magnetic fields, cosmic rays, and solar plasma was returned. Unfortunately, controllers were unable to regain contact during a subsequent communications attempt on 22 February. A later investigation indicated that the spacecraft had lost its "permanent" solar orientation due to an optical sensor (that was not pressurized) that malfunctioned because of excess heat after the spacecraft's thermal control system failed. The inert spacecraft eventually passed by Venus on 19–20 May 1961 at a distance of about 100,000 kilometers and entered heliocentric orbit. Data from Venera 1 helped detect plasma flow in deep space.

23

Ranger I

Nation: USA (10)
Objective(s): highly elliptical Earth orbit
Spacecraft: P-32
Spacecraft Mass: 306.18 kg
Mission Design and Management: NASA / JPL
Launch Vehicle: Atlas Agena B (Atlas Agena B no. 1 / Atlas D no. 111 / Agena B no. 6001)

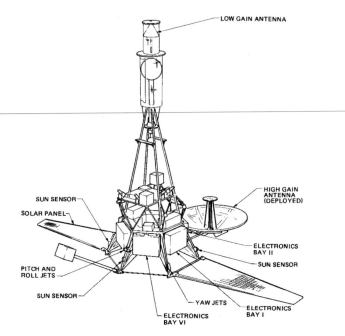

Ranger I and II design. *Credit: NASA*

Launch Date and Time: 23 August 1961 / 10:04 UT
Launch Site: Cape Canaveral / Launch Complex 12
Scientific Instruments:

1. electrostatic analyzer
2. photoconductive particle detectors
3. Rubidium vapor magnetometer
4. triple coincidence cosmic ray telescope
5. cosmic ray integrating ionization chamber
6. x-ray scintillation detectors
7. micrometeoroid dust particle detectors
8. Lyman alpha scanning telescope

Results: Ranger I was the first in a series of standardized spacecraft designed to take photos of the lunar surface during its descent to the Moon and rough-land simple instrumented capsules on the surface. The spacecraft consisted of a tubular central body connected to a hexagonal base containing basic equipment required for control and communications. Power was provided by solar cells and a silver-zinc battery. Ranger I's specific mission was to test performance of the new technologies intended for operational Ranger missions and also to study the nature of particles and fields in interplanetary space. Its intended orbit was

60,000 × 1.1. million kilometers. Ranger I was the first American spacecraft to use a parking orbit around Earth prior to its deep space mission. In this case, the Agena B upper stage cut off almost immediately after its ignition for trans-lunar injection (instead of firing for 90 seconds). The probe remained stranded in low Earth orbit (501 × 168 kilometers) and telemetry ceased by 27 August when the main battery went dead. The spacecraft reentered Earth's atmosphere three days later. The cause of the Agena failure was traced to a malfunctioning switch which had prematurely choked the flow of red fuming nitric acid to the rocket engine.

24

Ranger II

Nation: USA (11)
Objective(s): highly elliptical Earth orbit
Spacecraft: P-33
Spacecraft Mass: 306.18 kg
Mission Design and Management: NASA / JPL
Launch Vehicle: Atlas Agena B (Atlas Agena B no. 2 / Atlas D no. 117 / Agena B no. 6002)
Launch Date and Time: 18 November 1961 / 08:12 UT
Launch Site: Cape Canaveral / Launch Complex 12

Scientific Instruments:
1. electrostatic analyzer for solar plasma
2. photoconductive particle detectors
3. Rubidium vapor magnetometer
4. triple coincidence cosmic ray telescope
5. cosmic ray integrating ionization chamber
6. x-ray scintillation detectors
7. micrometeoroid dust particle detectors
8. Lyman alpha scanning telescope

Results: Like its predecessor, Ranger II was designed to operate in a highly elliptical Earth orbit that would take it into deep space beyond the Moon. Mission planners expected that during five months of operation, they could verify both the technical design of the vehicle and conduct key scientific experiments to study the space environment over a prolonged period. Since the Block I Rangers (Ranger I and II) carried no rocket engine, they could not alter their trajectories. On this attempt, Ranger II, like its predecessor, failed to leave low Earth orbit, the Agena B stage having failed to fire. In its low orbit, Ranger II lost its solar orientation and then eventually lost power, and reentered on 19 November 1961. The most probable cause of the failure was inoperation of the roll control gyroscope on the Agena B guidance system. As a result, the stage used up all attitude control propellant for its first orbit insertion burn. At the time of the second burn, without proper attitude, the engine failed to fire.

1962

Ranger III

Nation: USA (12)

Objective(s): lunar impact

Spacecraft: P-34

Spacecraft Mass: 330 kg

Mission Design and Management: NASA / JPL

Launch Vehicle: Atlas Agena B (Atlas Agena B no. 3 / Atlas D no. 121 / Agena B no. 6003)

Launch Date and Time: 26 January 1962 / 20:30 UT

Launch Site: Cape Canaveral / Launch Complex 12

Scientific Instruments:

1. vidicon TV camera
2. gamma-ray spectrometer
3. radar altimeter
4. single-axis seismometer

Results: This was the first U.S. attempt to impact a probe on the lunar surface. The Block II Ranger spacecraft carried a TV camera that used an optical telescope that would allow imaging during descent down to about 24 kilometers above the lunar surface. The main bus also carried a 42.6-kilogram instrument capsule that would separate at 21.4 kilometers altitude and then independently impact on the Moon. Protected by a balsa-wood outer casing, the capsule was designed to bounce several times on the lunar surface before coming to rest. The primary onboard instrument was a seismometer. Because of a malfunction in the Atlas guidance system (due to faulty transistors), the probe was inserted into a lunar transfer trajectory with an excessive velocity. A subsequent incorrect course change ensured that the spacecraft reached the Moon 14 hours early and missed it by 36,793 kilometers on January 28. The central computer and sequencer failed and the spacecraft returned no TV images. The probe did, however, provide scientists with the first measurements of interplanetary gamma ray flux. Ranger III eventually entered heliocentric orbit.

Ranger IV

Nation: USA (13)

Objective(s): lunar impact

Spacecraft: P-35

Spacecraft Mass: 331.12 kg

Mission Design and Management: NASA / JPL

Launch Vehicle: Atlas Agena B (Atlas Agena B no. 4 / Atlas D no. 133 / Agena B no. 6004)

Launch Date and Time: 23 April 1962 / 20:50 UT

Launch Site: Cape Canaveral / Launch Complex 12

Scientific Instruments:

1. vidicon TV camera
2. gamma-ray spectrometer
3. radar altimeter
4. single-axis seismometer

Results: Ranger IV was the first American spacecraft to reach another celestial body, in this case, the Moon. Like its predecessor, also a Block II spacecraft, it was designed to transmit pictures in the final 10 minutes of its descent to the Moon and rough-land on the lunar surface a balsawood capsule (about 65 centimeters in diameter) that, among other instruments, carried a seismometer. A power failure in the central computer and sequencer stopped the spacecraft's master clock, preventing the vehicle from performing any of its pre-planned operations, such as opening its solar panels. Drifting aimlessly and without any mid-course corrections, Ranger IV impacted the Moon on its far side at a velocity of about 9,600 kilometers/hour at 12:49:53 UT on 26 April 1962. Impact coordinates were 15.5° S / 229.3° E. Although the spacecraft did not achieve its primary objective, the Atlas Agena-Ranger combination performed without fault for the first time.

Mariner I

Nation: USA (14)

Objective(s): Venus flyby

Spacecraft: P-37 / Mariner R-1

Spacecraft Mass: 202.8 kg

Mission Design and Management: NASA / JPL

Launch Vehicle: Atlas Agena B (Atlas Agena B no. 5 / Atlas D no. 145 / Agena B no. 6901)

Launch Date and Time: 22 July 1962 / 09:21:23 UT

Launch Site: Cape Canaveral / Launch Complex 12

Scientific Instruments:

1. microwave radiometer
2. infrared radiometer
3. fluxgate magnetometer
4. cosmic dust detector
5. solar plasma spectrometer
6. energetic particle detectors
7. ionization chamber

Results: In formulating a series of early scientific missions to Venus, in early 1961, NASA originally planned two missions, P-37 and P-38, to be launched on Atlas Centaur rockets, each spacecraft weighing about 565 kilograms. By the time NASA Headquarters formally approved the plan in September 1961, problems with the Atlas Centaur necessitated a switch to the Atlas Agena B with a reduced payload. By that time, JPL prepared three spacecraft based on the design of the Ranger Block I series (therefore named Mariner R) to fly by Venus in late 1962. Each spacecraft carried a modest suite (9 kilograms) of scientific instrumentation,

but had no imaging capability. The spacecraft included 54,000 components and was designed to maintain contact with Earth for 2,500 hours—an ambitious goal given that the (still unsuccessful) Ranger was designed for only 65 hours contact. Mariner I would have flown by Venus at range of 29,000 kilometers on 8 December 1962 but due to an incorrect trajectory during launch, at T+294.5 seconds, range safety sent a signal to destroy the Atlas Centaur booster and its payload. The failure was traced to a guidance antenna on the Atlas as well as faulty software in its onboard guidance program, which was missing a single superscript bar. The press described it as "the most expensive hyphen in history."

[Venera, 2MV-1 no. 3]

Nation: USSR (14)

Objective(s): Venus impact

Spacecraft: 2MV-1 (no. 3)

Spacecraft Mass: 1,097 kg (350 kg impact capsule)

Mission Design and Management: OKB-1

Launch Vehicle: Molniya + Blok L (8K78 no. T103-12)

Launch Date and Time: 25 August 1962 / 02:56:06 UT

Launch Site: NIIP-5 / Site 1/5

Scientific Instruments:

Spacecraft Bus:

1. magnetometer
2. scintillation counter
3. gas discharge Geiger counters
4. Cherenkov detector
5. ion traps
6. cosmic wave detector
7. micrometeoroid detector

Impact Probe:

1. temperature, pressure, and density sensors
2. chemical gas analyzer
3. gamma-ray detector
4. Mercury level wave motion detector

Results: This was the first of a second generation of Soviet deep space probes based on a unified platform called 2MV ("2" for the second generation, "MV" for Mars and Venera) designed to study Mars and Venus. The series included four variants with the same bus but with different payload complements: 2MV-1 (for Venus impact), 2MV-2 (for Venus flyby), 2MV-3 (for Mars impact), and 2MV-4 (for Mars flyby). The buses were basically similar in design carrying all the essential systems to support the mission as well as a main engine, the S5.17 on the two Venus probes, and the S5.19 on the Mars probes. Both had a thrust of 200 kgf but the former was capable of one firing while the latter was designed for two. The payloads were designed in two variants depending on whether the mission was a flyby mission or an impact mission. In the former, there was an instrument module, and in the latter, it carried a 90-centimeter diameter spherical pressurized lander covered by thermal coating; the Venus landers were cooled with an ammonia-based system, while the Mars landers used a system of air conditioners. Both landers were sterilized with a special substance on recommendation from the Academy of Sciences' Institute of Microbiology. The buses were powered by solar panels with an area of 2.5 m² capable of providing 2.6 A. The Venus impact probes were to use a three-stage parachute system to descend through the atmosphere. For Venus, the Soviets prepared three spacecraft for the August–September 1962 launch period, one flyby spacecraft and two landers. This first spacecraft—a flyby plus lander combination—was successfully launched into Earth orbit, but the Blok L upper stage cut off its interplanetary burn after only 45 seconds (instead of the planned 240 seconds). Later investigation showed that the stage had been set on a tumbling motion prior to main engine ignition due to asymmetrical firing of the solid propellant stabilizing motors. The spacecraft remained in Earth orbit for three days before reentering Earth's atmosphere.

29

Mariner II

Nation: USA (15)
Objective(s): Venus flyby
Spacecraft: P-38 / Mariner R-2
Spacecraft Mass: 203.6 kg
Mission Design and Management: NASA / JPL
Launch Vehicle: Atlas Agena B (Atlas Agena B no. 6 / Atlas D no. 179 / Agena B no. 6902)
Launch Date and Time: 27 August 1962 / 06:53:14 UT
Launch Site: Cape Canaveral / Launch Complex 12
Scientific Instruments:

1. microwave radiometer
2. infrared radiometer
3. fluxgate magnetometer
4. cosmic dust detector

NASA image showing mission planners receiving data from Mariner II in January 1963, about five months after its launch. Note the reference to current tracking by the NASA station in South Africa. *Credit: NASA*

This 1961 photo shows Dr. William H. Pickering, (center) JPL Director, presenting a Mariner spacecraft model to President John F. Kennedy, (right). NASA Administrator James Webb is standing directly behind the Mariner model. *Credit: NASA*

5. solar plasma spectrometer
6. energetic particle detectors
7. ionization chamber

Results: NASA brought the Mariner R-2 spacecraft out of storage and launched it just 36 days after the failure of Mariner I. Mariner II was equipped with an identical complement of instrumentation as its predecessor (see Mariner I). The mission proved to be the first fully successful interplanetary mission performed by any nation. After a mid-course correction on 4 September, the spacecraft flew by Venus at a range of 34,854 kilometers at 19:59:28 UT on 14 December 1962. During a 42-minute scan of the planet, Mariner II gathered significant data on the Venusian atmosphere and surface before continuing on to heliocentric orbit. The radiometers, in particular, were able to conduct five scans of the nightside of the planet, eight across the terminator, and five on the daylight side. NASA maintained

contact until 07:00 UT on 3 January 1963 when the spacecraft was 86.68 million kilometers from Earth, a new distance record for a deep space probe. The data returned implied that there was no significant difference in temperature across Venus: readings from Mariner II's microwave radiometer indicated temperatures of 216°C (on the dark side) to 237°C (on the day side). Mariner II also found that there was a dense cloud layer that extended from 56 to 80 kilometers above the surface. The spacecraft detected no discernable planetary magnetic field, partly explained by the great distance between the spacecraft and the planet. If in terms of scientific results, Mariner II was only a modest success, it still retains the honor of being the very first successful planetary science mission in history. NASA elected to stand down the third spacecraft in the series (Mariner R-3) scheduled for the 1964 launch period.

30

[Venera, 2MV-1 no. 4]

Nation: USSR (15)
Objective(s): Venus impact
Spacecraft: 2MV-1 (no. 4)
Spacecraft Mass: c. 1,100 kg (370 kg impact capsule)
Mission Design and Management: OKB-1
Launch Vehicle: Molniya + Blok L (8K78 no. T103-13)
Launch Date and Time: 1 September 1962 / 02:12:33 UT
Launch Site: NIIP-5 / Site 1/5
Scientific Instruments:

Spacecraft Bus:
1. magnetometer
2. scintillation counter
3. gas discharge Geiger counters
4. Cherenkov detector
5. ion traps
6. cosmic wave detector
7. micrometeoroid detector

Impact Probe:
1. temperature, pressure, and density sensors
2. chemical gas analyzer
3. gamma-ray detector
4. Mercury level wave motion detector

Results: This was the second of three Venus spacecraft launched by the Soviets in 1962. Like its predecessor launched in August 1962 (also a Venus impact probe), the spacecraft never left parking orbit around Earth due to a malfunction in the Blok L upper stage designed to send the probe out of Earth orbit towards Venus. The valve that controlled the delivery of fuel into the combustion chamber of the Blok L engine (the S1.5400) never opened. As a result, the engine did not fire. The payload decayed within five days of launch.

31

[Venera, 2MV-2 no. 1]

Nation: USSR (16)
Objective(s): Venus flyby
Spacecraft: 2MV-2 (no. 1)
Spacecraft Mass: [unknown]
Mission Design and Management: OKB-1
Launch Vehicle: Molniya + Blok L (8K78 no. T103-14)
Launch Date and Time: 12 September 1962 / 00:59:13 UT
Launch Site: NIIP-5 / Site 1/5
Scientific Instruments:

Spacecraft Bus:
1. magnetometer
2. scintillation counter
3. gas discharge Geiger counters
4. Cherenkov detector
5. ion traps
6. cosmic wave detector
7. micrometeoroid detector

Instrument Module:
1. imaging system
2. ultraviolet spectrometer
3. infrared spectrometer

Results: Like its two predecessors (launched on 25 August and 1 September 1962), this Soviet Venus probe never left parking orbit around Earth. The Blok L upper stage designed to send the spacecraft towards Venus fired for only 0.8 seconds before shutting down due to unstable attitude. Later investigation indicated that the upper stage had been put into a tumble due to the violent shutdown (and destruction) of the third stage (Blok I) between T+530.95 and T+531.03 seconds. The tumble mixed air bubbles within the propellant tanks preventing a clean firing of the engine. Unlike its predecessors, this probe was designed for a Venus flyby rather than atmospheric entry and impact. The payload reentered two days after launch.

Ranger V

Nation: USA (16)
Objective(s): lunar impact
Spacecraft: P-36
Spacecraft Mass: 342.46 kg
Mission Design and Management: NASA / JPL
Launch Vehicle: Atlas Agena B (Atlas Agena B no. 7 / Atlas D no. 215 / Agena no. 6005)
Launch Date and Time: 18 October 1962 / 16:59:00 UT
Launch Site: Cape Canaveral / Launch Complex 12
Scientific Instruments:

1. imaging system
2. gamma-ray spectrometer
3. single-axis seismometer
4. surface-scanning pulse radio experiment

Results: This was the third attempt to impact the lunar surface with a Block II Ranger spacecraft. On this mission, just 15 minutes after normal operation, a malfunction led to the transfer of power from solar to battery power. Normal operation never resumed, and battery power was depleted after 8 hours, following which all spacecraft systems

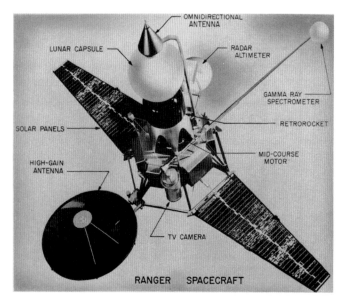

Scientific experiments on the Ranger Block II spacecraft. *Credit: NASA*

died. The first mid-course correction was never implemented, and Ranger V passed the Moon at a range of 724 kilometers on October 21 and entered heliocentric orbit. It was tracked to a distance of 1,271,381 kilometers. Before loss of signal, the spacecraft sent back about 4 hours of data from the gamma-ray experiment.

[Mars, 2MV-4 no. 3]

Nation: USSR (17)
Objective(s): Mars flyby
Spacecraft: 2MV-4 (no. 3 or no. 1)
Spacecraft Mass: c. 900 kg
Mission Design and Management: OKB-1
Launch Vehicle: Molniya + Blok L (8K78 no. T103-15)
Launch Date and Time: 24 October 1962 / 17:55:04 UT
Launch Site: NIIP-5 / Site 1/5
Scientific Instruments:

1. magnetometer
2. 2 scintillation counters
3. 2 gas discharge Geiger counters
4. Cherenkov counter
5. 2 ion traps
6. infrared spectrometer
7. micrometeoroid sensor
8. imaging system
9. ultraviolet spectrograph

Results: This was the first of three "second generation" interplanetary probes (two flyby probes and one impact probe) designed to reach Mars prepared by the Soviets for the late 1962 launch period. Because of the repeated failures of the Blok L upper stage during deep space missions, engineers elected to outfit the stage for the Mars missions with supplementary control and measurement equipment. As a result, most of the scientific instruments were deleted from the Mars spacecraft. The three missions were primarily technological test

A mostly accurate model of the Mars 1 spacecraft (of the 2MV-4 type) shown here at the Memorial Museum of Cosmonautics in Moscow. The main course correction engine, known as S5.19, can be seen on left. *Credit: Asif Siddiqi*

flights rather than scientific missions. In this case, the Blok L interplanetary stage failed again. Just 17 seconds after trans-Mars injection ignition, the turbopump of the main engine (the S1.5400A1) exploded, destroying the payload. The problem was traced to leaking lubricant. As many as 24 fragments were later tracked in 1,485 × 180-kilometer orbit at 64.8° inclination, the largest of which reentered on 29 October. The original probe was designed to fly by Mars on 17 June 1963.

34

Mars 1

Nation: USSR (18)

Objective(s): Mars flyby
Spacecraft: 2MV-4 (no. 4 or no. 1)
Spacecraft Mass: 893.5 kilograms
Mission Design and Management: OKB-1
Launch Vehicle: Molniya + Blok L (8K78 no. T103-16)
Launch Date and Time: 1 November 1962 / 16:14:06 UT
Launch Site: NIIP-5 / Site 1/5
Scientific Instruments:

1. magnetometer
2. 2 scintillation counters
3. 2 gas discharge Geiger counters
4. Cherenkov counter
5. 2 ion traps
6. infrared spectrometer
7. micrometeoroid sensor
8. imaging system
9. ultraviolet spectrograph

Results: The second of three Soviet spacecraft intended for the 1962 Mars launch period, Mars 1 was the first spacecraft sent by any nation to fly past Mars. Its primary mission was to photograph the surface during a flyby from a range of between 1,000 to 11,000 kilometers. In comparison to its predecessor, the probe had a slightly different main engine (the S5.19) than the Venus probes, with a reduced propellant tank mass. The camera system, weighing 32 kilograms, included both 35 and 750 mm lenses and used 70 mm film. It could take up to 112 frames, stored on film and then to be scanned for playback to Earth. After successful insertion into Earth orbit, the Blok L upper stage successfully fired the probe towards Mars, but immediately after engine cutoff, controllers discovered that pressure in one of the nitrogen gas bottles for the spacecraft's attitude control system dropped to zero (due to incomplete closure of a valve). Before all the compressed nitrogen was lost, on 6–7 November, controllers were able to spin the vehicle around the axis perpendicular to the plane of the solar panels to enable a backup gyroscope system to keep the solar panels constantly exposed to the Sun during the coast phase. Further mid-course corrections, however, proved impossible. Controllers maintained contact with the vehicle until 21 March 1963 when the probe was 106 million kilometers from Earth. According to TASS (on 16 May), because of the failure of orientation, "the direction of the station's antennas toward Earth was upset." This anomaly prevented further radio contact after 21 March. Mars 1 silently flew by Mars at a distance of 197,000 kilometers on 19 June 1963. Prior to loss of contact, scientists were able to collect data on interplanetary space (on cosmic ray intensity, Earth's magnetic fields, ionized gases from the Sun, and meteoroid impact densities) up to a distance of 1.24 AUs. The data from Mars 1 (from 20 November 1962 to 25 January 1963) showed that once past 0.24 AUs, i.e., Earth's radiation belts, the intensity of cosmic radiation was virtually constant.

35

[Mars, 2MV-3 no. 1]

Nation: USSR (19)
Objective(s): Mars impact
Spacecraft: 2MV-3 (no. 1)
Spacecraft Mass: [unknown]
Mission Design and Management: OKB-1
Launch Vehicle: Molniya + Blok L (8K78 no. T103-17)
Launch Date and Time: 4 November 1962 / 15:35:14 UT
Launch Site: NIIP-5 / Site 1/5
Scientific Instruments:

Spacecraft Bus:
1. magnetometer
2. scintillation counter
3. gas discharge Geiger counters
4. Cherenkov detector
5. ion traps
6. cosmic wave detector
7. micrometeoroid detector

Impact Probe:
1. temperature, pressure, and density sensors
2. chemical gas analyzer
3. gamma-ray detector
4. Mercury level movement detector

Results: This was the third and last of the Soviet "second generation" Mars attempts in 1962 and also the only impact lander in the series. During the trans-Mars injection firing of the Blok L upper stage, the main engine (the S1.5400A1) prematurely shut down after 33 seconds due to a malfunction in the programmed timer for the stage. The problem was later traced to excessive vibrations of the second stage during liftoff. These vibrations also jarred loose a pyrotechnic igniter from its support, preventing the Blok L from firing. The spacecraft remained stranded in Earth orbit and reentered the atmosphere on 5 November. The probe had been intended to fly by Mars on 21 June 1963.

1963

36

[Luna, Ye-6 no. 2]

Nation: USSR (20)
Objective(s): lunar soft-landing
Spacecraft: Ye-6 (no. 2)
Spacecraft Mass: 1,420 kg
Mission Design and Management: OKB-1
Launch Vehicle: Molniya + Blok L (8K78 no. T103-09)
Launch Date and Time: 4 January 1963 / 08:48:58 UT
Launch Site: NIIP-5 / Site 1/5
Scientific Instruments:
1. imaging system
2. gas-discharge counter

Results: This spacecraft was the first "second generation" Soviet lunar probe (known as Ye-6), designed to accomplish a survivable landing on the surface of the Moon. The Ye-6 probes were equipped with simple 100-kilogram lander capsules (called the Automatic Lunar Station or *Avtomaticheskaya lunnaya stantsiya*, ALS) whose primary objective was to send back photographs from the lunar surface. Each egg-shaped ALS was installed on a roughly cylindrical-shaped main bus. Like the Mars and Venera deep space probes, the Ye-6 Luna spacecraft were also launched by the four-stage 8K78 (Molniya) booster but modified for lunar missions. This first Ye-6 probe was designed to cruise for about three days before landing on the Moon on 7 January at 19:55:10 UT. Like many of its deep space predecessors, the probe failed to escape Earth orbit because of a failure in the Blok L trans-lunar injection stage. There was apparently a failure in a current converter in the power system of the I-100 instrument container (which controlled both the Blok L and the spacecraft), which as a result, failed to issue a command to fire the Blok L engine. The spacecraft remained in Earth orbit, unacknowledged by the Soviets until 11 January 1963.

37

[Luna, Ye-6 no. 3]

Nation: USSR (21)
Objective(s): lunar soft-landing
Spacecraft: Ye-6 (no. 3)
Spacecraft Mass: 1,420 kg
Mission Design and Management: OKB-1
Launch Vehicle: Molniya + Blok L (8K78 no. T103-10)
Launch Date and Time: 3 February 1963 / 09:29:14 UT
Launch Site: NIIP-5 / Site 1/5
Scientific Instruments:
1. imaging system
2. gas-discharge counter

Results: This was the second Soviet attempt to accomplish a soft-landing on the Moon (planned for 20:34:04 UT on 6 February). This time, the spacecraft failed to reach Earth orbit. After launch, at around T+105.5 seconds, the rocket began to lose attitude control along the pitch axis, which spread to the yaw axis after separation from the core booster. The third and fourth stages (along with payload) traced an arc and reentered over the Pacific Ocean near Midway Island. Later investigation indicated that the I-100 control system provided incorrect information to the booster's trajectory control system.

Luna 4

Nation: USSR (22)
Objective(s): lunar soft-landing
Spacecraft: Ye-6 (no. 4)
Spacecraft Mass: 1,422 kilograms
Mission Design and Management: OKB-1
Launch Vehicle: Molniya + Blok L (8K78 no. T103-11)
Launch Date and Time: 2 April 1963 / 08:16:38 UT
Launch Site: NIIP-5 / Site 1/5
Scientific Instruments:

1. imaging system
2. gas-discharge counter

Results: The third Soviet attempt to perform a lunar soft-landing (planned for 19:42:37 UT on 5 April) was the first in which the spacecraft actually left Earth orbit. During the coast to the Moon, the spacecraft's Yupiter-M astronavigation system suffered a major failure (probably related to its thermal control system) and left the probe in an incorrect attitude. As a result, Luna 4 was unable to perform its planned mid-course correction. Although communications were maintained with the spacecraft, it passed by the Moon at a range of 8,500 kilometers at 01:24 UT on 6 April and eventually entered heliocentric orbit from its intermediate barycentric orbit. Data from the gas-discharge counter was compared with data from Mars 1 to provide further clarification to a radiation map of Earth up to lunar distance. The data showed that the intensity of cosmic radiation remained "virtually constant" up to 0.24 AU from the Earth.

Kosmos 21 [Zond]

Nation: USSR (23)
Objective(s): deep space and return to Earth
Spacecraft: 3MV-1A (no. 2, also no. 1)
Spacecraft Mass: c. 800 kg
Mission Design and Management: OKB-1
Launch Vehicle: Molniya + Blok L (8K78 no. G103-18, also G15000-17)
Launch Date and Time: 11 November 1963 / 06:23:34 UT
Launch Site: NIIP-5 / Site 1/5
Scientific Instruments:

Spacecraft Bus:
1. radiation detector
2. charged particle detector
3. magnetometer
4. piezoelectric detector
5. LA-2 atomic hydrogen detector
6. Kassiopeya radio telescope
7. RSK-2M ultraviolet and Roentgen solar radiation experiment
8. VIKT-2 vapor friction technology experiment
9. plasma engines

Results: This was the first of the Soviet Union's "third generation" deep space planetary probes of the 3MV series. Like the second generation, Soviet engineers planned four types of the 3MV, the 3MV-1 (for Venus impact), 3MV-2 (for Venus flyby), 3MV-3 (for Mars impact), and 3MV-4 (for Mars flyby). The primary difference over the second generation was vastly improved (and in many cases doubled) orientation system elements as well as improved on board propulsion systems. While these four versions were meant to study Mars and Venus, the Soviets conceived of two additional variants of the series, similar but not identical to the 3MV-1 and 3MV-4 versions, with the designations 3MV-1A and 3MV-4A. These

"Object-Probes" (*ob'yekt-zond*) were designed to verify key technological systems during simpler missions into deep space and back to Earth. A government decree on March 21, 1963 had approved two to three such "object-probe" missions, one of which (a 3MV-1A) was designed to depart from Earth's ecliptic (the orbital plane of Earth around the Sun) out to 12–16 million kilometers from Earth and then return back to Earth after about six months when its orbit intersected with that of Earth again, aided by two mid-course corrections using its S5.45 main engine. The latter, capable of two firings, was a lighter version of that used on the 2MV model with higher specific impulse and a longer burn time. During this mission, the third and fourth stages separated abnormally, and after reaching Earth orbit, ground control lost telemetry (at about 06:45:44 UT) from the Blok L upper stage designed to send the vehicle past the Moon. As a result, the spacecraft remained stranded in Earth orbit. The stage's main engine turbopump probably exploded upon ignition destroying the spacecraft. With this mission, the Soviets began the practice of giving "Kosmos" designations to obscure the failure of lunar and planetary probes that remained stranded in Earth orbit. If the spacecraft had successfully departed from Earth orbit, it would probably have been called "Zond 1."

1964

40

Ranger VI

Nation: USA (17)

Objective(s): lunar impact

Spacecraft: P-53 / Ranger-A

Spacecraft Mass: 364.69 kg

Mission Design and Management: NASA / JPL

Launch Vehicle: Atlas Agena B (Atlas Agena B no. 8 / Atlas D no. 199 / Agena B no. 6008)

Launch Date and Time: 30 January 1964 / 15:49:09 UT

Launch Site: Cape Kennedy / Launch Complex 12

Scientific Instruments:

1. imaging system (six TV cameras)

Results: This fourth American attempt to lunar impact was the closest success so far. The spacecraft, the first Block III type vehicle with a suite of six TV cameras, was sterilized to avoid contaminating the lunar surface. The series would also serve as a testbed for future interplanetary spacecraft by deploying systems (such as solar panels) that could be used for more ambitious missions. The Block III spacecraft carried a 173-kilogram TV unit (replacing the impact capsule carried on the Block II Ranger spacecraft). The six cameras included two full-scan and four partial-scan cameras, capable of shooting 300 pictures a minute. Ranger VI flew to the Moon successfully and impacted precisely on schedule at 09:24:32 UT on 2 February. Unfortunately, the power supply for the TV camera package had short-circuited during Atlas booster separation three days previously and left the system inoperable. The cameras were to have transmitted high-resolution photos of the lunar approach from 1,448 kilometers to 6.4 kilometers range in support of Project Apollo. Impact coordinates were 9° 24′ N / 21° 30′ E.

41

[Zond, 3MV-1A no. 4A]

Nation: USSR (24)

Objective(s): Venus flyby

Spacecraft: 3MV-1A (no. 4A, also no. 2)

Spacecraft Mass: 800 kg

Mission Design and Management: OKB-1

Launch Vehicle: Molniya + Blok L (8K78 no. G15000-26)

Launch Date and Time: 19 February 1964 / 05:47:40 UT

Launch Site: NIIP-5 / Site 1/5

Scientific Instruments:

1. radiation detector
2. charged particle detector
3. magnetometer
4. piezoelectric detector
5. LA-2 atomic hydrogen detector
6. Kassiopeya radio telescope
7. RSK-2M ultraviolet and Roentgen solar radiation experiment
8. VIKT-2 technology experiment
9. plasma engines

Results: This was the second Soviet "Object-Probe" whose goal was to test systems in interplanetary space in preparation for actual landings and flybys of Venus and Mars. Unlike its predecessor (see Kosmos 21), it appears not to have carried a lander. Its mission was to test its propulsion, thermal, and communications systems during a four-month flight in the direction of Venus to a distance of about 40 million kilometers. In the event, the spacecraft failed to reach Earth orbit due to a malfunction in the launch vehicle's third stage. Later investigation indicated that a liquid oxygen leak through an unpressurized valve (the B4311-O) seal froze propellant in the main pipeline while the rocket was still on the launch pad. As a result, the pipeline cracked, leading to an explosion in the third stage.

42

[Luna, Ye-6 no. 6]

Nation: USSR (25)
Objective(s): lunar soft-landing
Spacecraft: Ye-6 (no. 6)
Spacecraft Mass: c. 1,420 kg
Mission Design and Management: OKB-1
Launch Vehicle: Molniya-M (8K78M no. T15000-20)
Launch Date and Time: 21 March 1964 / 08:14:33 UT
Launch Site: NIIP-5 / Site 1/5
Scientific Instruments:
1. imaging system
2. gas-discharge counter

Results: This fourth Soviet attempt to achieve a soft-landing on the Moon ended in failure when the spacecraft failed to reach an intermediate orbit around Earth. During the boost phase, the launcher's third stage engine's (8D715) main liquid oxygen valve failed to open when the valve rod broke. As a result, the engine never achieved full thrust and eventually cut off prematurely at T+487 seconds. The spacecraft never reached Earth orbit.

43

Kosmos 27 [Venera]

Nation: USSR (26)
Objective(s): Venus impact
Spacecraft: 3MV-1 (no. 5)
Spacecraft Mass: 948 kg (lander: 285 kg)

Mission Design and Management: OKB-1
Launch Vehicle: Molniya + Blok L (8K78 no. T15000-27)
Launch Date and Time: 27 March 1964 / 03:24:43 UT
Launch Site: NIIP-5 / Site 1/5
Scientific Instruments:
Spacecraft Bus:
1. STS-5 gas-discharge counter
2. scintillation counter
3. micrometeoroid detector
4. magnetometer
5. ion traps
6. LA-2 atomic hydrogen spectrometer
Impact Probe:
1. RMV barometer
2. TIPG thermometer
3. L-1A radiation detector
4. microorganism detection experiment
5. atmospheric composition experiment
6. R-3 acidity measurement experiment
7. K-2 electro-conductivity experiment
8. DAS-2 luminosity experiment

Results: The probe was the first dedicated 3MV spacecraft that the Soviets launched (earlier missions had been of the test "Object-Probe" versions). It was designed to accomplish atmospheric entry into Venus followed by descent and impact. The spacecraft successfully reached Earth orbit but failed to leave for Venus when the Blok L upper stage malfunctioned. The upper stage lost stable attitude due to a failure in the circuit of the power supply circuit that powered the valves for the attitude control system; hence, the stage remained uncontrollable and not ready to initiate a burn to leave Earth orbit. The problem was traced to a design error rather than one related to quality control. The spacecraft burned up in Earth's atmosphere the following day. If successful, this mission would probably have been given a "Venera" designation.

44

Zond 1 [Venera]

Nation: USSR (27)
Objective(s): Venus impact
Spacecraft: 3MV-1 (no. 4)
Spacecraft Mass: 948 kg
Mission Design and Management: OKB-1
Launch Vehicle: Molniya + Blok L (8K78 no. T15000-28)
Launch Date and Time: 2 April 1964 / 02:42:40 UT
Launch Site: NIIP-5 / Site 1/5
Scientific Instruments:

Spacecraft Bus:
1. STS-5 gas-discharge counter
2. scintillation counter
3. micrometeoroid detector
4. magnetometer
5. ion traps
6. LA-2 atomic hydrogen spectrometer

Impact Probe:
1. RMV barometer
2. TIPG thermometer
3. L-1A radiation detector
4. microorganism detection experiment
5. atmospheric composition experiment
6. R-3 acidity measurement experiment
7. K-2 electro-conductivity experiment
8. DAS-2 luminosity experiment

Results: This was the second dedicated launch of the 3MV series (not including two "Object-Probes"). Like its predecessor (see Kosmos 27), it was also designed for atmospheric entry and then impact on Venus. Although the probe was successfully sent towards Venus, ground controllers faced a series of major malfunctions in the spacecraft during its coast to the planet. These malfunctions included depressurization of the main spacecraft bus when the glass cover of a Sun-star attitude control sensor cracked. Additionally, the internal radio transmitters of the spacecraft were automatically switched on at precisely the wrong time, i.e., during depressurization, when the gas discharge created high-voltage currents that shorted out the communications system. As a result, communications had to be conducted through transmitters on the 290-kilogram pressurized descent module. Last contact was on 25 May 1964, by which time, controllers managed to conduct two major course corrections (at 560,000 and 13–14 million kilometers from Earth respectively), the first time such actions had been performed on a Soviet interplanetary spacecraft. The second correction, however, imparted 20 meters/second less velocity than required, ensuring that the vehicle would not intersect with Venus. The inert spacecraft eventually flew by Venus on 19 July 1964 at a range of 110,000 kilometers. The Soviets named the vehicle "Zond" (the Russian word for "probe") even though it was not one of the Object-Probe testbed spacecraft; this was done to disguise the fact that it was a failed Venus mission. If it had actually succeeded in its Venus mission, it probably would have been named "Venera 2." (Undoubtedly this has confused historians since this was not an Object-Probe mission). The Soviets later published some data on cosmic ray flux measured by Zond 1.

45

[Luna, Ye-6 no. 5]

Nation: USSR (28)
Objective(s): lunar soft-landing
Spacecraft: Ye-6 (no. 5)
Spacecraft Mass: c. 1,420 kg
Mission Design and Management: OKB-1
Launch Vehicle: Molniya-M + Blok L (8K78M no. T15000-21)
Launch Date and Time: 20 April 1964 / 08:08:28 UT
Launch Site: NIIP-5 / Site 1/5
Scientific Instruments:
1. imaging system
2. gas-discharge counter

First image of the Moon returned by a Ranger mission (Ranger VII in 1964). *Credit: NASA*

Results: This was the fifth Soviet attempt at a lunar soft-landing. The mission was aborted early, during the ascent to Earth orbit, when the launch vehicle's third stage engine (Blok I) prematurely shut down after 50 seconds of firing (at T+340 seconds). U.S. Air Force radars in Turkey apparently monitored the failed launch. A subsequent investigation indicated that the engine cut off due to loss of power when a circuit between a battery in the fourth stage (which powered the third stage engine) and the I-100 guidance unit was broken.

46

Ranger VII

Nation: USA (18)
Objective(s): lunar impact
Spacecraft: P-54 / Ranger-B
Spacecraft Mass: 365.6 kg
Mission Design and Management: NASA / JPL

Launch Vehicle: Atlas Agena B (Atlas Agena no. 9 / Atlas D no. 250 / Agena B no. 6009)

Launch Date and Time: 28 July 1964 / 16:50:07 UT

Launch Site: Cape Kennedy / Launch Complex 12

Scientific Instruments:

1. imaging system (six TV cameras)

Results: Ranger VII, the second of the Block III Ranger series, was, after 13 consecutive failures, the first unequivocal success in U.S. efforts to explore the Moon. In some ways, it marked a major milestone in American deep space exploration as the ratio in favor of successes increased dramatically after this point. After a nominal mid-course correction on 29 July, Ranger VII approached the Moon precisely on target two days later. Just fifteen minutes prior to impact, the suite of TV cameras began sending back spectacular photos of the approaching surface to JPL's Goldstone antenna in California. The last of 4,316 images was transmitted only 2.3 seconds prior to impact at 13:25:49 UT on 31 July 1964. The impact point was at 10° 38′ S / 20° 36′ W on the northern rim of the Sea of Clouds. Scientists on the ground were more than satisfied with results; image resolution was, in many cases, one thousand times better than photos taken from Earth. Scientists concluded that an Apollo crewed landing would be possible in the mare regions of the lunar surface, given their relative smoothness.

47

Mariner III

Nation: USA (19)

Objective(s): Mars flyby

Spacecraft: Mariner-64C / Mariner-C

Spacecraft Mass: 260.8 kg

Mission Design and Management: NASA / JPL

Launch Vehicle: Atlas Agena D (Atlas Agena D no. 11 / Atlas D no. 289 / Agena D no. AD68/6931)

Launch Date and Time: 5 November 1964 / 19:22:05 UT

Launch Site: Cape Kennedy / Launch Complex 13

Scientific Instruments:

1. imaging system
2. cosmic dust detector
3. cosmic ray telescope
4. ionization chamber
5. helium magnetometer
6. trapped radiation detector
7. solar plasma probe

Results: NASA approved two probes for the Mariner-Mars 1964 project in November 1962. The primary goal of the two spacecraft, code-named Mariner C, was to photograph the Martian surface using a single TV camera fixed on a scan platform that could return up to 22 frames after an eight-month journey. During the launch of Mariner III, the first of the two probes, the booster payload shroud failed to separate from the payload. Additionally, battery power spuriously dropped to zero (at T+8 hours 43 minutes) and the spacecraft's solar panels apparently never unfurled to replenish the power supply. Due to the incorrect mass of the spacecraft (since the payload shroud was still attached), it never entered a proper trans-Mars trajectory. The probe ended up in an unanticipated heliocentric orbit of 0.983 × 1.311 AU. A later investigation indicated that the shroud's inner fiberglass layer had separated from the shroud's outer skin, thus preventing jettisoning.

48

Mariner IV

Nation: USA (20)

Objective(s): Mars flyby

Spacecraft: Mariner-64D / Mariner-D

Spacecraft Mass: 260.8 kg

Mission Design and Management: NASA / JPL

Launch Vehicle: Atlas Agena D (Atlas Agena D no. 12 / Atlas D no. 288 / Agena D no. AD69/6932)

Launch Date and Time: 28 November 1964 / 14:22:01 UT

Launch Site: Cape Kennedy / Launch Complex 12

On 15 July 1965, Mariner IV transmitted this image of the Martian surface from 12,600 kilometers away. The photograph shows a 150-kilometer diameter crater. *Credit: NASA/JPL*

Scientific Instruments:

1. imaging system
2. cosmic dust detector
3. cosmic ray telescope
4. ionization chamber
5. helium magnetometer
6. trapped radiation detector
7. solar plasma probe

Results: The Mariner IV mission, the second of two Mars flyby attempts in 1964 by NASA, was one of the great early successes of the Agency, and indeed the Space Age, returning the very first photos of another planet from deep space. Using a new all-metal shroud, the spacecraft lifted off without any problems and was successfully boosted towards Mars by the Agena D upper stage. A single mid-course correction on December 5 ensured that the spacecraft would fly between 8,000 and 9,660

The Mariner IV spacecraft was assembled by engineers and technicians at the Jet Propulsion Laboratory in Pasadena, California. It is seen here being prepared for a weight test on 1 November 1963. *Credit: NASA/JPL*

kilometers from the Martian surface. On one of the scientific instruments, the plasma probe, there was a component failure making its readings unintelligible although due to a better telemetry rate, some data was received between January and May 1965. Additionally, a Geiger tube, one of the two sensors in the ionization chamber experiment, failed and stopped returning data in March 1965. Approximately 40 minutes prior to closest approach (which was at 01:00:57 UT on 15 July 1965 at a range of 9,846 kilometers), the TV camera began taking the first of 21 images (plus 22 lines of a 22nd) through red and green filters. About 1.25 hours after the encounter, Mariner IV dipped behind the right-hand side of Mars (as viewed from Earth) in an occultation experiment in order to refract its radio signals through the Martian atmosphere. Data indicated that surface pressure was quite low, i.e., future Mars landers would have to be equipped with retro-rocket engines in addition to parachutes. The images as well as the occultation experiment fundamentally transformed the scientific view of the Red Planet, providing hard data where speculation had previously dominated. The probe detected daytime surface temperatures at about −100°C. A very weak radiation belt, about 0.1% of that of Earth's, was also detected. The day after the closest encounter, Mariner IV began transmitting its photos back to Earth. The images clearly showed Mars to be an ancient Moon-like body with widespread cratering, thus incontrovertibly quashing any expectations of lost civilizations on the planet. Given the thin atmosphere, scientists believed that it was unlikely that Mars harbored any life. NASA maintained contact with the spacecraft until 1 October 1965 when the probe was 309 million kilometers from Earth. Two years later, in October 1967, the spacecraft was reactivated for attitude control tests in support of the Mariner V mission to Venus that used a similar spacecraft bus. Contact was maintained until 31 December 1967, over three years after launch.

49

Zond 2

Nation: USSR (29)
Objective(s): Mars flyby
Spacecraft: 3MV-4A (no. 2)
Spacecraft Mass: 996 kg
Mission Design and Management: OKB-1
Launch Vehicle: Molniya + Blok L (8K78 no. G15000-29)
Launch Date and Time: 30 November 1964 / 13:25 UT
Launch Site: NIIP-5 / Site 1/5
Scientific Instruments:

1. radiation detector (STS-5 scintillation and gas-discharge counters)
2. charged particle detector
3. magnetometer
4. piezoelectric detector?
5. Kassiopeya radio telescope
6. TyaMV nuclear component of cosmic rays experiment
7. RSK-2M ultraviolet and Roentgen solar radiation experiment
8. imaging system

Results: This was the last of three "Object-Probe" test vehicles launched as part of the third generation ("3MV") Soviet interplanetary probes, and the first intended towards Mars. These were designed to test out key technologies during deep space missions. Originally intended to fly in the April-May 1964 time period, this launch was constantly delayed and then ultimately timed for the late 1964 Mars opportunity. Besides carrying out long-distance communications tests and imaging Earth on the way out into deep space, this vehicle's trajectory was designed to allow it to intercept Mars (on approximately 6 August 1965) and become the first probe to enter its atmosphere and impact on its surface. After successfully entering a trans-Mars trajectory, ground controllers discovered that the probe's solar panels had not completely unfurled, depriving the vehicle of full power. Later investigation indicated that a tug cord designed to pull the panels free at the moment of separation from the Block L upper stage had broken off. Controllers were able to fully open the panel only on 15 December 1964, after "carrying out a number of dynamic operations on the station" according to the official institutional history, but by then the time for the first mid-course correction to fly by Mars had already passed. (Other reports suggest that even after the panels opened, they were partially obscured by radiators installed at the end of the solar panels which had not properly deployed). Additionally, there had been a failure in the onboard programmed timer immediately after trans-interplanetary injection that led to inappropriate thermal conditions for the spacecraft

between communications sessions. Before loss of contact, on 14 December, Zond 2 successfully fired six plasma electric rocket engines (twice) at a distance of 5.37 million kilometers from Earth. They were left "on" for 70 minutes and successfully maintained orientation of the spacecraft. These were technology demonstrators for future deep space missions. (These were actually carried on an earlier 3MV-1A model, launched on 11 November 1963, but that spacecraft failed to leave Earth orbit and was named Kosmos 21.) While some Western sources suggest that contact was maintained with Zond 2 until 4–5 May 1965, this is highly unlikely. The silent probe passed by Mars at a range of 650,000 kilometers on 6 August 1965 and entered heliocentric orbit. The spacecraft returned usable data on cosmic radio emissions at 210 and 2200 kc/second, up to 8 Earth radii distance from Earth.

1965

50

Ranger VIII

Nation: USA (21)
Objective(s): lunar impact
Spacecraft: Ranger-C
Spacecraft Mass: 366.87 kg
Mission Design and Management: NASA / JPL
Launch Vehicle: Atlas Agena B (Atlas Agena B no. 13 / Atlas D no. 196 / Agena B no. 6006)
Launch Date and Time: 17 February 1965 / 17:05:00 UT
Launch Site: Cape Kennedy / Launch Complex 12
Scientific Instruments:

    1. imaging system (six TV cameras)

Results: As successful as its predecessor, Ranger VIII returned 7,137 high resolution photographs of the lunar surface prior to lunar impact at 09:57:37 UT on 20 February. Unlike Ranger VII, however, Ranger VIII turned on its cameras about eight minutes earlier to return pictures with resolution comparable to Earth-based telescopes (for calibration and comparison purposes). Controllers attempted to align the cameras along the main velocity vector (to reduce imagine smear) but abandoned this maneuver to allow greater area coverage. There had also been a spurious loss of telemetry during a midcourse correction on 18 February that gave rise for concern, although the mission was completed successfully. Ranger VIII impacted at 2° 43′ N / 24° 38′ E, just 24 kilometers from its intended target point in the equatorial region of the Sea of Tranquility, an area that Apollo mission planners were particularly interested in studying as a possible landing site for future crewed missions.

51

[Atlas Centaur 5]

Nation: USA (22)
Objective(s): highly elliptical orbit
Spacecraft: SD-1
Spacecraft Mass: 951 kg
Mission Design and Management: NASA / JPL
Launch Vehicle: Atlas Centaur (AC-5 / Atlas C no. 156D / Centaur C)
Launch Date and Time: 2 March 1965 / 13:25 UT
Launch Site: Cape Kennedy / Launch Complex 36A
Scientific Instruments: [none]

The unsuccessful Atlas Centaur 5 just after launch on 2 March 1965. The rocket carried a Surveyor dynamic test model known as SD-1. *Credit: NASA*

Results: This mission was designed to rehearse a complete Centaur upper stage burn in support of the Surveyor lunar lander program. On a nominal mission, the Centaur would boost its payload on a direct ascent trajectory to the Moon. On this test flight, NASA planned to deliver the payload, a non-functional dynamic model known as SD-1, into an orbit of 167 × 926,625 kilometers that simulated a lunar transfer trajectory. During the actual launch, less than 2 seconds after liftoff, a faulty valve that incorrectly closed caused both Atlas main engines to shut down. As a result, the booster fell back onto the pad and exploded.

52

Kosmos 60 [Luna]

Nation: USSR (30)
Objective(s): lunar soft-landing
Spacecraft: Ye-6 (no. 9)
Spacecraft Mass: c. 1,470 kg
Mission Design and Management: OKB-1
Launch Vehicle: Molniya + Blok L (8K78 no. G15000-24)
Launch Date and Time: 12 March 1965 / 09:25 UT
Launch Site: NIIP-5 / Site 1/5
Scientific Instruments:
  1. imaging system
  2. SBM-10 radiation detector

Results: Yet another Soviet attempt to soft-land a Ye-6 probe on the lunar surface ended in failure when the Blok L upper stage failed to fire for the trans-lunar injection burn. Instead, the spacecraft remained stranded in Earth orbit. A later investigation indicated that there might have been a short circuit in the electric converter within the I-100 control system of the spacecraft (which also controlled the Blok L stage) preventing engine ignition. The spacecraft decayed five days later.

The S1.5400 engine powered the 4th stage known as the Blok L on Soviet lunar and planetary missions in the 1960s. This engine, capable of firing in vacuum, was the cause of numerous failures that left Soviet probes stranded in Earth orbit. *Credit: T. Varfolomeyev*

53

Ranger IX

Nation: USA (23)
Objective(s): lunar impact
Spacecraft: Ranger-D
Spacecraft Mass: 366.87 kg
Mission Design and Management: NASA / JPL
Launch Vehicle: Atlas Agena B (Atlas Agena B no. 14 / Atlas D no. 204 / Agena B no. 6007)
Launch Date and Time: 21 March 1965 / 21:37:02 UT
Launch Site: Cape Kennedy / Launch Complex 12

Scientific Instruments:

1. imaging system (six TV cameras)

Results: Ranger IX was the final Ranger mission of the Block III series and closed out the program as a whole. Since both Ranger VII and Ranger VIII had provided sufficient photographs of the mare regions (potential landing sites for the early Apollo missions), Ranger IX was targeted to the more geologically interesting Alphonsus crater in the lunar highlands, a possible site for recent volcanic activity. Following a mid-course correction on 23 March, the spacecraft headed directly to its impact point. Only 20 minutes prior to impact, Ranger IX began taking the first of 5,814 pictures from an altitude of 2,100 kilometers. Unlike its predecessors, the cameras this time were aimed directly in the direction of travel and provided some spectacular shots as the spacecraft approached the lunar surface. These pictures were converted for live viewing on commercial TV. Best resolution was up 25–30 centimeters just prior to impact. The spacecraft crashed onto the Moon at 14:08:20 UT on 24 March at 12.83° S / 357.63° E, about six-and-a-half kilometers from its scheduled target at a velocity of 2.67 kilometers/second.

54

[Luna, Ye-6 no. 8]

Nation: USSR (31)
Objective(s): lunar soft-landing
Spacecraft: Ye-6 (no. 8)
Spacecraft Mass: c. 1,470 kg
Mission Design and Management: OKB-1
Launch Vehicle: Molniya-M + Blok L (8K78M no. R103-26, also U15000-22)
Launch Date and Time: 10 April 1965 / 08:39 UT
Launch Site: NIIP-5 / Site 1/5
Scientific Instruments:

1. imaging system
2. SBM-10 radiation detector

Results: This was the seventh consecutive failure to accomplish a lunar soft-landing by the Soviets. On this mission, engineers redesigned the problematic I-100 control system that had caused most of the previous failures. Previously the I-100 unit had controlled both the Blok L upper stage and the spacecraft itself. On this mission (and subsequent Lunas), the fourth stage and the Ye-6 spacecraft had separate systems. Unfortunately, this probe never reached Earth orbit. During the launch, depressurization of a nitrogen pipe for the liquid oxygen tank on the third stage (Blok I) prevented third stage engine ignition. The spacecraft thus broke up over the Pacific without reaching Earth orbit.

55

Luna 5

Nation: USSR (32)
Objective(s): lunar soft-landing
Spacecraft: Ye-6 (no. 10)
Spacecraft Mass: 1,476 kg
Mission Design and Management: OKB-1
Launch Vehicle: Molniya-M + Blok L (8K78M no. U103-30, also U15000-24)
Launch Date and Time: 9 May 1965 / 07:49:37 UT
Launch Site: NIIP-5 / Site 1/5
Scientific Instruments:

1. imaging system
2. SBM-10 radiation detector

Results: Luna 5 became the first Soviet probe to head for the Moon in two years. Following a mid-course correction on 10 May, the spacecraft began spinning around its main axis due to a problem in a floatation gyroscope in the I-100 control system unit—the gyroscopes apparently had had too little time to warm up before being used for attitude control. A subsequent attempt to fire the main engine failed due to ground control error. A third attempt also failed, and having lost control of the spacecraft due to the gyroscope problem, controllers stood by helplessly as Luna 5 crashed on to the surface of the Moon at 19:10 UT on 12 May in the Sea of Clouds, about 700 kilometers from its planned landing point. Landing coordinates were 31° S / 8° W. It was the second Soviet spacecraft to impact on the Moon (following Luna 2 in 1959).

56

Luna 6

Nation: USSR (33)
Objective(s): lunar soft-landing
Spacecraft: Ye-6 (no. 7)
Spacecraft Mass: 1,442 kg
Mission Design and Management: OKB-1
Launch Vehicle: Molniya-M + Blok L (8K78M no. U103-31, also U15000-33)
Launch Date and Time: 8 June 1965 / 07:40 UT
Launch Site: NIIP-5 / Site 1/5
Scientific Instruments:
1. imaging system
2. SBM-10 radiation detector

Results: On this ninth Soviet attempt at a lunar soft-landing, the mission proceeded as planned until a major mid-course correction late on 9 June. Although the main retro-rocket engine (the S5.5A) ignited on time, it failed to cut off and continued to fire until propellant supply was exhausted. An investigation later indicated that the problem had been due to human error; a command had been mistakenly sent to the timer that ordered the main engine to shut down. Although the spacecraft was sent on a completely off-nominal trajectory, ground controllers put the spacecraft through a series of steps to practice an actual landing (such as inflating the airbags, separating the lander, etc.), all of which were satisfactorily accomplished. Luna 6 passed by the Moon late on 11 June at a range of 161,000 kilometers and eventually entered heliocentric orbit. Contact was maintained to a distance of 600,000 kilometers from Earth.

The Zond 3 spacecraft (of the 3MV-4 type) returned higher resolution pictures, as compared to Luna 3, of the farside of the Moon in 1965 during a flyby. *Credit: Don Mitchell*

57

Zond 3

Nation: USSR (34)
Objective(s): lunar flyby
Spacecraft: 3MV-4 (no. 3)
Spacecraft Mass: 950 kg
Mission Design and Management: OKB-1
Launch Vehicle: Molniya + Blok L (8K78 no. U103-32, also U15000-32)
Launch Date and Time: 18 July 1965 / 14:32 UT
Launch Site: NIIP-5 / Site 1/5
Scientific Instruments:
1. imaging system
2. ultraviolet spectrograph
3. ultraviolet and infrared spectrophotometer
4. meteoroid detectors
5. STS-5 scintillation and gas-discharge counters)

6. magnetometer
7. ion thrusters
8. radio telescope

Results: This "third generation" deep space probe had originally been slated for a Mars flyby in late 1964 but could not be prepared on time. Instead, Soviet designers diverted the mission for a simple lunar flyby in 1965 to test its basic systems and photograph the farside of the Moon. After a successful trans-lunar injection burn, Zond 3 approached the Moon after only a 33-hour flight. Its imaging mission began at 01:24 hours on 20 July at a range of 11,600 kilometers from the near side of the Moon and completed 68 minutes later. Zond 3 carried an imaging system somewhat similar to the one carried on Automatic Interplanetary Station ("Luna 3"), with onboard exposure, development, fixing, and drying prior to scanning for transmission to Earth. The new system, known as 15P52, weighed 6.5 kilograms, was developed by the Moscow-based NII-885 (as opposed to VNII-380, which developed the Luna-3 system). In total, the spacecraft took 25 visual and three ultraviolet images during its flyby. Closest approach was to 9,220 kilometers. These pictures were successfully transmitted back to Earth on July 29, nine days after its lunar encounter when it was 2.2 million kilometers from Earth. Further communications sessions occurred on 23 October (involving photo transmissions) when Zond 3 was 31.5 million kilometers from Earth. Last contact was sometime in early March 1966 when the spacecraft was 153.5 million kilometers away. During the mission, it photographed the unseen 30% of the farside of the Moon. Zond 3 also demonstrated successful course corrections using both solar and stellar orientation, a first for a Soviet spacecraft.

58

Surveyor Model 1

Nation: USA (24)
Objective(s): highly elliptical orbit
Spacecraft: SD-2
Spacecraft Mass: 950 kg
Mission Design and Management: NASA / JPL
Launch Vehicle: Atlas Centaur (AC-6 / Atlas D no. 151D / Centaur D)
Launch Date and Time: 11 August 1965 / 14:31:04 UT
Launch Site: Cape Kennedy / Launch Complex 36B
Scientific Instruments: [none]
Results: This was the second attempt to launch a dummy Surveyor lunar lander spacecraft into a barycentric orbit towards a simulated Moon. Unlike the previous attempt (in March 1965), this time all systems worked without fault; the Centaur fired flawlessly and put the Surveyor dynamic model on a simulated lunar trajectory so precise that it would have impacted on the Moon without a trajectory correction. Orbital parameters were $164 \times 822{,}135$ kilometers at 28.6° inclination. The payload reentered after 31 days.

59

Luna 7

Nation: USSR (35)
Objective(s): lunar soft-landing
Spacecraft: Ye-6 (no. 11)
Spacecraft Mass: 1,506 kg
Mission Design and Management: OKB-1
Launch Vehicle: Molniya-M + Blok L (8K78M no. U103-27, also U15000-54)
Launch Date and Time: 4 October 1965 / 07:56:40 UT
Launch Site: NIIP-5 / Site 1/5

Scientific Instruments:

1. imaging system
2. SBM-10 radiation detector

Results: The first attempt to launch this vehicle took place at 07:05:36 UT on 4 September 1965 but the launch was aborted due to a problem in a sensor that measured relative velocity on the rocket. The subsequent launch a month later went off successfully and the Blok L upper stage successfully sent Luna 7 on its way to the Moon. Unlike its predecessors, Luna 7 successfully carried out its mid-course correction on October 5 on the way to the Moon, in anticipation of a soft-landing two days later (at the time planned for 2208 UT on 7 October). Unfortunately, immediately prior to planned retro-fire during the approach to the lunar surface, the spacecraft suddenly lost attitude control and failed to regain it. One of its attitude control sensors—the one designed to lock on to Earth—stopped functioning, preventing it from reaching the desired orientation for firing its retro-engine. Automatic programmed systems then prevented the main engine from firing. As controllers observed helplessly, Luna 7 plummeted to the lunar surface at a very high speed, crashing at 22:08:24 UT on 7 October west of the Kepler crater, relatively near to the actual intended target. Impact coordinates were 9° N / 49° W. Later investigation indicated that the optical sensor of the Yupiter-M astronavigation system had been set at the wrong angle and had lost sight of Earth during the critical attitude control maneuver. It was the tenth consecutive failure in the Ye-6 program.

60

Venera 2

Nation: USSR (36)

Objective(s): Venus flyby

Spacecraft: 3MV-4 (no. 4)

Spacecraft Mass: 963 kg

Mission Design and Management: OKB-1

Launch Vehicle: Molniya-M + Blok L (8K78M no. U103-42, also U15000-42)

Launch Date and Time: 12 November 1965 / 04:46 UT

Launch Site: NIIP-5 / Site 31/6

Scientific Instruments:

1. triaxial fluxgate magnetometer
2. spectrometers
3. micrometeoroid detectors
4. ion traps
5. cosmic radio emission receiver
6. radio detector
7. STS-5 gas-discharge and solid-state detectors
8. imaging system
9. infrared spectrometer
10. 2 other spectrometers

Results: Although the 3MV-3 and 3MV-4 type spacecraft were originally intended for Mars exploration, the Soviets re-equipped three of the series, left over from the 1964 Mars launch periods, for Venus exploration in 1965. This particular vehicle was scheduled to fly past the sunlit side of Venus at no more than 40,000 kilometers range and take photographs. About 3 hours after injection into heliocentric orbit, contact was temporarily lost with the spacecraft, and although it was regained soon after, the event was symptomatic of the mission in general during which communications were generally poor. Before closest approach in late February 1966, ground control switched on all the onboard scientific instrumentation. Closest approach to the planet was at 02:52 UT on 27 February 1966 at about 24,000 kilometers range. After its flyby, when the spacecraft was supposed to relay back the collected information, ground control was unable to regain contact. Controllers finally gave up all attempts at communication on 4 March. Venera 2 eventually entered heliocentric orbit. Later investigation indicated that improper functioning of 40 thermal radiator elements caused a sharp increase in gas temperatures in the spacecraft. As a result, elements of the receiving and decoding units failed (and the solar panels overheated), and contact was lost. Ironically, the scientific instruments may have collected valuable data, but none of it was ever transmitted back to Earth.

61

Venera 3

Nation: USSR (37)
Objective(s): Venus impact
Spacecraft: 3MV-3 (no. 1)
Spacecraft Mass: 969 kg
Mission Design and Management: OKB-1
Launch Vehicle: Molniya + Blok L (8K78 no. U103-31, also U15000-31)
Launch Date and Time: 16 November 1965 / 04:13 UT
Launch Site: NIIP-5 / Site 31/6
Scientific Instruments:

Spacecraft Bus:
1. triaxial fluxgate magnetometer
2. spectrometers
3. SBT-9 cosmic ray sensor
4. ion traps
5. STS-5 gas-discharge and solid-state detectors

Impact Probe:
1. temperature, pressure, and density sensors
2. chemical gas analysis
3. photometer
4. gamma-ray counter

Results: This was the second of three 3MV spacecraft the Soviets attempted to launch towards Venus in late 1965. Venera 3 successfully left Earth orbit carrying a small 0.9-meter diameter 310-kilogram landing capsule to explore the Venusian atmosphere and transmit data on pressure, temperature, and composition of the Venusian atmosphere back to Earth during the descent by parachute. During the outbound trajectory, ground controllers successfully performed a mid-course correction on 26 December 1965 and completed 63 communications sessions during which scientists on the ground recorded valuable information. For example, between 16 November 1965 and 7 January 1966, a modulation charged particle trap (of the same type carried on Zond 2), provided valuable data on the energy spectra of solar wind ion streams beyond the Earth's magnetosphere. Contact was, however, lost on 16 February 1966, shortly before the Venusian encounter. The spacecraft subsequently automatically released its lander probe which impacted inertly onto the Venusian surface at 06:56:26 UT on 1 March 1966, only 4 minutes earlier than planned. It was the first time a human-made object made physical contact with another planetary body other than the Moon. The impact location was on the night side of Venus, near the terminator, in the range of $-20°$ to $20°$ N / $60°$ to $80°$ E. In response to concern from some American and British scientists, the Soviet press emphasized that "prior to liftoff, the descent module was subject to careful sterilization, required to dispose of all microorganisms of terrestrial origin and thus prevent the possibility of contamination." Later investigation confirmed that Venera 3 suffered many of the same failures as Venera 2, i.e., overheating of internal components and the solar panels.

62

Kosmos 96 [Venera]

Nation: USSR (38)
Objective(s): Venus flyby
Spacecraft: 3MV-4 (no. 6)
Spacecraft Mass: c. 950 kg
Mission Design and Management: OKB-1
Launch Vehicle: Molniya + Blok L (8K78 no. U103-30, also U15000-30)
Launch Date and Time: 23 November 1965 / 03:14 UT
Launch Site: NIIP-5 / Site 31/6
Scientific Instruments:
1. three-component magnetometer
2. imaging system
3. solar x-radiation detector
4. cosmic ray gas-discharge counters
5. piezoelectric detectors
6. ion traps
7. photon Geiger counter
8. cosmic radio emission receivers

A model of the Ye-6-type lunar probe on display at the Memorial Museum of Cosmonauts in Moscow. The package on 'right' is the lunar lander (the ALS). The three silver balls visible in the foreground are three of the four gas storage bottles for the attitude control system. The two (of four total) black-and-white nozzles facing left are verniers. The main S5.5A engine is at the left or aft end of the entire spacecraft. *Credit: Asif Siddiqi*

Results: This was the third and last spacecraft prepared for a Venus encounter by the Soviets in 1965. All three spacecraft had originally been intended for Mars exploration in 1964–1965. In this case, during coast to Earth orbit, a combustion chamber in the booster's third stage engine exploded due to a crack in the fuel pipeline. Although the payload reached Earth orbit, the Blok L upper stage was tumbling and was unable to fire for trans-Venus trajectory injection. The probe remained stranded in Earth orbit and the Soviets named it Kosmos 96 to disguise its true mission. The probe decayed on 9 December 1965.

63

Luna 8

Nation: USSR (39)
Objective(s): lunar soft-landing
Spacecraft: Ye-6 (no. 12)
Spacecraft Mass: 1,552 kg
Mission Design and Management: OKB-1
Launch Vehicle: Molniya-M + Blok L (8K78M no. U103-28, also U15000-48)
Launch Date and Time: 3 December 1965 / 10:46:14 UT
Launch Site: NIIP-5 / Site 31/6
Scientific Instruments:
1. imaging system
2. SBM-10 radiation detector

Results: This, the tenth Soviet attempt to achieve a lunar soft-landing, nearly succeeded. The Blok L upper stage successfully dispatched the probe towards the Moon. After a successful mid-course correction at 19:00 UT on 4 December, the spacecraft headed towards its targeted landing site on the Moon without any apparent problems. Just prior to the planned retro-fire burn, a command was sent to inflate cushioning airbags around the ALS lander probe. Unfortunately, a plastic mounting bracket appears to have pierced one of the two bags. The resulting expulsion of air put the spacecraft into a spin (of 12°/second). The vehicle momentarily regained attitude, long enough for a 9-second retro-engine firing, but then lost it again. Without a full retro-burn to reduce approach velocity sufficient for a survivable landing, Luna 8 plummeted to the lunar surface and crashed at 21:51:30 UT on 6 December just west of the Kepler crater. Impact coordinates were 9° 8′ N / 63° 18′ W. The Soviet news agency TASS merely reported that "the station's systems functioned normally at all stages of the landing except the final one."

Pioneer VI was the first in a series of solar-orbiting spacecraft designed to obtain measurements on a continuing basis of interplanetary phenomena from widely separated points in space. *Credit: NASA*

64

Pioneer VI

Nation: USA (25)
Objective(s): heliocentric orbit
Spacecraft: Pioneer A
Spacecraft Mass: 62.14 kg
Mission Design and Management: NASA / ARC
Launch Vehicle: Thrust Augmented Delta (Thor Delta E no. 35 / Thor no. 460/DSV-3E)
Launch Date and Time: 16 December 1965 / 07:31:21 UT
Launch Site: Cape Kennedy / Launch Complex 17A
Scientific Instruments:
1. single-axis fluxgate magnetometer
2. plasma Faraday cup
3. electrostatic analyzer
4. cosmic ray telescope
5. cosmic ray anisotropy detector
6. two-frequency beacon receiver

Results: Pioneer VI was the first of several NASA spacecraft designed for launch at six-month intervals to study interplanetary phenomena in space in heliocentric orbits similar to that of Earth. These spacecraft successfully provided simultaneous scientific measurements at widely dispersed locations in heliocentric orbit. The so-called Improved Thrust Augmented Delta launch vehicle's third stage burned for 23 seconds to boost Pioneer VI into heliocentric orbit. Initial solar orbit for the spacecraft ranged from 0.814 AU (perihelion) to 0.985 AU (aphelion) with a period of 311.3 days. By 2 March 1966, Pioneer VI had transmitted about 250 million readings from its six scientific instruments. In the fall of 1969, JPL and UCLA scientists reported the results of a solar occultation performed from 21–24 November 1968. This was the first time that a spacecraft had been tracked while passing behind the Sun, allowing scientists, despite an unfavorable signal-to-noise ratio of data transmission, to examine the solar corona during this passage. Scientists used instruments on Pioneer VI in coordination with those on Pioneer VII, in November and December 1969, to measure solar wind particles and carry out long-distance communications experiments. Five years after launch, at the end of 1970, Pioneer VI had orbited the Sun six times, and had passed by the Sun's far side (relative to Earth), sending back new information on the solar atmosphere and regular solar weather reports. One of the two radio receivers was still operational, and although some of the solar cells had been damaged by solar flares, the spacecraft was still getting sufficient power to operate satisfactorily. Pioneer VI returned the first data on the tenuous solar atmosphere and later recorded the passage of Comet Kohoutek's tail in 1974 (in conjunction with Pioneer VIII) from a range of about 100 million kilometers from the comet's nucleus. Along with Pioneers VII, VIII, and IX, the spacecraft formed a ring of solar weather stations spaced along Earth's orbit. Measurements by the four Pioneers were used to predict solar storms for approximately 1,000 primary users including the Federal Aviation Agency, commercial airlines, power companies, communication companies, military organizations, and entities involved in surveying, navigation, and electronic prospecting. By December 1990, Pioneer VI had circled

the Sun 29 times (travelling 24.8 billion kilometers) and had been operational for twenty straight years—a record for a deep space probe. Its original slated lifetime had been only six months, achieved on 16 June 1966. Of the spacecraft's six scientific instruments, two (the plasma Faraday cup and the cosmic ray detector) functioned well into the 1990s. NASA maintained contact with the spacecraft once or twice each year during the 1990s. For example, one hour's worth of scientific data was collected on 29 July and 15 December 1995, although the primary transmitter failed the following year. Soon after, on 31 March 1997, NASA officially declared the mission complete largely due to the costs associated with continuing communications sessions. Despite the decision, contact was established with the backup transmitter on 6 October 1997 as part of a training exercise for the Lunar Prospector spacecraft. By this point, the probe's solar arrays had deteriorated although the transmitter could still be turned on at perihelion when the solar flux was strong enough to provide sufficient power. On 8 December 2000, ground controllers established successful contact for 2 hours to commemorate the 35th year of operation. This, however, proved to be the very last contact made with the probe.

65

Luna 9

Nation: USSR (40)
Objective(s): lunar soft-landing
Spacecraft: Ye-6M (no. 202)
Spacecraft Mass: 1,583.7 kg
Mission Design and Management: GSMZ imeni Lavochkina
Launch Vehicle: Molniya-M + Blok L (8K78M no. U103-32, also U15000-49)
Launch Date and Time: 31 January 1966 / 11:41:37 UT

Launch Site: NIIP-5 / Site 31/6
Scientific Instruments:
1. imaging system
2. gamma-ray spectrometer
3. KS-17M radiation detector

Results: With this mission, the Soviets accomplished another spectacular first in the space race, the first survivable landing of a human-made object on another celestial body and the transmission of photographs from its surface. Luna 9 was the twelfth attempt at a soft-landing by the Soviets; it was also the first deep space probe built by the Lavochkin design bureau that would design and build all future Soviet (and Russian) lunar and interplanetary

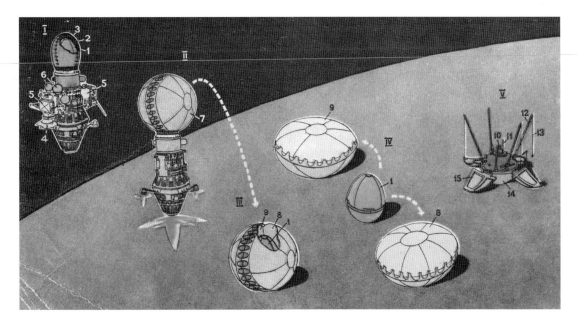

A model of the Ye-6-type lunar lander (the ALS) shows its ingenuous "petal" design. In scene I, we see the Luna spacecraft with (1) Automatic Lunar Station (ALS) which is covered by (3) a thermal covering. Ideally, at an altitude of 75 kilometers, the padding (2) around the ALS would be pressurized as we see in scene II. At a command from the radio-altimeter located at (4), two side packages (5) from the main bus are jettisoned and the main S5.5A retro engine is turned on. At a given altitude from the lunar surface, the pressurized ball (7) would be separated from the main bus and would impact on the surface as shown in scene III. The pressurized covering surrounding the ALS would then separate into two parts (8 and 9) revealing the inner ALS. As shown in scene V, the petals of the lander would then unfurl, stabilizing the main body of the ALS (14), allowing a small suite of scientific instruments to operate. These included antennae (12 and 15) and a camera (10). *Credit: Nauka i zhizn*

spacecraft. All operations prior to landing occurred without fault. A 48-second mid-course correction at 19:29 UT on 1 February some 233,000 kilometers from the Moon directed the probe directly to its target in the Ocean of Storms. About one hour before touchdown at a distance of 8,300 kilometers, Luna 9 was put in proper attitude for retro-fire. Just prior to engine ignition, two side compartments were jettisoned, followed by inflation of two shock-absorbing airbags covering the lander to a pressure of 1 atmosphere. Its main S5.5A engine ignited at an altitude of 74.885 kilometers above the surface and fired for 48 seconds until the probe was just 260–265 meters above ground, thus decelerating Luna 9 from 2,600 meters/second to a few meters/second. Just above the surface, a long boom sensor made contact with the lunar surface, thus issuing a command to eject the 58-centimeter spheroid ALS capsule weighing 99.8 kilograms from the main bus. The ALS (still enclosed in surrounded airbags) landed a few meters away. The impact time was recorded as 18:45:30 UT on 3 February 1966 west of the Reiner and Marius craters in the Ocean of Storms (reported as 7° 8′ N / 64° 32′ W but closer to 8° N / 64° W). About four minutes after landing, the airbags split open, and the petals covering the top of the ALS were deployed. Precisely four minutes and 10 seconds after touchdown, Luna 9 began transmitting initial telemetry data back to Earth, although it would be another 7 hours (at 01:50 UT on 4 February, after the Sun climbed from 3° to 7° elevation) before the probe began sending back the first of nine images (including five panoramas) of the surface of the Moon. The first panoramic images arrived very early in the morning in Moscow, and because officials were afraid to wake up Soviet space program curator Dmitriy Ustinov (1908–1984) (whose permission was required for publication in the Soviet media), the first panoramic images were actually published in the British media courtesy of Sir Bernard Lovell (1913–2012) at the Jodrell Bank Observatory who had intercepted and analyzed the same data. The later images had the Sun much high up, up to 41°, thus causing the shadow relief of the images to change. These were the first images sent back from the surface of another planetary body. Controllers noticed at one point that Luna 9's vantage point had slightly shifted over the sequence of images, possibly caused by the diminishing water supply of its thermal control system which changed its weight distribution. This change in perspective (of about 100 mm) opened up the possibility of stereo photography of the surface. The KS-17M radiation detector measured a dosage of 30 millirads per day. Perhaps the most important discovery from the mission was determining that a foreign object would not simply sink into the lunar dust, i.e., that the ground could support a heavy lander. Mission controllers expected that the last communications session would be on 5 February (from 16:00 to 17:41 UT) but were pleased to have an additional one, on 6 February (from 20:37 to 22:55 UT). By the time contact was lost, controllers had communicated with Luna 9 over seven communications sessions lasting 8 hours and 5 minutes.

66

Kosmos 111 [Luna]

Nation: USSR (41)
Objective(s): lunar orbit
Spacecraft: Ye-6S (no. 204)
Spacecraft Mass: c. 1,580 kg
Mission Design and Management: GSMZ imeni Lavochkina
Launch Vehicle: Molniya-M + Blok L (8K78M (no. N103-41, also U15000-50)
Launch Date and Time: 1 March 1966 / 11:03:49 UT
Launch Site: NIIP-5 / Site 31/6
Scientific Instruments:
1. magnetometer
2. gamma-ray spectrometer
3. five gas-discharge counters
4. two ion traps and a charged particle trap
5. piezoelectric micrometer detector
6. infrared detector
7. low-energy x-ray photon counters

Results: In early 1966, the Soviets hastily put together an interim lunar orbiter program, the Ye-6S, partly to upstage the American Lunar Orbiter project, and partly to commemorate the 23rd Congress of the Communist Party held in March 1966. Engineers quickly designed a set of two rudimentary probes using the old Ye-6 (lander) buses for these missions. The first of these was prepared in less than a month but failed to leave Earth orbit. During Earth orbit operations, the Blok L upper stage lost roll control and failed to fire to send the probe towards the Moon. The official Soviet media named the stranded satellite Kosmos 111 which reentered two days after launch.

67

Luna 10

Nation: USSR (42)
Objective(s): lunar orbit
Spacecraft: Ye-6S (no. 206)
Spacecraft Mass: 1,583.7 kg
Mission Design and Management: GSMZ imeni Lavochkina
Launch Vehicle: Molniya-M + Blok L (8K78M no. N103-42, also U15000-51)
Launch Date and Time: 31 March 1966 / 10:46:59 UT
Launch Site: NIIP-5 / Site 31/6
Scientific Instruments:
1. SG-59M 3-component magnetometer
2. gamma-ray spectrometer
3. SL-1 radiometer for detecting radiation near the Moon
4. D-153 solar plasma detector
5. RMCh-1 meteorite detector
6. ID-1 infrared radiation detector
7. RFL-1 x-ray fluorescence detector

Results: After a mid-course correction on 1 April, Luna 10, the second and backup model of two hastily prepared Soviet Ye-6S probes, successfully entered lunar orbit two days later at 18:44 UT, thus becoming the first human-made object to go into orbit around another planetary body. A 248.5-kilogram instrument compartment separated from the main bus at 18:45:39 UT which was in a 350 × 1,016-kilometer orbit inclined at 71.9° to the lunar equator. The spacecraft carried a set of solid-state oscillators which had been programmed to reproduce the notes of the *Internationale* so that it could be broadcast live to the 23rd Communist Party Congress. During a rehearsal on the night of 3 April, the playback went well, but the following morning, controllers discovered a missing note and so played the previous night's tape to the assembled gathering at the Congress, claiming it was a live broadcast from the Moon. Luna 10 conducted extensive research in lunar orbit, gathering important data on the weakness of the Moon's magnetic field, radiation belts, and micrometeoroid density. In 2012, a Soviet scientist made the claim that Luna 10's achievements included "the first gamma-ray spectrometer used in the history of space research to first define the contents of natural radio nuclides in the lunar soil." Data from Luna 10 suggested that Moon rocks were comparable to terrestrial basalt rocks. Based on data collected by the probe, Efraim Akim (1929–2010) at the USSR Academy of Sciences identified the "noncentrality of the gravitational field of the Moon" which he argued was "the essential fact determining the evolution of the orbit of Luna 10," as reported in an Academy journal in 1966. Based on Akim's claim, some sources incorrectly credit Luna 10 with discovery of mass concentrations (called "mascons")—areas of high density below the mare basins that distort lunar orbital trajectories. Their discovery, however, came much later, after the creation of a gravimetric map of the Moon (albeit at the time, only of the near side of the Moon). Paul Muller and William Sjogren published their conclusions in 1968 based on data from the Lunar Orbiters showing that very large positive gravity anomalies exist in all of the circular ringed sea basins on the Moon. The discovery of mascons thus should be credited to them. Controllers made last contact with Luna 10 on 30 May 1966.

68

Surveyor Model 2

Nation: USA (26)
Objective(s): highly elliptical orbit
Spacecraft: SD-3
Spacecraft Mass: 784 kg
Mission Design and Management: NASA / JPL
Launch Vehicle: Atlas Centaur (AC-8 / Atlas no. 184D / Centaur D)
Launch Date and Time: 8 April 1966 / 01:00:02 UT
Launch Site: Cape Kennedy / Launch Complex 36B
Scientific Instruments: [none]
Results: This was a test to launch a dummy Surveyor lunar lander spacecraft into a barycentric orbit towards a simulated Moon. Unlike the two previous Surveyor mass model tests, this flight was supposed to demonstrate a restart capability for the Centaur upper stage. The Centaur-Surveyor combination successfully achieved parking orbit around Earth (with a first firing), but at the desired time, when it came time for the second firing, the Centaur RL-10 engines fired for only a few seconds (instead of 107 seconds). A thrust imbalance left the payload tumbling and in an incorrect orbit of 182 × 335 kilometers at 30.7° inclination. The problem was later traced to a hydrogen peroxide leak in the ullage motors of the Centaur stage. With no hope of reaching its ultimate orbit (planned for 167 × 380,000 km), the payload reentered on 5 May 1966.

69

Surveyor I

Nation: USA (27)
Objective(s): lunar soft-landing
Spacecraft: Surveyor-A
Spacecraft Mass: 995.2 kg
Mission Design and Management: NASA / JPL

Launch Vehicle: Atlas Centaur (AC-10 / Atlas D no. 290 / Centaur D)
Launch Date and Time: 30 May 1966 / 14:41:01 UT
Launch Site: Cape Kennedy / Launch Complex 36A
Scientific Instruments:
1. TV camera
Results: In January 1961, NASA selected Hughes Aircraft Company to build a series of seven soft-landing vehicles, each weighing about 340 kilograms, to "land gently on the moon, perform chemical analyses of the lunar surface and subsurface and relay back to Earth television pictures of lunar features." These vehicles were to be designed and built under the technical direction of JPL and launched in the period of 1963–1966. Unlike the Soviet Luna landers, Surveyor was a true soft-lander, comprising a three-meter tall vehicle based on a 27-kilogram thin-walled aluminum triangular structure with one of three legs at each corner and a large solid-propellant retro-rocket engine (that comprised over 60% of the spacecraft's overall mass) in the center. The spacecraft was equipped with a Doppler velocity-sensing system that fed information into the spacecraft computer to implement a controllable descent to the surface. Each of the three landing pads also carried aircraft-type shock absorbers and strain gauges to provide data on landing characteristics, important for future Apollo missions. Surveyor I, the first in the series, was an unprecedented success. NASA accomplished the first true soft-landing on the Moon on its very first try when the probe landed in the southwestern region of the Ocean of Storms at 06:17:36 UT on 2 June 1966, just 63.6 hours after launch from Cape Canaveral. Touchdown coordinates were announced as 2.46° S / 43.32° W, just 14 kilometers from the planned target. At landing, the spacecraft weighed 294.3 kilograms. The initial panoramic views from the lunar surface indicated that Surveyor I was resting in a 100-kilometer diameter crater that contained boulders more than one meter scattered all around. The photos showed crestlines of low mountains in the distant horizon. The lander transmitted 11,240 high-resolution

images over two separate communications sessions by 6 July. Although the primary mission was completed by 14 July, NASA maintained contact until 7 January 1967. Without doubt, Surveyor I was one of the great successes of NASA's early lunar and interplanetary program.

70

Explorer XXXIII

Nation: USA (28)
Objective(s): lunar orbit
Spacecraft: AIMP-D
Spacecraft Mass: 93.4 kg
Mission Design and Management: NASA / GSFC
Launch Vehicle: Thor Delta E-1 (Thor Delta E-1 no. 39 / Thor no. 467/DSV-3E)
Launch Date and Time: 1 July 1966 / 16:02:25 UT
Launch Site: Cape Kennedy / Launch Complex 17A
Scientific Instruments:
1. ionizing radiation experiment
2. 3-grid Faraday cup / thermal ion experiment
3. 3 GM tubes and a PN junction semiconductor / energetic particle experiment
4. plasma probe experiment
5. magnetometer (from Ames)
6. magnetometer (from GSFC)
7. solar cell damage experiment

Results: It was hoped that Explorer XXXIII (33), also known as the Anchored Interplanetary Monitoring Platform (AIMP), would become the first U.S. spacecraft to enter lunar orbit (planned parameters were 1,300 × 6,440 kilometers at 175° inclination), but the Thor Delta E-1 second stage accelerated too rapidly, ensuring that lunar orbit would not be possible. Instead mission planners adopted an alternate mission which required the probe's (Thiokol TE-M-458) 35.8 kgf thrust retro-rocket to fire about 6.5 hours after launch. Under the new plan, the spacecraft was put into a highly elliptical Earth orbit of 449,174 × 30,550 kilometers at 28.9° inclination. In this orbit, Explorer XXXIII was high

enough to be perturbed by the Moon's gravitation. In its new orbit, Explorer XXXIII approached the Moon (on its very first orbit) to a distance of 35,000 kilometers with subsequent close approaches in September, November, and December 1966 varying from 40,000 to 60,000 kilometers. During this time, the probe returned key data on Earth's magnetic tail, the interplanetary magnetic field, and radiation. In January 1967, Goddard Space Flight Center engineers used an electric "screwdriver" to restore power to the spacecraft after a temporary blackout. The emergency repair was the most remote repair carried out on a spacecraft to date. The mission was declared a success by 1 January 1967, although the spacecraft continued to return useful data until 15 September 1971.

71

Lunar Orbiter I

Nation: USA (29)
Objective(s): lunar orbit
Spacecraft: LO-A
Spacecraft Mass: 386.9 kg
Mission Design and Management: NASA / LaRC
Launch Vehicle: Atlas Agena D (Atlas Agena D no. 17 / Atlas D no. 5801 / Agena D no. AD121/6630)
Launch Date and Time: 10 August 1966 / 19:26:00 UT
Launch Site: Cape Kennedy / Launch Complex 13
Scientific Instruments:
1. imaging system
2. micrometeoroid detectors
3. radiation dosimeters

Results: The Lunar Orbiter program originated as a response to the need to obtain one-meter resolution photographs of potential Apollo landing sites. NASA planned launches of a series of three-axis stabilized spacecraft with four solar panels and a main engine (derived from an Apollo attitude control thruster) for lunar orbit insertion. The primary instrument on board was a 68-kilogram

Eastman Kodak imaging system (using wide and narrow-angle lenses) that could develop exposed film, scan them, and send them back to Earth. In a twist that was not known until after the end of the Cold War, the Eastman Kodak camera flown on the Lunar Orbiters was originally developed by the National Reconnaissance Office (NRO) and flown on the Samos E-1 spy satellite. The narrow-angle pictures taken by this system provided resolution up to 60 to 80 meters, while the wide-angle photos showed resolutions up to 0.5 kilometers. Lunar Orbiter I was launched into a parking orbit around Earth before its Agena upper stage fired at 20:04 UT to insert it on a translunar trajectory. On the way to the Moon, the spacecraft's Canopus star tracker failed to acquire its target, probably because the spacecraft's structure was reflecting too much light. Flight controllers used a backup method by using the same sensor, but with the Moon to orient the vehicle. The vehicle also displayed higher than expected temperatures but successfully entered a 1,866.8 × 189.1-kilometer orbit around the Moon on 24 August, thus becoming the first U.S. spacecraft to do so. The spacecraft's primary mission was to photograph nine potential Apollo landing sites, seven secondary areas, and the Surveyor I landing site. By 28 August, Lunar Orbiter I had completed its main photography mission, having exposed a total of 205 frames, of which 38 were taken in the initial orbit and 167 in lower orbits, covering an area of 5.18 million km². As planned, it photographed all 9 potential Apollo landing sites as well as 11 sites on the far side of the Moon. Some of the high-resolution photos were blurred due to smearing (stemming from problems in the imaging system), but the medium resolution images were the best lunar surface images returned to date. One of the images returned, unplanned but taken on 23 August, was the first picture of Earth taken from the vicinity of the Moon. Lunar Orbiter I began an extended non-photography phase of its mission on 16 September that was focused on engineering goals, but by 28 October, the spacecraft's condition had deteriorated. As such, the day after,

on its 577th orbit, ground controllers commanded the orbiter to crash onto the lunar surface (at 13:30 UT) to prevent its transmissions from interfering with future Lunar Orbiters. Impact coordinates were 6° 42′ N / 162° E.

Pioneer VII

Nation: USA (30)
Objective(s): heliocentric orbit
Spacecraft: Pioneer-B
Spacecraft Mass: 62.75 kg
Mission Design and Management: NASA / ARC
Launch Vehicle: Thrust Augmented Improved Delta (Thor Delta E-1 no. 40 / Thor no. 462/DSV-3E)
Launch Date and Time: 17 August 1966 / 15:20:17 UT
Launch Site: Cape Kennedy / Launch Complex 17A
Scientific Instruments:
1. single-axis fluxgate magnetometer
2. plasma Faraday cup
3. electrostatic analyzer
4. cosmic ray telescope
5. cosmic ray anisotropy detector
6. two-frequency beacon receiver

Results: Identical to Pioneer VI, Pioneer VII was put into heliocentric orbit at 1.01 × 1.125 AU with a period of 402.95 days to study magnetic fields of solar origin, measure various characteristics of the solar wind, and distinguish between solar and galactic cosmic rays. On 17 August 1966, Pioneer VII flew through Earth's magnetic tail region at 5.6 million kilometer range from Earth, and discovered long periods when the solar wind was completely or partially blocked out, suggesting that its instruments were monitoring the end of an organized tail region. On 7 September 1968, the spacecraft was correctly aligned with the Sun and Earth to begin studying Earth's magnetic tail. In 1977, 11 years after its launch, Pioneer VII registered the magnetic tail 19.3 million kilometers out, three times further into space than recorded prior. At 23:36 UT, on 20 March 1986, the spacecraft flew within 12.1 million kilometers of Halley's Comet—the closest a U.S. spacecraft approached the comet—and monitored the interaction between the cometary hydrogen tail and the solar wind. Like Pioneer VI and Pioneer VIII, NASA maintained intermittent contact with Pioneer VII in the 1990s, more than 30 years after its mission began (with data returned from its cosmic ray detector and plasma analyzer in February 1991, for example). On 31 March 1995, the plasma analyzer was turned on during 2 hours of contact with the ground, this being the final contact made with the spacecraft.

73

Luna 11

Nation: USSR (43)
Objective(s): lunar orbit
Spacecraft: Ye-6LF (no. 101)
Spacecraft Mass: 1,640 kg
Mission Design and Management: GSMZ imeni Lavochkina
Launch Vehicle: Molniya-M + Blok L (8K78M no. N103-43, also N15000-52)
Launch Date and Time: 24 August 1966 / 08:03:21 UT
Launch Site: NIIP-5 / Site 31/6
Scientific Instruments:
1. gamma-ray spectrometer
2. RMCh-1 meteorite detector
3. SL-1 radiometer for measuring radiation near the Moon
4. RFL-F instrument for detecting x-ray fluorescence
5. Kassiopeya KYa-4 instrument for measuring intensity of longwave radio-radiation
6. US-3 spectro-photometer
7. 2 cameras (high and low-resolution)
8. R-1 gear transmission experiment

Results: This subset of the "second generation" Luna spacecraft, the Ye-6LF, was designed to take the first photographs of the surface of the Moon from lunar orbit. A secondary objective was to obtain data on gravitational anomalies on the Moon (later identified by U.S. researchers as "mascons") also detected by Luna 10. Using the basic Ye-6 bus, a suite of scientific instruments included an imaging system similar to the one used on Zond 3, which was capable of high- and low-resolution imaging and whose lenses faced the direction of the S5.5 main engine. This package replaced the small lander capsule used on the soft-landing flights. The resolution of the photos was reportedly 15 to 20 meters. A technological experiment included testing the efficiency of gear transmission in vacuum for the future Ye-8 lunar rover (which worked successfully for 5 hours in vacuum). Luna 11, launched only two weeks after the U.S. Lunar Orbiter, successfully entered lunar orbit at 21:49 UT on 27 August about 5 minutes earlier than planned. Parameters were 163.5 × 1,193.6 kilometers. Within 3 hours of lunar orbit insertion, the spacecraft was to be stabilized for its imaging mission which would include taking 42 frames taken over a 64-minute session. Due to an off-nominal position of the vehicle, the camera only took images (64 of them) of blank space. Investigators determined that a foreign object had probably been dislodged in the nozzle of one of the attitude control thrusters. The other instruments functioned without fault (although no data was returned by the spectrophotometer) before the mission formally ended on 1 October 1966 after power supply had been depleted.

74

Surveyor II

Nation: USA (31)
Objective(s): lunar soft-landing
Spacecraft: Surveyor-B
Spacecraft Mass: 995.2 kg
Mission Design and Management: NASA / JPL
Launch Vehicle: Atlas Centaur (AC-7 / Atlas D no. 194 / Centaur D)

Launch Date and Time: 20 September 1966 / 12:32:00 UT

Launch Site: Cape Kennedy / Launch Complex 36A

Scientific Instruments:

1. imaging system

Results: Surveyor II, similar in design to its predecessor, was aimed for a lunar soft-landing in Sinus Medii. During the coast to the Moon, at 05:00 UT on 21 September, one of three thrusters failed to ignite for a 9.8-second mid-course correction, and as a result, put the spacecraft into an unwanted spin. Despite as many as 39 repeated attempts to fire the recalcitrant thruster, the engine failed to ignite, and Surveyor II headed to the Moon without proper control. Just 30 seconds after retro-fire ignition at 09:34 UT on 22 September, communications ceased, and the lander crashed on to the surface of the Moon at 5° 30′ N / 12° W, just southeast of Copernicus crater.

75

Luna 12

Nation: USSR (44)

Objective(s): lunar orbit

Spacecraft: Ye-6LF (no. 102)

Spacecraft Mass: 1,640 kg

Mission Design and Management: GSMZ imeni Lavochkina

Launch Vehicle: Molniya-M + Blok L (8K78M no. N103-44, also N15000-53)

Launch Date and Time: 22 October 1966 / 08:42:26 UT

Launch Site: NIIP-5 / Site 31/6

Scientific Instruments:

1. gamma-ray spectrometer
2. RMCh-1 meteorite detector
3. SL-1 radiometer for measuring radiation near the Moon
4. RFL-F instrument for detecting x-ray fluorescence
5. Kassiopeya KYa-4 instrument for measuring intensity of longwave radio-radiation

6. US-3 spectro-photometer
7. 2 cameras (high- and low-resolution)
8. R-1 gear transmission experiment

Results: Luna 12 was launched to complete the mission that Luna 11 had failed to accomplish, i.e., take high resolution photos of the Moon's surface from lunar orbit. The propulsion system, now called S5.5A, was redesigned to account for the failure of Luna 11 but otherwise was almost identical. Luna 12 successfully reached the Moon on 25 October 1966 and entered a 103 × 1,742-kilometer orbit. About 2 hours later, the imaging system was turned on and worked for 64 minutes, returning 28 high resolution and 14 panoramic images. Film was developed, fixed, and dried automatically and scanned for transmission to Earth. The Soviet press released the first photos taken of the surface on 29 October, pictures that showed the Sea of Rains and the Aristarchus crater. Resolution was as high as 15–20 meters. No further photos were released at the time, although apparently 42 total images were obtained. After completing its main imaging mission, Luna 12 was put into a spin-stabilized roll to carry out its scientific mission which was fulfilled quite successfully—the only major failure was of the US-3 spectro-photometer. Contact was finally lost on 19 January 1967 after 302 communications sessions.

76

Lunar Orbiter II

Nation: USA (32)

Objective(s): lunar orbit

Spacecraft: LO-B (Spacecraft 5)

Spacecraft Mass: 385.6 kg

Mission Design and Management: NASA / LaRC

Launch Vehicle: Atlas Agena D (Atlas Agena D no. 18 / Atlas D no. 5802 / Agena D no. AD122/6631)

Launch Date and Time: 6 November 1966 / 23:21:00 UT

Launch Site: Cape Kennedy / Launch Complex 13

Scientific Instruments:

1. imaging system
2. micrometeoroid detectors
3. radiation dosimeters

Results: Lunar Orbiter II's mission was to photograph 13 primary and 17 secondary landings sites for the Apollo program in the northern region of the Moon's near side equatorial area. After a single mid-course correction on the way to the Moon, on 10 November 1966, the spacecraft entered a 196 × 1,850 kilometer orbit around the Moon. After 33 orbits, Lunar Orbiter II was moved to its photographic orbit with a perilune of 49.7 kilometers; on 18 November, it began its photography mission, returning excellent quality medium and high-resolution photographs, including the impact point of Ranger VIII. The spacecraft ended its photography mission on 26 November and transmission of the images was concluded by 7 December, by which time the probe had transmitted back 211 pictures of both the near side and large areas of the farside. These photos covered nearly four million km² of the lunar surface. The high-gain transmitter failed during this time, but did not significantly affect the coverage afforded by the photos. On 23 November, Lunar Orbiter II took perhaps the most memorable photo of any in the series, a spectacular shot looking across the Copernicus crater from an altitude of only 45 kilometers that vividly emphasized the three-dimensional nature of the lunar surface. On 8 December, after the main photographic mission was over, Lunar Orbiter II fired its main engine to

This image was taken on 24 November 1966 by Lunar Orbiter II from an altitude of 45.7 kilometers from the lunar surface. It shows very vividly the striking topography within the crater Copernicus. *Credit: NASA / LOIRP*

change its orbital plane in order to provide tracking data of the Moon's gravitational field over a wider swath. Finally, on 11 October 1967, when attitude control gas was almost depleted, a retro-burn deliberately crashed the spacecraft onto the lunar surface at 4° S / 98° E on the farside to prevent communications interference on future missions.

77

Luna 13

Nation: USSR (45)
Objective(s): lunar soft-landing
Spacecraft: Ye-6M (no. 205)
Spacecraft Mass: c. 1,620 kg
Mission Design and Management: GSMZ imeni Lavochkina
Launch Vehicle: Molniya-M + Blok L (8K78M no. N103-45, also N15000-55)
Launch Date and Time: 21 December 1966 / 10:17:08 UT
Launch Site: NIIP-5 / Site 1/5
Scientific Instruments:
1. Two TV cameras
2. ID-3 instrument to measure heat stream from surface
3. GR-1 penetrometer
4. RP radiation densitometer
5. KS-17MA instrument for corpuscular radiation
6. DS-1 Yastreb instrument to register loads on landing

Results: Luna 13 became the second Soviet spacecraft to successfully soft-land on the surface of the Moon. It began its mission by entering an initial Earth orbit of 223 × 171 kilometers at 51.8° inclination. The Blok L upper stage soon fired to send the spacecraft on a trajectory to the Moon. After a routine course correction on 22 December, Luna 13 began its approach to our only natural satellite. The retro-rocket engine fired at 17:59 UT on 24 December about 70 kilometers above the surface. Within two minutes, by 18:01 UT, Luna 13 was safely on the lunar surface, having landed in the Ocean of Storms between the Krafft and Seleucus craters at 18° 52′ N / 62° 04′ W, some 440 kilometers from Luna 9. The first signal from the probe was received at 18:05:30 UT. Unlike its predecessor, the heavier Luna 13 lander (113 kilograms) carried a suite of scientific instruments in addition to the usual imaging system. A three-axis accelerometer within the pressurized frame of the lander recorded the landing forces during impact to determine the regolith structure down to a depth of 20–30 centimeters. A pair of spring-loaded booms (not carried on Luna 9) capable of extending 1.5 meters beyond the lander, was also deployed. One of these was equipped with the RP radiation densitometer for determining the composition of the lunar surface, and the other with the GR-1 penetrometer to investigate the mechanical strength of the soil. At 18:06 UT, a small solid-propellant motor fired and forced the GR-1 instrument, a 3.5-centimeter diameter titanium-tipped rod, into the lunar surface. The instrument recorded how fast and how far into the soil (4.5 centimeters) the probe penetrated, thus helping scientists gain valuable information for future landers. At the same time, the RP densitometer emitted gamma quanta from a cesium-137 sample into the soil. The resulting scattering was then recorded by three independent pickups to provide an estimate of the density of the soil, found to be about 0.8 g/cm³. In addition, four radiometers recorded infrared radiation from the surface indicating a noon temperature of 117±3°C while a radiation detector indicated that radiation levels would be less than hazardous for humans. The lander returned a total of five panoramas of the lunar surface, showing a terrain smoother than that seen by Luna 9. One of the two cameras (intended to help return stereo images) failed but this did not diminish the quality of the photographs. A fully successful mission concluded with a last communications session between 04:05 UT and 06:13 UT on 28 December when the onboard batteries were exhausted.

78

Lunar Orbiter III

Nation: USA (33)

Objective(s): lunar orbit

Spacecraft: LO-C (Spacecraft 6)

Spacecraft Mass: 385.6 kg

Mission Design and Management: NASA / LaRC

Launch Vehicle: Atlas Agena D (Atlas Agena D no. 20 / Atlas D no. 5803 / Agena D no. AD128/6632)

Launch Date and Time: 5 February 1967 / 01:17:01 UT

Launch Site: Cape Kennedy / Launch Complex 13

Scientific Instruments:

1. imaging system
2. micrometeoroid detectors
3. radiation dosimeters

Results: Lunar Orbiter III was the final Lunar Orbiter mission to study potential Apollo landing sites although its mission was focused on "site confirmation" rather than "site search." For the mission, the spacecraft's orbital inclination was increased to 21° to ensure photography both north and south of the lunar equator. Building on the stereo photography taken by Lunar Orbiter II, the third mission was focused on making two "footprints" of the same area on two successive orbits. Lunar Orbiter III was also designed to obtain precision trajectory information for defining the lunar gravitational field, measure micrometeoroid flux, and measure radiation dosage levels around the Moon. The spacecraft arrived in lunar orbit after a 9 minute 2.5 second engine burn on 8 February 1967 after a single mid-course correction. Initial orbital parameters were 210.2 × 1,801.9 kilometers at 20.93° inclination. About four days later, the spacecraft entered its operational 55 × 1,847-kilometer orbit at 20.9° inclination. At the time, Lunar Orbiter II was still in operation around the Moon, thus providing key experience for NASA's ability to track and communicate with two simultaneous spacecraft around the Moon. Lunar Orbiter III began its photographic mission on 15 February. The spacecraft exposed 211 (out of a possible 212) frames of pictures by the time that imaging concluded on 23 February. Soon, the spacecraft began to "read out" the images to the ground but this activity suddenly stopped on 4 March due to a problem with the film advance mechanism in the readout section of the imaging system. As such, only 182 images were returned—72 others were never read out to the ground. Despite the glitch, Lunar Orbiter wholly fulfilled its original mission objectives, returning images of 15.5 million km² of the near side and 650,000 km² of the farside. One of its images showed the Surveyor II lander on the lunar surface. On 30 August 1967, ground controllers commanded the vehicle to circularize its orbit to 160 kilometers in order to simulate an Apollo trajectory. Later, on 9 October 1967, the probe was intentionally crashed onto the lunar surface at 14° 36′ N / 91° 42′ W. The photographs from the first three Lunar Orbiters allowed NASA scientists to pick eight preliminary landing sites for Apollo by early April 1967, including site 2 in the Sea of Tranquility where Apollo 11 would land and site 5 in the Ocean of Storms where Apollo 12 (and also Surveyor III) would disembark.

Astronaut Alan L. Bean of Apollo 12 inspecting the remains of the Surveyor III craft in November 1969. Surveyor landed on the Moon on 20 April 1967. In the background, the Apollo 12 Lunar Module Intrepid is visible. *Credit: NASA*

79

Surveyor III

Nation: USA (34)

Objective(s): lunar soft-landing

Spacecraft: Surveyor-C

Spacecraft Mass: 997.9 kg

Mission Design and Management: NASA / JPL

Launch Vehicle: Atlas Centaur (AC-12 / Atlas D no. 292 / Centaur D)

Launch Date and Time: 17 April 1967 / 07:05:01 UT

Launch Site: Cape Kennedy / Launch Complex 36B

Scientific Instruments:

1. TV camera
2. surface sampler

Results: Surveyor III was the third engineering flight of the series, but for the first time carried a surface-sampling instrument that could reach up to 1.5 meters from the lander and dig up to 18 centimeters deep. Unlike the previous Surveyors, Surveyor III began its mission from parking orbit around Earth with a burn from the Centaur upper stage, now capable of multiple firings. During the descent to the lunar surface, highly reflective rocks confused the lander's descent radar, and the main engine failed to cut off at the correct moment at about 4.3-meters altitude. As a result, Surveyor III bounced off the lunar surface twice, the first time to a height of 10 meters and the second time, to 3 meters. The third time, the lander settled down to a soft-landing at 00:04:17 UT on 20 April 1967 in the south-eastern region of Oceanus Procellarum at 3.0° S / 23.41° W. Less than an hour after landing, the spacecraft began transmitting the first of 6,326 TV pictures of the surrounding areas. The most exciting experiments of the mission included deployment of the surface sampler for digging trenches, making bearing tests, and manipulating lunar material in the view of the TV system. Via commands from Earth, it dug four trenches, and performed four bearing tests and thirteen impact tests. Based on these experiments, scientists concluded that lunar soil had a consistency similar to wet sand, with a bearing strength of 0.7 kilograms/cm², i.e., solid enough for an Apollo Lunar Module. The lander's TV camera consisted of two 25 and 100 mm focal length lenses and was mounted under a mirror that could be moved horizontally and vertically. It took about 20 seconds to transmit a single 200-line picture of the surface. The camera was also capable of 600-line images which used digital picture transmission. Last contact was made on 4 May 1967, two days after the lunar night began. More than three years later, on 18 November 1969, Apollo 12 astronauts Charles Conrad, Jr. and Alan L. Bean landed the Intrepid Lunar Module approximately 180 meters from the inactive Surveyor III lander. During their second extra-vehicular activity (EVA) on 19 November, the astronauts walked over to Surveyor III and recovered parts, including the soil scoop and camera system, to allow scientists to evaluate the effects of nearly two-and-a-half years of exposure on the Moon's surface.

80

Lunar Orbiter IV

Nation: USA (35)
Objective(s): lunar orbit
Spacecraft: LO-D (Spacecraft 7)
Spacecraft Mass: 385.6 kg
Mission Design and Management: NASA / LaRC
Launch Vehicle: Atlas Agena D (Atlas Agena D no. 22 / Atlas D no. 5804 / Agena D no. AD131/6633)
Launch Date and Time: 4 May 1967 / 22:25:00 UT
Launch Site: Cape Kennedy / Launch Complex 13
Scientific Instruments:

1. imaging system
2. micrometeoroid detectors
3. radiation dosimeters

Results: Lunar Orbiter IV was the first in the series dedicated to scientific surveys of the Moon. Its goal was to acquire contiguous photographic coverage of the lunar surface of at least 80% of the near side at 50–100-meter resolution. After a midcourse correction on 5 May 1967, Lunar Orbiter IV fired its engine at 15:08 UT on 8 May to insert the spacecraft into an initial lunar polar orbit of 6,111 × 2,706 kilometers at 85.5° inclination, thus becoming the first spacecraft to go into polar orbit around the Moon. Orbital period was about 12 hours. The spacecraft began its photographic mission at 15:46 UT on 11 May. A potentially serious problem threatened the mission when on 13 May, controllers found a problem with a camera thermal door that failed to close, leaking light onto exposed images. They were able to devise a fix that worked and the spacecraft continued its imaging mission. During its two-month mission, Lunar Orbiter IV took pictures of 99% of the near side and 75% of the farside of the Moon in a total of 163 frames. The imaging mission ended on the orbiter's 34th orbit due to worsening readout difficulties. Fortunately, all but 30 of the 163 images collected, many with a resolution down to 60 meters, were successfully transmitted to Earth by 1 June. In early June, controllers lowered the spacecraft's orbit to match that of Lunar Orbiter V so that scientists could collect gravitational data in support of the latter mission. Before losing contact on 17 July, Lunar Orbiter IV took the first photos of the lunar south pole and discovered a 240-kilometer long crustal fault on the farside. Since contact was lost before controlled impact, the spacecraft naturally crashed onto the Moon on 6 October 1967 due to gravitational anomalies.

81

Kosmos 159 [Luna]

Nation: USSR (46)
Objective(s): highly-elliptical orbit around Earth
Spacecraft: Ye-6LS (no. 111)
Spacecraft Mass: 1,640 kg
Mission Design and Management: GSMZ imeni Lavochkina
Launch Vehicle: Molniya-M + Blok L (8K78M no. Ya716-56, also N15000-56)
Launch Date and Time: 16 May 1967 / 21:43:57 UT
Launch Site: NIIP-5 / Site 1/5
Scientific Instruments: [none]
Results: This spacecraft was a one-off high apogee Earth satellite developed to acquire data on new telecommunications systems for upcoming crewed missions to the Moon. Besides a usual complement of telemetry and communications equipment, the vehicle also carried a transceiver as part of the long-range communications system (*Dal'nyy radiokompleks*, DRK) and the BR-9-7 telemetry system, equipment designed to work with the new Saturn-MS-DRK ground station located near the village of Saburovo, about 10 kilometers from NIP-14, a station, close to Moscow, belonging to the Soviet ground-based tracking network. The spacecraft was similar to Luna 11 but had a slightly lengthened (by 15 cm) instrument container so as to accommodate the modified DRK and new BR-9-7

telemetry systems. Mission designers had planned to send the probe into a highly elliptical orbit with an apogee of 250,000 kilometers, but the Blok L upper stage cut off too early. Instead, the spacecraft, named Kosmos 159, entered a lower orbit of 380 × 60,600 kilometers at 51.5° inclination. Despite the incorrect orbit, controllers were able to accomplish the original mission, carried out over a period of nine days during which it was discovered that the energy potential of the UHF downlink from the spacecraft to the ground was 1–2 orders magnitude below the calculated value. Kosmos 159 reentered Earth's atmosphere on 11 November 1967.

82

Venera 4

Nation: USSR (47)
Objective(s): Venus impact
Spacecraft: V-67 (1V no. 310)
Spacecraft Mass: 1,106 kg
Mission Design and Management: GSMZ imeni Lavochkina
Launch Vehicle: Molniya-M + Blok VL (8K78M no. Ya716-70, also Ya15000-70)
Launch Date and Time: 12 June 1967 / 02:39:45 UT
Launch Site: NIIP-5 / Site 1/5

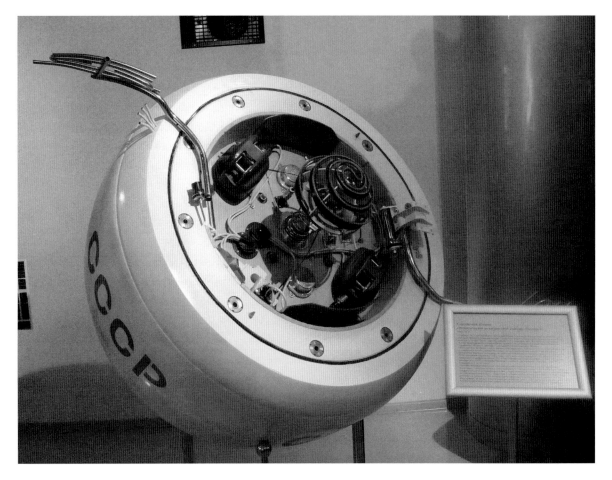

A model of the 1-meter diameter landing capsule of Venera 4 on display at the Memorial Museum of Cosmonautics in Moscow. *Credit: Asif Siddiqi*

Scientific Instruments:

Lander:

1. MDDA altimeter
2. G-8 and G-10 gas analyzers
3. TPV densitometer and thermometer
4. barometer

Bus:

1. SG-59M triaxial magnetometer
2. 4 ion traps
3. STS-5 gas discharge counter
4. radiation detectors
5. SBT-9 gas discharge counter
6. KS-18-2M cosmic ray particle counters
7. LA-2 spectrometer

Results: Venera 4 was the first spacecraft to transmit data from a planet's atmosphere. It was also the first Venus probe built by the Lavochkin design bureau, although Lavochkin engineers retained the basic design layout of the earlier 3MV probes built under Chief Designer Sergey Korolev (1907–1966). The spacecraft consisted of a main bus about 3.5 meters high and a 383-kilogram lander probe designed to transmit data as it descended through the Venusian atmosphere. This capsule could endure loads as high as 300 g's and land on both land and liquid. For atmospheric entry it was equipped with a thick ablative heatshield. Launch, for the first time, used an upgraded Blok L (4th stage), the Blok VL. After a mid-course correction on 29 July 1967, Venera 4 approached Venus on 18 October. About 2 hours before arrival at Venus, at a distance of 45,000 kilometers, on command from Earth, the spacecraft was properly oriented for entry. The bus released the lander at 04:34 UT, and the latter entered the Venusian atmosphere at a velocity of 11 kilometers/second. The bus returned data for some time before being destroyed in the atmosphere. The lander, meanwhile experienced as much as 300 g's and reduced its speed to 210 meters/second at which point the parachute system was deployed. Simultaneously, the lander began to transmit information back to Earth. Because the altimeter was designed to record pressures only up to 7.3 atmospheres, it went off-scale rather quickly. Temperature measurements (from 33°C to 262°C) continued to come back for 93 minutes as the probe slowly descended through the atmosphere. Initially, Soviet scientists believed that the probe transmitted until contact with the surface, announcing that "a calculation of the rate of descent of the station until termination of transmission of data indicates that it continued transmitting until it actually touched the surface of the planet." Later analysis of data showed that transmissions ceased at an altitude of 28 kilometers when the high atmospheric pressure and temperatures damaged the probe. The inert probe impacted the surface near 19° N / 36° E. The data implied that surface temperatures and pressure were 500°C and 75 atmospheres respectively. Venera 4's gas analyzers also found that the planet's atmosphere was composed of 90–95% carbon dioxide (with a sensitivity of ±7%) with no nitrogen, which had previously been assumed would constitute most of the atmosphere. Data from the ionizing densitometer showed that cloud cover in the Venusian atmosphere exists at altitudes below 52 kilometers with the lower boundary at roughly 35 kilometers. The spacecraft bus measured the planet's weak magnetic field and found no ring of radiation belts. It detected a very weak atmosphere of atomic hydrogen about 9,900 kilometers above the planet. Noteworthy was Venera 4's detection of the planet's bow shock, identified a day before by Mariner V (and later confirmed by Venera 5). The mission's importance was underscored when NASA Administrator James E. Webb (1906–1992) issued a statement on 18 October 1967, noting that the landing "represents an accomplishment any nation can be proud of." Researchers at NASA Ames Research Center were particularly interested in data from Venera 4, especially the effect of planetary atmosphere on the propagation of radio signals from the spacecraft, in anticipation of future NASA missions to Venus.

Mariner V

Nation: USA (36)

Objective(s): Venus flyby

Spacecraft: Mariner-67E / Mariner-E

Spacecraft Mass: 244.9 kg

Mission Design and Management: NASA / JPL

Launch Vehicle: Atlas Agena D (Atlas Agena D no. 23 / Atlas D no. 5401 / Agena D no. AD157/6933)

Launch Date and Time: 14 June 1967 / 06:01:00 UT

Launch Site: Cape Kennedy / Launch Complex 12

Scientific Instruments:

1. ultraviolet photometer
2. S-band occultation experiment
3. dual frequency occultation experiment
4. solar plasma probe
5. magnetometer
6. trapped radiation detector
7. celestial mechanics experiment

Results: In December 1965, NASA approved a project to modify the Mariner IV backup spacecraft to conduct a closer flyby of Venus than the only other NASA probe to fly past Venus, Mariner II. The primary goal of the mission was to conduct a radio-occultation experiment (much like Mariner IV at Mars) to determine atmospheric properties of Venus. Unlike Mariner IV, however, Mariner V did not carry an imaging instrument. Initially, NASA had planned to send Mariner V on a flyby at a miss distance of 8,165 kilometers to the surface, but the Agency altered its plan in favor of a more modest 75,000-kilometer flyby in order to preclude the non-sterilized vehicle from crashing into the planet. After a mid-course correction on 19 June, Mariner V began transmitting data on Venus on 19 October during its encounter. Closest approach was at 17:34:56 UT at a range of 4,094 kilometers, much closer than expected due to the course correction. Mariner V found that although Venus does not have a magnetic field, the dense daylight ionosphere produces a bow shock that deflects the solar wind around the planet. The ultraviolet photometer detected a hydrogen corona (as found by the Soviet Venera 4) but no oxygen emission. Mariner V's instruments indicated that the planet's surface temperature and pressure were 527°C and 75 to 100 atmospheres, respectively—which countered the Soviet claim that its Venera 4 spacecraft had managed to transmit from the planet's surface. The encounter with Venus deflected the spacecraft towards the Sun, and Mariner V entered solar orbit with parameters ranging from 0.579 AU and 0.735 AU. On 4 December 1967, NASA lost contact with the spacecraft although controllers briefly regained contact on 14 October 1968. The spacecraft did not transmit any further telemetry and NASA eventually stopped attempts (on 5 November 1968) to communicate with the vehicle, now in heliocentric orbit. Planetary scientists reviewed data from both Mariner V and the Soviet Venera 4 at a conference at Kitt Peak National Observatory in March 1968, one of the first major international meetings to discuss the results of planetary exploration. Scientists concluded that neither Mariner V or Venera 4 had been entirely successful in communicating data about conditions at the planet's surface.

Kosmos 167 [Venera]

Nation: USSR (48)

Objective(s): Venus impact

Spacecraft: V-67 (1V no. 311)

Spacecraft Mass: c. 1,100 kg

Mission Design and Management: GSMZ imeni Lavochkina

Launch Vehicle: Molniya-M + Blok VL (8K78M no. Ya716-71, also Ya15000-71)

Launch Date and Time: 17 June 1967 / 02:36:38 UT

Launch Site: NIIP-5 / Site 1/5

Scientific Instruments:

Lander:

1. MDDA altimeter
2. G-8 and G-10 gas analyzers
3. TPV densitometer and thermometer
4. barometer

Bus:

1. SG-59M triaxial magnetometer
2. 4 ion traps
3. STS-5 gas discharge counter
4. radiation detectors
5. SBT-9 gas discharge counter
6. KS-18-2M cosmic ray particle counters
7. LA-2 spectrometer

Results: This identical twin craft to Venera 4 failed to leave Earth orbit when its Blok VL trans-interplanetary stage failed to fire, because the engine's turbopump had not been cooled prior to ignition. The spacecraft remained stranded in Earth orbit and reentered on 25 June 1967.

85

Surveyor IV

Nation: USA (37)

Objective(s): lunar soft-landing

Spacecraft: Surveyor-D

Spacecraft Mass: 1,037.4 kg

Mission Design and Management: NASA / JPL

Launch Vehicle: Atlas Centaur (AC-11 / Atlas D no. 291 / Centaur D)

Launch Date and Time: 14 July 1967 / 11:53:29 UT

Launch Site: Cape Kennedy / Launch Complex 36A

Scientific Instruments:

1. TV camera
2. surface sampler

Results: Like Surveyor III, Surveyor IV was equipped with a surface claw (with a magnet in the claw) to detect and measure ferrous elements in the lunar surface. The mission appeared successful until all communications were abruptly lost 2 seconds prior to retro-rocket cutoff at 02:03 UT on 17 July

1967 just two-and-a-half minutes before landing on the Moon. The landing target was Sinus Medii (Central Bay) at 0.4° N / 1.33° W. NASA believed that the solid propellant retro-rocket might have exploded, destroying the vehicle.

86

Explorer XXXV / Anchored International Monitoring Platform 6

Nation: USA (38)

Objective(s): lunar orbit

Spacecraft: AIMP-E

Spacecraft Mass: 104.3 kg

Mission Design and Management: NASA / GSFC

Launch Vehicle: Thor Delta E-1 (Thor Delta E-1 no. 50 / Thor no. 488/DSV-3E)

Launch Date and Time: 19 July 1967 / 14:19:02 UT

Launch Site: Cape Kennedy / Launch Complex 17B

Scientific Instruments:

1. magnetometers
2. thermal ion detector
3. ion chambers and Geiger tubes
4. Geiger tubes and p-on-n junction
5. micrometeoroid detector
6. Faraday cup

Results: Explorer XXXV (35) also known as the Anchored Interplanetary Monitoring Platform (AIMP-E or AIMP-6) was designed to study interplanetary space phenomena, particularly the solar wind, the interplanetary magnetic field, dust distribution near the Moon, the lunar gravitational field, the weak lunar ionosphere, and the radiation environment. The spacecraft left Earth on a direct ascent trajectory and entered lunar orbit on 21 July 1967 after a 23-second burn. The main engine separated 2 hours later, the spacecraft having entered an initial elliptical orbit of 800 × 7,692 kilometers at 147° inclination. The spacecraft, similar to Explorer XXXIII, which had failed to achieve lunar orbit, found that the Moon has no magnetosphere, that solar wind particles impact directly onto the surface, and that the Moon creates a "cavity" in the solar wind stream. After six years of successful operation, the satellite was turned off on 24 June 1973. The lunar satellite later impacted on the surface, although the precise location remains unknown. Explorer XXXV was launched by the fiftieth Thor Delta booster, of which only three had failed to date, giving it a success rating of 94%.

87

Lunar Orbiter V

Nation: USA (39)
Objective(s): lunar orbit
Spacecraft: LO-E (Spacecraft 3)
Spacecraft Mass: 385.6 kg
Mission Design and Management: NASA / LaRC
Launch Vehicle: Atlas Agena D (Atlas Agena D no. 24 / Atlas D no. 5805 / Agena D no. AD159/6634)
Launch Date and Time: 1 August 1967 / 22:33:00 UT
Launch Site: Cape Kennedy / Launch Complex 13
Scientific Instruments:

1. imaging system
2. micrometeoroid detectors
3. radiation dosimeters

Results: Lunar Orbiter V was the last in a series of highly successful missions to map the Moon for potential landing sites and conduct general observational surveys. This last mission's objectives were both photographic (the primary mission) and non-photographic (the secondary). The former involved taking additional pictures of Apollo sites, broad surveys of unphotographed areas of the farside, imaging the Surveyor landing sites, and photographic areas of scientific value. The secondary goals included acquisition of precise trajectory information for improving the definition of the lunar gravitational field, measurement of the micrometeoroid flux and radiation dose in the lunar environment, and helping to prepare the Manned Space Flight Network for the Apollo missions.

After a single mid-course correction on 3 August, Lunar Orbiter V entered lunar polar orbit two days later after an engine firing at 16:48 UT that lasted 8 minutes, 28 seconds. Initial orbital parameters were 194.5 × 6,023 kilometers at 85.01° inclination. The orbital period was 8.5 hours. Lunar Orbiter V began its photography mission at 01:22 UT on 7 August before executing a maneuver to bring it to its operational orbit at 100 × 6,023 kilometers. The spacecraft photographed 36 different areas on the near side and mapped most of the farside via a set of 212 frames until the photography mission ended on 18 August. These included five potential Apollo landing sites, 36 science sites, and 23 previously unphotographed areas of the farside, as well possible targets for future Surveyor missions. Controllers also extensively used the spacecraft to map the Moon's gravitational field in order to predict orbital perturbations on future lunar orbital missions. The probe also obtained a spectacular high-quality photo of Earth showing Africa, the Middle East, and parts of Africa at 09:05 UT on 8 August 1967. A further orbital change on 9 August brought the orbit down to 1,499.37 × 98.93 kilometers at 84.76° inclination. Lunar Orbiter V was commanded to impact on the lunar surface on 31 January 1968, which it did at 0° N / 70° W. In total, the five Lunar Orbiters photographed 99% of the lunar surface. Perhaps the most important finding credited to data from the Lunar Orbiters (and Lunar Orbiter V, in particular) is the discovery of "mascons" or lunar mass concentrations under the lunar ringed maria, first published by JPL researchers P. M. Muller and W. L. Sjogren in mid-1968.

88

Surveyor V

Nation: USA (40)
Objective(s): lunar soft-landing
Spacecraft: Surveyor-E
Spacecraft Mass: 1,006 kg

Mission Design and Management: NASA JPL

Launch Vehicle: Atlas Centaur (AC-13 / Atlas 3C no. 5901C / Centaur D-1A)

Launch Date and Time: 8 September 1967 / 07:57:01 UT

Launch Site: Cape Kennedy / Launch Complex 36B

Scientific Instruments:

1. TV camera
2. alpha-scattering instrument
3. footpad magnet

Results: Surveyor V was similar to its predecessor but the surface sampler from the earlier vehicle was replaced by an alpha-backscatter instrument to determine the relative abundance of the chemical elements in lunar material. In addition, a small bar magnet was installed on one of the lander's footpads to indicate whether the lunar soil had magnetic properties. Overcoming a near-fatal helium leak in a pressure regulator, engineers from JPL and Hughes Aircraft Company (the prime contractor for the spacecraft) expertly managed to safely land Surveyor V on the surface of the Moon in the south-eastern region of Mare Tranquilitatis at 1.42° N / 23.20° E at 00:46:42 UT on 11 September 1967. The malfunction put the lander about 29 kilometers away from its target in an angular incline within the slope of rimless crater. Surveyor V was, however, the most successful of the series so far. The lander returned 18,006 photos before lunar night descended on 24 September. Controllers successfully commanded the vehicle to take a further 1,000 photographs during the second lunar day between 15 and 24 October 1967 and the fourth lunar day in December. In total, 20,018 pictures were transmitted. In a new experiment, on 13 September controllers fired the main engine for 0.55 seconds to examine the effects of disturbing the lunar surface. NASA announced that no new craters were created, nor was there any significant dust cloud due to the firing. The alpha-scattering instrument had earlier been activated and found the soil to be composed of more than half oxygen with various amounts of silicon and aluminum. Contact was lost with the lander on 16 December 1967.

89

[Zond, 7K-L1 no. 4L]

Nation: USSR (49)

Objective(s): circumlunar flight

Spacecraft: 7K-L1 (no. 4L)

Spacecraft Mass: c. 5,375 kg

Mission Design and Management: TsKBEM

Launch Vehicle: Proton-K + Blok D (8K82K no. 229-01 / 11S824 no. 12L)

Launch Date and Time: 27 September 1967 / 22:11:54 UT

Launch Site: NIIP-5 / Site 81/23

Scientific Instruments: [unknown]

Results: This spacecraft, a 7K-L1 type, was the first of a series of spacecraft that the Soviets tried to send on circumlunar missions as part of a larger project to send two cosmonauts around the Moon. The program, which was officially approved in October 1965, involved combining forces between two of the leading space organizations in the Soviet Union, those headed by Vasiliy Mishin (1917–2001) (who provided the 7K-L1 spaceship and the Blok D translunar injection stage) and Vladimir Chelomey (1914–1984) (who provided the 3-stage Proton-K launch vehicle). The 7K-L1 spacecraft was a stripped-down version of the larger 7K-OK Soyuz spacecraft intended for Earth-orbital operations. The proximate goal was to send two cosmonauts around the Moon by the fiftieth anniversary of the Bolshevik Revolution, in November 1967. The project moved into a new stage with two technological flights in Earth orbit in March and April 1967, which involved high-speed reentries. During this particular launch, the ascent was steady for 61 seconds before diverting from a nominal path, which activated the emergency rescue system into action. The 7K-LI descent module immediately separated from the wandering launch vehicle, and although the capsule was destabilized at the moment of separation because of an unexpected pressure shock, it landed safely in one piece not far from wreckage of the booster which was destroyed at T+97.4 seconds.

90

Surveyor VI

Nation: USA (41)

Objective(s): lunar soft-landing

Spacecraft: Surveyor-F

Spacecraft Mass: 1,008.3 kg

Mission Design and Management: NASA / JPL

Launch Vehicle: Atlas Centaur (AC-14 / Atlas 3C no. 5902C / Centaur D-1A)

Launch Date and Time: 7 November 1967 / 07:39:01 UT

Launch Site: Cape Kennedy / Launch Complex 36B

Scientific Instruments:

1. TV camera
2. alpha-scattering instrument
3. footpad magnets

Results: Surveyor VI landed safely on the Moon at 01:01:04 UT on 10 November 1967 in the Sinus Medii (Central Bay) at coordinates announced as 0.46° N / 1.37° W (but probably closer to 0.473° N / 1.427° W). The spacecraft returned 29,952 images of the lunar surface during less than two weeks of operation before the onset of lunar night on 24 November, because of which the spacecraft was placed in hibernation mode on 26 November. During its initial operations, the alpha-scattering instrument acquired about 30 hours of data on the chemical composition of the lunar surface. Although controllers regained contact briefly on 14 December 1967, primary landing operations had ceased by this time. Before termination of operations, on 17 November 1967, Surveyor VI was commanded to fire its three main liquid propellant thrusters for 2.5 seconds. As a result, the lander became the first spacecraft to be launched from the lunar surface. Surveyor VI lifted up to about three meters before landing two-and-a-half meters west of its original landing point. Cameras then studied the original landing footprints in order to determine the soil's mechanical properties and also accomplish some stereo imaging, now that the source point had been displaced. Surveyor VI also sent back pictures of the bar magnet at the footpad allowing investigators to determine the concentration of magnetic material in the lunar surface.

91

[Zond, 7K-L1 no. 5L]

Nation: USSR (50)

Objective(s): circumlunar flight

Spacecraft: 7K-L1 (no. 5L)

Spacecraft Mass: c. 5,375 kg

Mission Design and Management: TsKBEM

Launch Vehicle: Proton-K + Blok D (8K82K no. 230-01 / 11S824 no. 13L)

Launch Date and Time: 22 November 1967 / 19:07:59 UT

Launch Site: NIIP-5 / Site 81/24

Scientific Instruments: [unknown]

Results: This was the second Soviet attempt at a robotic circumlunar mission as part of a larger project to send cosmonauts around the Moon. On this launch, one of the four second stage engines of the Proton-K rocket failed to ignite at T+125.5 seconds due to a break in the engine nozzle. The wayward booster was then destroyed on command from the ground at T+129.9 seconds. Once again, the emergency rescue system was activated, shooting the 7K-L1 descent module away from the rocket. The capsule landed by parachute about 80 kilometers southwest of the town of Dzhezkazgan. The actual impact was a hard one because of a spurious command from the altimeter which fired the capsule's soft-landing engines at an altitude of 4.5 kilometers instead of a few meters above the ground.

92

Pioneer VIII

Nation: USA (42)

Objective(s): heliocentric orbit

Spacecraft: Pioneer-C

Spacecraft Mass: 65.36 kg

Mission Design and Management: NASA ARC

Launch Vehicle: Thrust-Augmented Thor-Delta (Thor Delta E-1 no. 55 / Thor no. 489 / DSV-3E)

Launch Date and Time: 13 December 1967 / 14:08 UT

Launch Site: Cape Kennedy / Launch Complex 17B

Scientific Instruments:

1. single-axis fluxgate magnetometer
2. plasma analyzer
3. cosmic ray telescope
4. radio-wave propagation experiment
5. cosmic ray gradient detector
6. electric field detector
7. cosmic dust detector
8. celestial mechanics experiment

Results: Pioneer VIII, like its two predecessors, was sent into heliocentric orbit to study interplanetary space, particularly to carry collect information on magnetic fields, plasma, and cosmic rays for two or more passages of solar activity. Although the spacecraft carried a different complement of scientific instruments than Pioneers VI and VII, its findings were correlated with the other two probes. The spacecraft was launched into a path ahead of Earth to provide the vehicle with added velocity in solar orbit to move out beyond Earth's orbit at 1.0080 × 0.9892. AU. Pioneer VIII arrived at Earth's magnetospheric bounds at 19:00 UT on 15 December 1967. Later, on 18 January 1968, the probe, the Sun, and Earth were perfectly aligned to allow investigation of Earth's magnetic tail in detail, first performed by Pioneer VII in 1968. Already by June 1968, data from the probe led scientists to speculate that Earth's magnetic tail might be shorter than the 320 million kilometers suggested by theoretical calculations. In August of the same year, NASA scientists at the Deep Space Network (DSN) announced that they had managed to quadruple the distance over which signals from the three solar Pioneers so far launched could be heard. These improvements were enabled by changes in the DSN receivers. In October 1982, Pioneers VIII and IX came within 2.4 million kilometers of each other, an "encounter" used to recalibrate Pioneer VIII's damaged plasma instrument. Controllers intermittently maintained contact with the spacecraft for nearly 30 years although only one instrument, the electric field detector, remained operational past 1982. During tracking on 23 July 1995, NASA was unable to switch on Pioneer VIII's transmitter, probably because the spacecraft was too far away from the Sun to charge the solar panels. On 22 August 1996, contact was reacquired via a backup transmitter. Although there were no further plans to contact the spacecraft, the on-board electric field detector remained at least hypothetically functional in 2015, nearly 48 years after launch. The original launch vehicle also carried a second payload, the Tts I (Test and Training Satellite I, later renamed Tetr I), fixed to the second stage, which was successfully ejected after the third stage finished firing, and entered an Earth orbit of 488 × 301 kilometers at 33° inclination. It reentered on 14 May 1968.

1968

Surveyor VII

Nation: USA (43)

Objective(s): lunar soft-landing

Spacecraft: Surveyor-G

Spacecraft Mass: 1,040.1 kg

Mission Design and Management: NASA / JPL

Launch Vehicle: Atlas Centaur (AC-15 / Atlas 3C no. 5903C / Centaur D-1A)

Launch Date and Time: 7 January 1968 / 06:30:00 UT

Launch Site: Cape Kennedy / Launch Complex 36A

Scientific Instruments:

1. TV camera
2. alpha-scattering instrument
3. surface sampler
4. footpad magnets
5. stereoscopic and dust detection mirrors

Results: Since Surveyors I, III, V, and VI successfully fulfilled requirements in support of Apollo, NASA opted to use the last remaining Surveyor for a purely scientific mission outside of a potential landing site for the early Apollo flights. After an uneventful coast to the Moon (including one of two planned mid-course corrections), Surveyor VII successfully set down at 01:05:36.3 UT on 10 January 1968 on the ejecta blanket emanating from the bright Tycho crater in the south of the nearside. Landing coordinates were 40.97° S / 11.44° W, about 29 kilometers north of Tycho's rim, and 2.4 kilometers from its target. Initial photos from the surface showed surprisingly few craters, much like the mare sites, although the general area was rougher. About 21 hours after landing, ground controllers fired pyrotechnic charges to drop the alpha-scattering instrument on the lunar surface. When the instrument failed to move beyond an intermediate position, controllers used the surface sampler (robot arm) to force it down. The sampler was then used to pick up the alpha-scattering instrument after its first chemical analysis and move it to two additional locations. About 66 hours of alpha-scattering data were obtained during the first lunar day on three samples: the undisturbed lunar surface, a Moon rock, and an area dug up by the surface sampler. The alpha-scattering instrument collected 34 more hours of data during the second lunar day. The scoop on the sampler's arm was used numerous times for picking up soil and digging trenches, and for conducting at least 16 surface-bearing tests. Apart from taking 21,274 photographs (20,993 on the first lunar day and 45 during the second), Surveyor VII also served as a target for Earth-based lasers (of 1 watt power) to accurately measure the distance between Earth and the Moon. Communications were paused with Surveyor VII at 14:12 UT on 26 January 1968, about 80 hours after sunset. Second lunar day operations began at 19:01 UT on 12 February 1968, and extended to 00:24 UT on 21 February effectively ending the mission. In total, the five successful Surveyors returned more than 87,000 photos of the lunar surface and operated for about 17 months total on the lunar surface, and most important, demonstrated the feasibility of soft-landing a spacecraft on the lunar surface. Originally planned as a 7-spacecraft series, in 1963, NASA added 3 more missions (Surveyors H, I, and J) for a total of 10 missions. These later Surveyors, each weighing 1,134 kilograms (as opposed to the 998-kilogram model), were, however, canceled in December 1966 based on the successful results of the Ranger missions, Lunar Orbiters, and Surveyors already launched by then.

94

[Luna, Ye-6LS no. 112]

Nation: USSR (51)

Objective(s): lunar orbit

Spacecraft: Ye-6LS (no. 112)

Spacecraft Mass: 1,640 kg

Mission Design and Management: GSMZ imeni Lavochkina

Launch Vehicle: Molniya-M + Blok L (8K78M no. Ya716-57, also Ya15000-57)

Launch Date and Time: 7 February 1968 / 10:43:54 UT

Launch Site: NIIP-5 / Site 1/5

Scientific Instruments:

1. SL-2 radiometer
2. IK-2 dosimeter

Results: During launch to Earth orbit, the third stage (Blok I) engine cut off prematurely at T+524.6 seconds because of an excessive propellant consumption rate via the gas generator. The spacecraft never reached Earth orbit. The goal of the mission was to test communications systems in support of the N1-L3 human lunar landing program. (For more details, see Kosmos 159 on p. 67).

95

Zond 4

Nation: USSR (52)

Objective(s): deep space mission

Spacecraft: 7K-L1 (no. 6L)

Spacecraft Mass: c. 5,375 kg

Mission Design and Management: TsKBEM

Launch Vehicle: Proton-K + Blok D (8K82K no. 231-01 + 11S824 no. 14L)

Launch Date and Time: 2 March 1968 / 18:29:23 UT

Launch Site: NIIP-5 / Site 81/23

Scientific Instruments: [unknown]

Results: The Soviets decided to send this next 7K-L1 spacecraft not on a circumlunar flight, but to about 330,000 kilometers into deep space in the opposite direction of the Moon in order to test the main spacecraft systems without the perturbing effects of the Moon, much like the Surveyor model test flights in 1965–1966. After returning from its high apogee, the spacecraft would carry out a high-speed reentry into Earth's atmosphere and be recovered and investigated for the effects of reentry. After launch into Earth orbit, at T+71 minutes and 56 seconds, the Blok D upper stage fired a second time (for 459 seconds) to send the 7K-L1 on a highly elliptical Earth orbit with an apogee of 354,000 kilometers. The Soviet news agency TASS publicly named the spacecraft "Zond 4," thus connecting the mission with a series of completely unrelated deep space probes—a typically obfuscating maneuver from the Soviet media. Controllers were unable to carry out a mid-course correction at 04:53 UT on 4 March when a star sensor (the 100K) of the attitude control system failed. A second attempt also failed the following day and the main omni-directional antenna also did not deploy fully, compromising communications. A third attempt at a correction proved successful, by using a special filter on the sensor to read signals accurately. After the spacecraft separated into its two constituent parts, however, the descent module was unable to maintain a stable and proper attitude for a guided reentry, instead moving into a ballistic reentry trajectory, entering the atmosphere at 18:18:58 UT on 9 March. A crew on board would have experienced about 20 g's but probably would have survived. However, because the descent module was falling into an unanticipated area, to prevent "foreign" observers from recovering the wayward spacecraft, an automatic emergency destruct system destroyed the returning capsule at an altitude of 10–15 kilometers over the Gulf of Guinea. For some years, the official Soviet press claimed that Zond 4 had entered heliocentric orbit.

96

Luna 14

Nation: USSR (53)
Objective(s): lunar orbit
Spacecraft: Ye-6LS (no. 113)
Spacecraft Mass: 1,640 kg
Mission Design and Management: GSMZ imeni Lavochkina
Launch Vehicle: Molniya-M + Blok L (8K78M no. Ya716-58, also Ya15000-58)
Launch Date and Time: 7 April 1968 / 10:09:32 UT
Launch Site: NIIP-5 / Site 1/5
Scientific Instruments:
1. SL-2 radiometer
2. IK-2 dosimeter

Results: Luna 14 successfully entered lunar orbit at 1925 UT on 10 April 1968. Initial orbital parameters were 160 × 270 kilometers at 42° inclination. The primary goal of the flight, like its predecessors (Kosmos 159 and the launch failure in February 1968), was to test communications systems in support of the N1-L3 piloted lunar landing project. In addition, ground tracking of the spacecraft's orbit allowed controllers to accurately map lunar gravitational anomalies to predict future trajectories of future lunar missions such as those of the LOK and LK lunar landing vehicles. Luna 14 also carried scientific instruments to measure radiation near the Moon, a "tissue dosimeter for studying doses of ionizing radiation," as well as technical elements of the future Ye-8 lunar rover, in particular, the R-1-I, R-1-II, and R-1-III gear pairs (both steel and ceramic), various types of ball bearings with lubrication to test drives, and the M-1 drive shaft. The mission was slated to last 30 days but spanned 75 days; the entire program was fulfilled as planned, and allowed designers to make the final selection for elements of the Ye-8 lunar rover chassis.

97

[Zond, 7K-L1 no. 7L]

Nation: USSR (54)
Objective(s): circumlunar flight
Spacecraft: 7K-L1 (no. 7L)
Spacecraft Mass: c. 5,375 kg
Mission Design and Management: TsKBEM
Launch Vehicle: Proton-K + Blok D (8K82K no. 232-01 + 11S824 no. 15L)
Launch Date and Time: 22 April 1968 / 23:01:27 UT
Launch Site: NIIP-5 / Site 81/24
Scientific Instruments: [unknown]

Results: During this third attempt at a circumlunar mission, the Proton rocket's second stage engine spuriously shut down at T+194.64 seconds due to a false signal from the payload which had erroneously detected a problem in the launch vehicle. The emergency rescue system was activated and the 7K-L1 capsule was later successfully recovered about 520 kilometers from the launch pad, approximately 110 kilometers east of the town of Dzhezkazgan in Kazakhstan.

98

Zond 5

Nation: USSR (55)
Objective(s): circumlunar flight
Spacecraft: 7K-L1 (no. 9L)
Spacecraft Mass: c. 5,375 kg
Mission Design and Management: TsKBEM
Launch Vehicle: Proton-K + Blok D (8K82K no. 234-01 + 11S824 no. 17L)
Launch Date and Time: 14 September 1968 / 21:42:11 UT
Launch Site: NIIP-5 / Site 81/23
Scientific Instruments:
1. biological payload
2. radiation detectors
3. imaging system

Results: After three failures, the Soviets tried again to accomplish a robotic circumlunar mission. Besides a complement of sensors to monitor its basic systems and cameras, the spacecraft also carried a large biological payload: two Steppe tortoises (*Testudo horsfieldi* or the Horsfield's tortoise), each with a mass of 0.34 to 0.40 kilograms, hundreds of *Drosophila* eggs of the Domodedovo-32 line; air-dried cells of wheat, barley, peas, carrots, and tomatoes; a flowering plant of *Tradescantia paludosa*, three different strains of chlorella, and a culture of lysogenic bacteria. The tortoises, each 6–7 years old and weighing 340–400 grams, were part of a group of eight, with the other six serving as controls on the ground. The two flight tortoises were placed in the Zond 6 spacecraft on September 2, 12 days before launch. From that moment on, they (and two of the control group) were deprived of food to ensure that the only effect on them was due to the space mission. The ascent to orbit was perfect: the 7K-L1 spacecraft + Blok D combination successfully entered a 191 × 219-kilometer Earth orbit. About 67 minutes after launch, the Blok D fired again for lunar injection, after which the Soviet press announced the mission as "Zond 5." Once again, the 100K star sensor of the spacecraft's attitude control system failed (due to contamination of its exposed surface). Controllers, however, managed to carry out a mid-course correction at 03:11 UT on 17 September using the less accurate solar and Earth-directed sensors. At the time, Zond 5 was 325,000 kilometers from Earth. The spacecraft successfully circled around the farside of the Moon at a range of 1,950 kilometers on 18 September, taking spectacular high-resolution photos of the Moon and Earth. On the return leg of the flight, a second attitude control sensor (the 101K that used Earth for attitude control reference) failed and the spacecraft's three-axis stabilization platform switched off the guided reentry system. As a result, controllers were forced to maintain reentry attitude using the one remaining sensor (the 99K that used the Sun); they alternately fired two attitude control jets on each side of the vehicle to swing the spacecraft into the proper reentry corridor for a direct ballistic reentry into the backup target area in the Indian Ocean. Zond 5's descent module successfully splashed down in the Indian Ocean at 32° 38′ S / 65° 33′ E, about 105 kilometers from the nearest Soviet tracking ship. Landing time was 16:08 UT on 21 September, making it a mission lasting 6 days, 18 hours, and 24 minutes. It was the first successful circumlunar mission carried out by any nation. The tortoises survived the trip and arrived back in Moscow on 7 October. The results of dissection, performed on 11 October after "a 39-day fast," showed that "the main structural changes in the tortoises were caused by starvation" rather than flight to lunar distance or the subsequent travel back from the Indian Ocean to Moscow.

99

Pioneer IX

Nation: USA (44)

Objective(s): solar orbit

Spacecraft: Pioneer-D

Spacecraft Mass: 65.36 kg

Mission Design and Management: NASA / ARC

Launch Vehicle: Thrust-Augmented Improved Thor-Delta (Thor Delta E-1 no. 60 / Thor no. 479 / DSV-3E)

Launch Date and Time: 8 November 1968 / 09:46:29 UT

Launch Site: Cape Kennedy / Launch Complex 17B

Scientific Instruments:

1. triaxial fluxgate magnetometer
2. plasma analyzer
3. cosmic ray-anesotropy detector
4. cosmic ray gradient detector
5. radio wave propagation experiment
6. electric field detector
7. cosmic dust detector
8. celestial mechanics experiment

Results: Pioneer IX was the fourth in a series of five probes designed to collect data on electromagnetic and plasma properties of interplanetary space from widely separated points in heliocentric orbit over at least six passages of solar activity centers. In its 297.55-day orbit at 0.75 × 0.99 AU, the cylindrical, spin-stabilized spacecraft obtained valuable data on the properties of the solar wind, cosmic rays, and interplanetary magnetic fields. The Delta launch vehicle also carried the Test and Training Satellite known as Tetr II (TETR-B) which was put into Earth orbit to test ground-based communications systems in support of the Apollo program. By May 1969, NASA announced that the mission had already achieved all its objectives, having transmitted more than 6 billion bits of data. NASA maintained contact with Pioneer IX until 19 May 1983. Subsequent attempts to use Search for Extraterrestrial Intelligence (SETI) equipment to establish contact with the probe on 3 March 1987 failed, and the Agency officially declared the spacecraft inactive.

100

Zond 6

Nation: USSR (56)
Objective(s): circumlunar flight
Spacecraft: 7K-L1 (no. 12L)
Spacecraft Mass: c. 5,375 kg
Mission Design and Management: TsKBEM
Launch Vehicle: Proton-K + Blok D (8K82K no. 235-01 / 11S824 no. 19L)
Launch Date and Time: 10 November 1968 / 19:11:31 UT
Launch Site: NIIP-5 / Site 81/23
Scientific Instruments:
1. biological payload
2. radiation detectors
3. imaging system
4. photo-emulsion camera
5. micrometeoroid detector

Results: Zond 6 was the second spacecraft that the Soviets sent around the Moon as part of the human circumlunar program. Soon after trans-lunar injection, which happened at 20:18:30 UT on launch day, ground controllers discovered that the vehicle's high-gain antenna had failed to deploy. Given that the main attitude control sensor was installed on the antenna boom, controllers had to make plans to use a backup sensor for further attitude control. After a mid-course correction at 05:41 UT on 12 November (at a distance of 246,000 kilometers), the spacecraft circled the farside of the Moon at a closest range of 2,420 kilometers, taking high resolution black-and-white photographs of the Moon at a range of 11,000 and 3,300 kilometers. During the return flight, temperatures in a hydrogen peroxide tank for the attitude control thrusters dropped far below acceptable levels. Engineers attempted to heat the tank by direct sunlight, but as they later discovered, such a procedure affected the weak pressurization seal of the main hatch and led to slow decompression of the main capsule, which would have undoubtedly killed a crew on board. Despite the failures, Zond 6 conducted two mid-course corrections (on 16 November at 06:40 UT and 17 November at 05:36 UT), and then successfully carried out a fully automated guided reentry, requiring two successive "dips" into the atmosphere, each reducing velocity significantly (the first from 11.2 kilometers/second to 7.6 kilometers/second, and the second down to just 200 meters/second) and headed for the primary landing zone in Kazakhstan. Each of the "dips" was automatically and expertly controlled by attitude control jets to vary roll control so as to provide lift and reduce g-loads. The Zond 6 descent module experienced a maximum of only 4 to 7 g's. After the successful reentry, a gamma-ray altimeter, detecting the now practically depressurized spacecraft (with pressure down to only 25 millibars), issued a command to jettison the main parachute at an altitude of about 5.3 kilometers instead of much lower. As a result, the spacecraft plummeted down to the ground and was destroyed, with impact at 14:10 UT, just 16 kilometers from where the spacecraft had been launched 6 days and 19 hours previously. Although the main biological payload—unspecified by the Soviets—was killed, rescuers salvaged film from the cameras and even managed to scavenge seedlings carried on board.

1969

101

Venera 5

Nation: USSR (57)
Objective(s): Venus landing
Spacecraft: V-69 (2V no. 330)
Spacecraft Mass: 1,130 kg
Mission Design and Management: GSMZ imeni Lavochkina
Launch Vehicle: Molniya-M + Blok VL (8K78M no. V716-72, also V15000-72)
Launch Date and Time: 5 January 1969 / 06:28:08 UT
Launch Site: NIIP-5 / Site 1/5
Scientific Instruments:

Lander:
1. radio altimeter
2. MDDA-A aneroid barometers
3. 11 G-8 and G-10 gas analyzer cartridges
4. FO-69 to measure luminosity
5. VIP to measure atmospheric density
6. IS-164D thermometers

Bus:
1. KS-18-3M cosmic ray particle counters
2. LA-2U spectrometer

Results: Venera 5 and 6 were two identical spacecraft designed to penetrate Venus' atmosphere and transmit a variety of scientific data back to Earth during descent. Both spacecraft were targeted to reach Venus only a day apart, thus allowing some cross-calibration of data. The bus was basically similar in design to the Venera 4 bus but the 410-kilogram lander had some significant alterations. Because of data from Venera 4, the new lander was designed to survive pressures as high as 25 atmospheres and temperatures as high as 320°C. These changes forced a fairly significant increase in the structural strength of the lander, increasing its mass by 27 kilograms. This is why the mass of the spacecraft bus was reduced by 14 kilograms by removing the SG 59 magnetometer). The main and drogue parachutes were also reduced in size, thus reducing the time to descend to the surface. The new lander weighted 410 kilograms and was designed to endure g-loads as high as 450 (as compared to 300 for their predecessors). After performing 73 communications sessions with ground control and completing one mid-course correction on 14 March 1966, Venera 5 approached the dark side of Venus on 16 May 1969 and detached its lander whose speed reduced from 11.17 kilometers/second to 210 meters/second after it hit the Venusian atmosphere at 06:02 UT. One minute later, controllers reestablished contact with the lander and began receiving data on pressure, temperature, and composition (sampled at least twice during descent) of the Venusian atmosphere for 52.5 minutes. Contact was lost at an altitude of about 18 kilometers when the pressure exceeded 27 atmospheres; in other words, the probe probably cracked and became inert. Impact coordinates were 3° S / 18° E. Information extrapolated from Venera 5's data suggested that ground temperature and pressure at the Venusian surface was 140 atmospheres and 530°C, respectively.

102

Venera 6

Nation: USSR (58)

Objective(s): Venus landing

Spacecraft: V-69 (2V no. 331)

Spacecraft Mass: 1,130 kg

Mission Design and Management: GSMZ imeni Lavochkina

Launch Vehicle: Molniya-M + Blok VL (8K78M no. V716-73, also V15000-73)

Launch Date and Time: 10 January 1969 / 05:51:52 UT

Launch Site: NIIP-5 / Site 1/5

Scientific Instruments:

Lander:

1. radio altimeter
2. MDDA-A aneroid barometers
3. 11 G-8 and G-10 gas analyzer cartridges
4. FO-69 to measure luminosity
5. VIP to measure atmospheric density
6. IS-164D thermometers

Bus:

1. KS-18-3M cosmic ray particle counters
2. LA-2U spectrometer

Results: Identical to Venera 5, Venera 6 reached Venus after performing 63 communications sessions with Earth and one mid-course correction at a range of 15.7 million kilometers from Earth on 16 March 1969. Its 405-kilogram lander separated from the main bus 25,000 kilometers from the planet and entered the Venusian atmosphere at a velocity of 11.17 kilometers/second at 06:05 UT on 17 May 1969. The Venera 6 capsule transmitted data for 51 minutes before contact was lost, probably at an altitude of about 18 kilometers. Pressure was 27 atmospheres at loss of contact, similar to that measured by Venera 5 at a much higher altitude indicating that Venera 6 may have come down over a mountain or high plateau. Landing coordinates were 5° S / 23° E. Results from the Venera 5 and 6 missions, published by the Soviets in March 1970, seemed to confirm and sharpen earlier findings from Venera 4, suggesting that the planetary atmosphere consisted of 97% carbon dioxide, <2% nitrogen, and <0.1% oxygen. Data from Venera 6 suggested the ground pressure was about 60 atmospheres and ground temperature was about 400°C. This data compared with Venera 4's readings which indicated pressure at 75 atmospheres and temperature at 500°C.

103

[Zond, 7K-L1 no. 13L]

Nation: USSR (59)

Objective(s): circumlunar flight

Spacecraft: 7K-L1 (no. 13L)

Spacecraft Mass: c. 5,375 kg

Mission Design and Management: TsKBEM

Launch Vehicle: Proton-K + Blok D (8K82K no. 237-01 + 11S824 no. 20L)

Launch Date and Time: 20 January 1969 / 04:14:36 UT

Launch Site: NIIP-5 / Site 81/23

Scientific Instruments: [unknown]

Results: This was the sixth attempt at a robotic circumlunar flight in support of the L1 piloted lunar program and the first after the resounding success of the American Apollo 7 in December 1968. The Proton launch vehicle lifted off on time and first stage operation was nominal. However, during second stage firing, one of the four engines of the stage spuriously switched off at T+313.66 seconds, about 25 seconds early. The other engines continued firing (and could have actually compensated for the loss of thrust), but subsequently, the primary third stage engine also switched off early during its firing sequence, at T+500.03 seconds, due to a breakdown in the main pipeline feeding fuel to the fuel gas generator. After a near-ballistic flight, the L1 payload landed (and was recovered) southeast of Irkutsk near the border between the USSR and Mongolia.

104

[Luna, Ye-8 no. 201]

Nation: USSR (60)
Objective(s): lunar roving operations
Spacecraft: Ye-8 (no. 201)
Spacecraft Mass: c. 5,700 kg
Mission Design and Management: GSMZ imeni Lavochkina
Launch Vehicle: Proton-K + Blok D (8K82K no. 239-01 + 11S824 no. 201L)
Launch Date and Time: 19 February 1969 / 06:48:48 UT
Launch Site: NIIP-5 / Site 81/24
Scientific Instruments:

1. imaging system (two low-resolution TV + four high-resolution photometers)
2. x-ray spectrometer
3. penetrometer
4. laser reflector
5. radiation detectors
6. x-ray telescope
7. odometer/speedometer

Results: The Ye-8 represented the "third generation" of Soviet robotic lunar probes. The basic Ye-8 comprised a lander stage (the "KT") topped off by an eight-wheeled remote-controlled lunar rover (the "8YeL") for exploring the Moon's surface. Essentially a pressurized magnesium alloy container on wheels, the 8YeL was designed to operate over a period of three lunar days (i.e., roughly three Earth months) and collect scientific data from various points on the lunar surface. This first attempt to put the rover on the Moon was a complete failure. At T+51.42 seconds, the payload stack disintegrated and the booster exploded at T+54 seconds. Debris landed about 15 kilometers from the launch site. Later investigation indicated that maximum dynamic pressure during the ascent trajectory tore the new and untested payload shroud off at its weakest tension points. Despite an intensive effort, searchers were unable to find the polonium-210 radioactive isotope heat source in the rover. Unconfirmed rumors still abound that soldiers at the launch site used the isotope to heat their barracks during the bitter winter of 1968–1969.

105

[N1 launch test, 7K-L1S no. 2]

Nation: USSR (61)
Objective(s): lunar orbit
Spacecraft: 7K-L1S (no. 2)
Spacecraft Mass: 6,900 kg
Mission Design and Management: TsKBEM
Launch Vehicle: N1 (no. 15003)
Launch Date and Time: 21 February 1969 / 09:18:07 UT
Launch Site: NIIP-5 / Site 110/38
Scientific Instruments: [unknown]
Results: This was the first attempted launch of the giant N1 booster as part of early test operations in the Soviet piloted lunar landing program. N1 development began in 1962 after two years of initial R&D on heavy booster designs. Although the first launch had been originally planned for 1965, a major redesign of the booster in 1964 and financial and organizational difficulties delayed the launch by four years. The Soviet Communist Party and government officially sanctioned the human lunar landing program in August 1964, more than three years after President John F. Kennedy's famous speech calling on the United States to land an American on the Moon before the end of the 1960s. Development of both the N1 rocket and the L3 payload was plagued by many delays. On this first launch, the N1 carried a basic 7K-L1 spacecraft modified for operations in lunar orbit (rather than for circumlunar flight). Known as the 7K-L1S, the spacecraft was equipped with an Engine Orientation Complex (DOK) for attitude control in lunar orbit. The plan was for the spacecraft to carry out a short mission in lunar orbit (during which time it would have tried to photograph the Ye-8 rover on

the surface, the one that had failed to reach orbit just two days earlier). During the launch of the N1, two first stage engines (of a total of 30 NK-15 engines) shut down, but the remaining 28 engines operated for over a minute despite a growing fire at the base of the rocket. The rocket's KORD control system effectively shut down all first state engines at T+70 seconds. The booster crashed about 50 kilometers from the launch site and the payload successfully used its launch escape system to descend without problem 32–35 kilometers from the pad. Investigators believed that booster failed due when a pipe for measuring fuel pressure broke at T+ 23.3 seconds that set in motion a sequence of events that led to a huge fire at T+54.5 seconds in the tail of the first stage. The fire short-circuited the control system and shut down all the engines at T+70 seconds.

The trajectory design model shown above allowed Mariner mission planners in 1967 to illustrate the orientation of the planet and calculate the expected path of the Mariner VI and VII spacecraft, as well as the window of opportunity for the instruments and television cameras to operate during the flyby. *Credit: NASA*

106

Mariner VI

Nation: USA (45)

Objective(s): Mars flyby

Spacecraft: Mariner-69F / Mariner-F

Spacecraft Mass: 381 kg

Mission Design and Management: NASA / JPL

Launch Vehicle: Atlas Centaur (AC-20 / Atlas 3C no. 5403C / Centaur D-1A)

Launch Date and Time: 25 February 1969 / 01:29:02 UT

Launch Site: Cape Kennedy / Launch Complex 36B

Scientific Instruments:

1. imaging system (two TV cameras)
2. infrared spectrometer
3. ultraviolet spectrometer
4. infrared radiometer
5. celestial mechanics experiment
6. S-band occultation experiment
7. conical radiometer

Results: Mariner VI and VII, identical spacecraft intended to fly by Mars, were the first Mariner spacecraft launched by the Atlas Centaur, permitting a heavier instrument suite. Both spacecraft were intended to study the surface and atmosphere of Mars during close flybys (approximately 3,200 kilometers) that would include a radio-occultation experiment. All onboard instrumentation was designed to collect data on Mars; there were no experiments for study of interplanetary space. The 3.35-meter tall spacecraft was built around an eight-sided magnesium framework with four rectangular solar panels for 449 W power. The heart of the spacecraft was the 11.8-kilogram Control Computer and Sequencer (CC&S) computer which was designed to independently operate Mariner without intervention from ground control. After a mid-course correction on 28 February 1969 and preliminary imaging sessions (50 photos) on 28 July Mariner VI flew by Mars at 05:19:07 UT on 31 July at a distance of 3,429 kilometers. Just 15 minutes prior to closest approach (south of the Martian equator), the two TV cameras on a scan platform began taking photos of the planet automatically every 42 seconds. Mariner VI took 24 near-encounter photos during a period of 17 minutes which were stored on a tape recorder and later transmitted back to Earth some 20 hours after the flyby at a rate of one frame every 5 minutes. The photos showed heavily cratered and chaotic areas not unlike parts of the Moon. Images of the south polar region showed intriguing detail of an irregular border. The scientific instruments indicated that the polar cap gave off infrared radiation consistent with solid carbon dioxide. Mariner VI found surface pressure to be equal to about 30.5 kilometers above Earth's surface. Atmospheric composition was about 98% carbon dioxide. Surface temperatures ranged from −73°C at night to −125°C at the south polar cap. Mariner VI eventually entered heliocentric orbit (1.14 × 1.75 AU) and NASA continued to receive data from the vehicle until mid-1971.

107

[Mars, M-69 no. 521]

Nation: USSR (62)
Objective(s): Mars orbit
Spacecraft: M-69 (no. 521)
Spacecraft Mass: 4,850 kg
Mission Design and Management: GSMZ imeni Lavochkina
Launch Vehicle: Proton-K + Blok D (8K82K no. 240-01 + 11S824 no. 521L)
Launch Date and Time: 27 March 1969 / 10:40:45 UT
Launch Site: NIIP-5 / Site 81/23
Scientific Instruments:
1. RA69 radiometer
2. IV1 instrument to measure water vapor levels
3. USZ ultraviolet spectrometer
4. UTV1 infrared Fourier spectrometer
5. KM69 cosmic ray detector

6. PL18M solar plasma spectrometer
7. RIP-803 low-energy spectrometer
8. GSZ gamma-ray spectrometer
9. UMR2M hydrogen/helium mass spectrometer
10. imaging system (3 cameras)
11. D-127 charged particle traps

Results: The M-69 series of Mars spacecraft was the first of a new generation of Mars probes designed by the Lavochkin design bureau for launch on the heavy Proton booster. Although the 1969 missions were originally meant for both Mars orbit and landing, weight constraints late in mission planning forced engineers to delete the lander and retain only the orbiter. These new probes were designed around a single large spherical tank to which three pressurized compartments were attached. After two en route mid-course corrections during a six-month flight to Mars, the spacecraft were intended to enter orbit around Mars at roughly 1,700 × 34,000 kilometers at 40° inclination. After an initial photography mission, the probes would lower their pericenter to about 500–700 kilometers for a second imaging mission. Total mission lifetime would be about three months. During the launch of the first M-69, the Proton's third stage stopped firing at T+438.46 seconds after its turbopump had caught on fire because of a faulty rotor bearing. The probe, scheduled to reach Mars orbit on 11 September 1969, never even reached Earth orbit.

108

Mariner VII

Nation: USA (46)
Objective(s): Mars flyby
Spacecraft: Mariner-69G
Spacecraft Mass: 381 kg
Mission Design and Management: NASA / JPL
Launch Vehicle: Atlas Centaur (AC-19 / Atlas 3C no. 5105C / Centaur D-1A)

Launch Date and Time: 27 March 1969 / 22:22:01 UT

Launch Site: Cape Kennedy / Launch Complex 36A

Scientific Instruments:

1. imaging system (two TV cameras)
2. infrared spectrometer
3. ultraviolet spectrometer
4. infrared radiometer
5. celestial mechanics experiment
6. S-band occultation experiment
7. conical radiometer

Results: Identical to Mariner VI, Mariner VII had a similar mission of flying by Mars. After Mariner VI had returned intriguing photos of Mars' south polar cap, controllers reprogrammed Mariner VII's control system to increase the number of scans of the south pole for the second spacecraft from 25 to 33. After a perfect mid-course correction on the way to Mars on 8 April 1969, on 30 July, just 7 hours before its twin was scheduled to fly by Mars, the deep space tracking station at Johannesburg, South Africa, lost contact with the spacecraft's high-gain antenna. One of two stations in Madrid, Spain was diverted from their original missions of tracking Pioneer VIII and joined the search for Mariner VII. Fortunately, the Pioneer station at Goldstone picked up faint signals from the spacecraft. Controllers sent commands to Mariner VII to switch to the low-gain antenna that worked well afterwards. Mission controllers later speculated that the spacecraft had been thrown out of alignment when struck by a micrometeoroid (although later speculations centered on the silver-zinc battery on board which might have exploded, with venting electrolytes acting like a thruster). As a result, 15 telemetry channels were lost. Despite problems with positional calibration, Mariner VII recorded 93 far-encounter and 33 near-encounter images of the planet, showing heavily cratered terrain very similar to Mariner VI. Closest approach to Mars was at 05:00:49 UT on 5 August 1969 at a distance of 3,430 kilometers. Oddly, despite the high resolution of 300 meters, Mariner VII found the center of Hellas to be devoid of craters. The spacecraft found a pressure of 3.5 millibars and a temperature of –90°F at 59° S / 28° E in the Hellespontus region, suggesting that this area was elevated about 6 kilometers above the average terrain. Post fight analysis showed that at least three photos from Mariner VII included the moon Phobos. Although surface features were not visible, the pictures clearly showed the moon to be irregularly shaped. Mariner VII entered heliocentric orbit (1.11 × 1.70 AU) and NASA maintained continued to receive data from the vehicle until mid-1971.

109

[Mars, M-69 no. 522]

Nation: USSR (63)

Objective(s): Mars orbit

Spacecraft: M-69 (no. 522)

Spacecraft Mass: 4,850 kg

Mission Design and Management: GSMZ imeni Lavochkina

Launch Vehicle: Proton-K + Blok D (8K82K no. 233-01 + 11S824 no. 522L)

Launch Date and Time: 2 April 1969 / 10:33:00 UT

Launch Site: NIIP-5 / Site 81/24

Scientific Instruments:

1. RA69 radiometer
2. IV1 instrument to measure water vapor levels
3. USZ ultraviolet spectrometer
4. UTV1 infrared Fourier spectrometer
5. KM69 cosmic ray detector
6. PL18M solar plasma spectrometer
7. RIP-803 low-energy spectrometer
8. GSZ gamma-ray spectrometer
9. UMR2M hydrogen/helium mass spectrometer
10. imaging system (3 cameras)
11. D-127 charged particle traps

Results: The second M-69 spacecraft was identical to its predecessor (launched six days before) and was intended to enter orbit around Mars on 15 September 1969. Like its twin, it also never reached intermediate Earth orbit. At launch, at T+0.02 seconds, one of the six first-stage engines of the Proton malfunctioned. Although the booster lifted off using the remaining five engines, it began veering off course and eventually turned with its nose toward the ground (at about 30° to the horizontal). At T+41 seconds, the booster impacted three kilometers from the launch site and exploded into a massive fireball. The launch complex was not affected although windows shattered in the Proton assembly building. Engineers believed that even if either or both of these M-69 Mars spacecraft had gotten off the ground, they probably had very little chance of success in their primary missions. Both the M-69 spacecraft were brought to launch in a period of immense stress and hurry for engineers at Lavochkin. In remembering the M-69 series, one leading designer, Vladimir Dolgopolov, later remembered that, "these were examples of how not to make a spacecraft."

110

[Luna, Ye-8-5 no. 402]

Nation: USSR (64)
Objective(s): lunar sample return
Spacecraft: Ye-8-5 (no. 402)
Spacecraft Mass: c. 5,700 kg
Mission Design and Management: GSMZ imeni Lavochkina
Launch Vehicle: Proton-K + Blok D (8K82K no. 238-01 + 11S824 no. 401L)
Launch Date and Time: 14 June 1969 / 04:00:48 UT
Launch Site: NIIP-5 / Site 81/24
Scientific Instruments:
1. stereo imaging system
2. remote arm for sample collection
3. radiation detector

Results: The Ye-8-5 was a variant of the basic Ye-8 lunar rover spacecraft developed by the Lavochkin design bureau. This particular version, whose development began in 1968, was designed to recover a

small portion of soil from the lunar surface and return it to Earth. It shared the basic lander stage ("KT") as the rover variant (built around a structure comprising four spherical propellant tanks linked together in a square), which was now installed with a robot arm to scoop up lunar soil. The rover was replaced by a new "ascent stage" that was built around three spherical propellant tanks that consisted of a main rocket engine (the S5.61) to lift off from the Moon, a pressurized compartment for electronics, and a small 39-kilogram spherical capsule which would detach from the stage and reenter Earth's atmosphere with its valuable payload of lunar dust. On the first launch attempt of the Ye-8-5 robot scooper, the first three stages of the Proton worked without fault, but the Blok D fourth stage, which was to fire to attain orbital velocity, failed to ignite due to a disruption in the circuit of its guidance system. The payload reentered over the Pacific Ocean without reaching Earth orbit.

111

[N1 test flight, 7K-L1S]

Nation: USSR (65)
Objective(s): lunar orbit
Spacecraft: 7K-L1S (no. 5)
Spacecraft Mass: c. 6,900 kg
Mission Design and Management: TsKBEM
Launch Vehicle: N1 (no. 15005)
Launch Date and Time: 3 July 1969 / 20:18:32 UT
Launch Site: NIIP-5 / Site 110/38
Scientific Instruments: [unknown]
Results: This was the second attempt to launch the giant N1 rocket. Like its predecessor, its payload consisted of a basic 7K-L1 spacecraft equipped with additional instrumentation and an attitude control block to enable operations in lunar orbit. Moments after launch, the first stage of the booster exploded in a massive inferno that engulfed the entire launch pad and damaged nearby buildings and structures for several kilometers around the

area. Amazingly, the payload's launch escape system fired successfully at T+14.5 seconds and the 7K-L1 descent module was recovered safely two kilometers from the pad. An investigation commission traced the cause of the failure to the entry of a foreign object into the oxidizer pump of one of the first stage engines at T-0.25 seconds. The ensuing explosion started a fire that began to engulf the first stage. The control system shut down all engines except one by T+10.15 seconds. The booster lifted about 200 meters off the pad and then came crashing down in a massive explosion. Some estimates suggest that the resulting explosion had a power of about 25 tons of TNT.

112

Luna 15

Nation: USSR (66)

Objective(s): lunar sample return

Spacecraft: Ye-8-5 (no. 401)

Spacecraft Mass: 5,667 kg

Mission Design and Management: GSMZ imeni Lavochkina

Launch Vehicle: Proton-K + Blok D (8K82K no. 242-01 + 11S824 no. 402L)

Launch Date and Time: 13 July 1969 / 02:54:42 UT

Launch Site: NIIP-5 / Site 81/24

Scientific Instruments:

1. stereo imaging system
2. remote arm for sample collection
3. radiation detector

Results: Luna 15, launched only three days before the historic Apollo 11 mission to the Moon, was the second Soviet attempt to recover and bring lunar soil back to Earth. In a race to reach the Moon and return to Earth, the parallel missions of Luna 15 and Apollo 11 were, in some ways, the culmination of the Moon race that defined the space programs of both the United States and the Soviet Union in the 1960s. Prior to launch, due to mass constraints, designers removed one of two

1.28-kilogram radio transmitters from the Rye-85 ascent stage, leaving only one for the entire return part of the mission. On the way to the Moon, controllers detected abnormally high temperatures in the propellant tanks feeding the S5.61 engine (to be used for takeoff from the Moon and return to Earth). By carefully keeping the tank in the Sun's shadow, controllers were able to reduce the temperature and the avoid the risk of an explosion en route. After a mid-course correction the day after launch, Luna 15 entered lunar orbit at 10:00 UT on 17 July 1969. Originally, plans were to carry out two orbital corrections, on 18 and 19 July, respectively, to put the vehicle on its landing track, but the ruggedness of the lunar terrain prompted a delay. Instead, controllers spent nearly four days studying data (over 20 communication sessions) to map out a plan of action to account for the rough geography. The two delayed corrections were eventually carried out on 19 July (at 13:08 UT) and 20 July (at 14:16 UT), putting Luna 15 into its planned 110 × 16-kilometer orbit at a retrograde inclination of 127°. Less than 6 hours after the second correction, Apollo 11 began its descent to the Moon, landing at 20:17 UT on 20 July. The original plan was for Luna 15 to embark on the Moon, less than 2 hours after Apollo 11 but it was not to be. Unsure of the terrain below, controllers delayed the landing by another 18 hours. During this critical period, Apollo 11 astronauts Neil A. Armstrong and Edwin E. "Buzz" Aldrin walked on the Moon. Finally, at 15:46:43 UT on 21 July, a little more than 2 hours prior to the Apollo 11 liftoff from the Moon, Luna 15, now on its 52nd orbit around the Moon, began its descent to the surface. Transmissions, however, abruptly ceased after four minutes instead of nearly five. According to the original plan, the main engine was to fire for 267.3 seconds and bring the vehicle down to about 2.5 kilometers altitude. During the descent, transmissions from the vehicle abruptly and suddenly ended 237 seconds into the engine firing at 15:50:40 UT. The data seemed to show that the spacecraft was 3 kilometers above the lunar surface. Later analysis indicated that Luna 15 had probably crashed onto the side of a mountain (at something like 480 kilometers/hour) as a result of incorrect attitude of the vehicle at the time of ignition of the descent engine—in other words, the spacecraft was probably descending not directly towards the surface but at a slight angle. Luna 15 crashed about 15 kilometers laterally away and 45 kilometers ahead of its assumed location. Impact was roughly at 17° N / 60° E in Mare Crisium.

113

Zond 7

Nation: USSR (67)
Objective(s): circumlunar flight
Spacecraft: 7K-L1 (no. 11)
Spacecraft Mass: c. 5,375 kg
Mission Design and Management: TsKBEM
Launch Vehicle: Proton-K + Blok D (8K82K no. 243-01 + 11S824 no. 18L)
Launch Date and Time: 7 August 1969 / 23:48:06 UT
Launch Site: NIIP-5 / Site 81/23
Scientific Instruments:
1. biological payload
2. radiation detectors
3. imaging system

Results: After a spate of partial successes and catastrophic failures, Zond 7 was the first fully successful Soviet circumlunar mission. The spacecraft was the last 7K-L1 vehicle manufactured explicitly for robotic flight—later models were equipped for crews. Like its predecessors, Zond 7 carried a set of biological specimens, including four male steppe tortoises that were part of a group of thirty selected for the experiment. After a mid-course correction on 8 August at a distance of 250,000 kilometers from Earth, the spacecraft successfully circled the farside of the Moon two days later at a range of 1,200 kilometers. Zond 7 performed color imaging sessions on 8 August (of Earth) and 11 August (two sessions of both Earth and the Moon). The only major malfunction during the

mission was the non-deployment of the main parabolic antenna (due to a problem in the securing cables), but this did not prevent a fulfillment of all the primary goals of the mission. Zond 7 successfully carried out a guided reentry into Earth's atmosphere and landed without problem south of Kustanay in Kazakhstan about 50 kilometers from the intended landing point after a 6-day, 18-hour, 25-minute flight. Zond 7 (and Zond 8) carried on board a full-size human mannequin known as FM-2 to help study the effects of radiation and gravitational loads on various parts of the body during lunar-distance flights.

114

Pioneer

Nation: USA (47)
Objective(s): solar orbit
Spacecraft: Pioneer-E
Spacecraft Mass: 65.4 kg
Mission Design and Management: NASA / ARC
Launch Vehicle: Thrust-Augmented Improved Thor-Delta (Thor Delta L no. D73 / Thor no. 540)
Launch Date and Time: 27 August 1969 / 21:59:00 UT
Launch Site: Cape Kennedy / Launch Complex 17A
Scientific Instruments:
1. three-axis magnetometer
2. cosmic ray telescope
3. radio propagation detector
4. electric field detector
5. quadrispherical plasma analyzer
6. cosmic ray anesotropy detector
7. cosmic dust detector
8. celestial mechanics experiment

Results: During launch of this Pioneer probe, at T+31 seconds, the hydraulics system of the first stage of the booster malfunctioned, eventually causing complete loss of pressure at T+213 seconds, only 4 seconds prior to main engine cutoff of the first stage. Although second stage performance was nominal, there was no way to compensate for the large pointing error introduced by the malfunctions in the first stage. With the booster veering off course, ground control sent a command to destroy the vehicle at T+484 seconds. Pioneer-E was the last in a series of probes intended for studying interplanetary space from heliocentric orbit. An additional payload on the Thor Delta L was a Test and Training Satellite (TETR C) to test the Apollo ground tracking network.

115

Kosmos 300 [Luna]

Nation: USSR (68)
Objective(s): lunar sample return
Spacecraft: Ye-8-5 (no. 403)
Spacecraft Mass: c. 5,700 kg
Mission Design and Management: GSMZ imeni Lavochkina
Launch Vehicle: Proton-K + Blok D (8K82K no. 244-01 + 11S824 no. 403L)
Launch Date and Time: 23 September 1969 / 14:07:37 UT
Launch Site: NIIP-5 / Site 81/24
Scientific Instruments:
1. stereo imaging system
2. remote arm for sample collection
3. radiation detector

Results: This was the third attempt to send a sample return spacecraft to the Moon (after failures in June and July 1969). On this attempt, the spacecraft successfully reached Earth orbit, but failed to inject itself on a trans-lunar trajectory. Later investigation indicated that the Blok D upper stage had failed to fire a second time for trans-lunar injection due to a problem with a fuel injection valve

that had become stuck during the first firing of the Blok D (for Earth orbital insertion). As a result, all the liquid oxygen in the Blok D had been depleted. The Soviet press named the vehicle Kosmos 300 without alluding to its lunar goal. The payload decayed from orbit about four days after launch.

116

Kosmos 305 [Luna]

Nation: USSR (69)
Objective(s): lunar sample return
Spacecraft: Ye-8-5 (no. 404)
Spacecraft Mass: c. 5,700 kg
Mission Design and Management: GSMZ imeni Lavochkina
Launch Vehicle: Proton-K + Blok D (8K82K no. 241-01 + 11S824 no. 404L)
Launch Date and Time: 22 October 1969 / 14:09:59 UT
Launch Site: NIIP-5 / Site 81/24
Scientific Instruments:
1. stereo imaging system
2. remote arm for sample collection
3. radiation detector

Results: Exactly one lunar month after the failure of Kosmos 300, the Soviets launched another Ye-8-5 lunar sample return spacecraft. Once again, the spacecraft failed to leave Earth orbit. When the Blok D upper stage was meant to fire for trans-lunar injection, telemetry readings went off scale and communications were lost. There was apparently a programming failure in one of the radio-command blocks designed to command the Blok D to fire. The Soviet press merely referred to the probe as Kosmos 305. The spacecraft decayed over Australia before completing a single orbit of Earth.

1970

117

[Luna, Ye-8-5 no. 405]

Nation: USSR (70)
Objective(s): lunar sample return
Spacecraft: Ye-8-5 (no. 405)
Spacecraft Mass: c. 5,700 kg
Mission Design and Management: GSMZ imeni Lavochkina
Launch Vehicle: Proton-K + Blok D (8K82K no. 247-01 + 11S824 no. 405L)
Launch Date and Time: 6 February 1970 / 04:16:05 UT
Launch Site: NIIP-5 / Site 81
Scientific Instruments:
1. stereo imaging system
2. remote arm for sample collection
3. radiation detector

Results: This launch continued the spate of failures in the robotic lunar sample return program. On this fifth attempt to recover soil from the Moon, the Proton booster failed to deposit its payload in Earth orbit. An erroneous command shut down the second stage at T+127 seconds and the booster was destroyed. Subsequently, the design organization responsible for the Proton, the Central Design Bureau of Machine Building (TsKBM) headed by General Designer Vladimir Chelomey, implemented a thorough review of the Proton's performance with a simple (and successful) suborbital diagnostic flight on 18 August 1970 to verify corrective measures.

118

Venera 7

Nation: USSR (71)
Objective(s): Venus landing
Spacecraft: V-70 (3V no. 630)
Spacecraft Mass: 1,180 kg
Mission Design and Management: GSMZ imeni Lavochkina
Launch Vehicle: Molniya-M + Blok NVL (8K78M no. Kh15000-62)
Launch Date and Time: 17 August 1970 / 05:38:22 UT
Launch Site: NIIP-5 / Site 31/6
Scientific Instruments:
Bus:
1. KS-18-4M cosmic ray detector
Lander:
1. GS-4 gamma-ray spectrometer
2. instrument for determining pressure and temperature (ITD)
3. DOU-1M instrument for measuring maximal acceleration during braking

Results: Venera 7 was one of a pair of spacecraft prepared by the Soviets in 1970 to make a survivable landing on the surface of Venus. The spacecraft were quite similar in design to Venera 4, 5, and 6 with a main bus and a spherical lander (now with a mass of 490 kilograms). Since the last mission, engineers had redesigned the landing capsule to withstand pressures up to 150–180 atmospheres and temperatures up to 540°C. Venera 7 successfully left Earth orbit using a slightly modified fourth stage (now called Blok NVL) and implemented two mid-course corrections on 2 October and 17 November. It began its Venus encounter operations on 12 December 1970 when the lander probe's batteries were charged up (using solar panels on the bus) and internal temperature lowered. At 04:58:38 UT on 15 December, the lander separated from the bus and entered the Venusian atmosphere at an altitude of 135 kilometers and a velocity of 11.5 kilometers/second. When aerodynamic drag had reduced velocity down to 200 meters/second at an altitude of 60 kilometers, the parachute system deployed (at 04:59:10 UT). The ride down was a bumpy one and it's possible that the parachute tore and ultimately collapsed before impact, which was at a velocity of 17 meters/second at 05:34:10 UT on the nightside of the Venusian landscape, about 2,000 kilometers from the morning terminator. Although transmissions appeared to have ended at the moment of landing, Soviet ground tracking stations recorded what at first proved to be unintelligible noise. After computer processing of the data, Soviet scientists discovered a valuable 22 minutes 58 seconds of information from the capsule—the first transmissions from the surface of another planet. Quite likely, the initial loss-of-signal occurred when the capsule tipped over on its side. Venera 7's data indicated a surface temperature of 475±20°C and a pressure of 90±15 atmospheres. The information was a good fit with previous Soviet and American estimates. Impact point was 5° S / 351° longitude at Tinatin Planitia.

119

Kosmos 359 [Venera]

Nation: USSR (72)
Objective(s): Venus landing
Spacecraft: V-70 (3V no. 631)
Spacecraft Mass: c. 1,200 kg
Mission Design and Management: GSMZ imeni Lavochkina
Launch Vehicle: Molniya-M + Blok NVL (8K78M no. Kh15000-61)

Launch Date and Time: 22 August 1970 / 05:06:08 UT
Launch Site: NIIP-5 / Site 31/6
Scientific Instruments:

Bus:

1. KS-18-4M cosmic ray detector

Lander:

1. GS-4 gamma-ray spectrometer
2. instrument for determining pressure and temperature (ITD)
3. DOU-1M instrument for measuring maximal acceleration during braking

Results: This was the second of a pair of probes designed to land on Venus and transmit information back to Earth. In this case, after the spacecraft had reached Earth orbit, the main engine of the Blok NVL upper stage was late in igniting and then cut off early (after only 25 seconds) due to incorrect operation of a sequencer and a failure in the DC transformer in the power supply system. The payload remained stranded in Earth orbit, eventually reentering on 6 November 1970. The spacecraft was named Kosmos 359 by the Soviet press to disguise the failure.

120

Luna 16

Nation: USSR (73)
Objective(s): lunar sample return
Spacecraft: Ye-8-5 (no. 406)
Spacecraft Mass: 5,725 kg
Mission Design and Management: GSMZ imeni Lavochkina
Launch Vehicle: Proton-K + Blok D (8K82K no. 248-01 + 11S824 no. 203L)
Launch Date and Time: 12 September 1970 / 13:25:52 UT
Launch Site: NIIP-5 / Site 81/23
Scientific Instruments:

1. stereo imaging system
2. remote arm for sample collection
3. radiation detector

A lifesize model of the Luna 16 lander at the Memorial Museum of Cosmonautics in Moscow. *Credit: Asif Siddiqi*

Results: Luna 16 was a landmark success for the Soviets in their deep space exploration program, being the first fully automatic recovery of lunar samples from the surface of the Moon. The success came after five consecutive failures. After an uneventful coast to the Moon (which included one mid-course correction), Luna 16 entered a roughly circular lunar orbit at 118.6 × 102.6 kilometers with a 70° inclination on 17 September, using its 11D417 propulsion unit (which consisted of a primary engine capable of 11 firings and a set of low-thrust engines). Two further orbital adjustments on 18 and 19 September altered both altitude and inclination in preparation for its descent to the Moon. The following day, when Luna 16 reached 13.28 kilometers altitude, close to perilune, it fired its main engine to begin its descent to the surface, using the DA-018 Planeta Doppler landing radar in conjunction with the Vega altimeter. After firing for about 270 seconds and descending down to 2.45 kilometers, Luna 16 went into freefall until the probe was at a height of 600 meters (falling at 700 meters/second) when the main engine fired again briefly, cutting off at 20 meters altitude. Two smaller engines then fired to reduce the landing velocity to a gentle 9 kilometers/hour. It had taken roughly 6 minutes from beginning of deorbit to landing, which occurred at 05:18 UT on 20 September, about 280 hours after launch. Coordinates were announced as 0°41′ S / 56°18′ E, in the northeast area of the Sea of Fertility approximately 100 kilometers west of Webb crater. Mass of the spacecraft at landing was 1,880 kilograms. Attempts to photograph possible sampling sites proved to be less than successful due to poor lighting, but less than an hour after landing, at 06:03 UT, an automatic drill penetrated into the lunar surface to collect a soil sample. After drilling for 7 minutes, the drill reached a stop at 35 centimeters depth and then withdrew its sample (largely dark grey loose rock or regolith) and lifted it in an arc to the top of the spacecraft, depositing the precious cargo in a small spherical capsule mounted on the main spacecraft bus. (Some of the soil apparently fell out during this procedure). Finally, at 07:43:21 UT on 21 September, the spacecraft's 512 kilogram upper stage lifted off from the Moon using the S5.61 engine developed by the Isayev design bureau. This ingeniously designed single-firing engine weighed 42 kilograms and generated 1,917 kgf thrust for a guaranteed 60 seconds. Three days later, after a direct ascent traverse with no mid-course corrections, the capsule with its 101 grams of lunar soil reentered Earth's atmosphere at a velocity of 10.95 kilometers/second, experiencing up to 350 g's deceleration. The capsule, weighing 34 kilograms, parachuted down 80 kilometers southeast of the town of Dzhezkazgan in Kazakhstan at 05:26 UT on 24 September 1970. Analysis of the dark basalt material indicated a close resemblance to soil recovered by the American Apollo 12 mission. The sample was found to be a mature mare regolith, with an abundance of fused soil and glass fragments, between 4.25 and 4.85 billion years old. The Soviets shared samples with representatives from France, the German Democratic Republic, and Iraq, among a number of nations. Based on images taken in 2009 and 2010, NASA's

Lunar Reconnaissance Orbiter (LRO) was able to more precisely identify the landing coordinates of Luna 16 as 0.5134° S / 56.3638° E.

121

Zond 8

Nation: USSR (74)
Objective(s): circumlunar flight
Spacecraft: 7K-L1 (no. 14)
Spacecraft Mass: c. 5,375 kg
Mission Design and Management: TsKBEM
Launch Vehicle: Proton-K + Blok D (8K82K no. 250-01 + 11S824 no. 21L)
Launch Date and Time: 20 October 1970 / 19:55:39 UT
Launch Site: NIIP-5 / Site 81/23
Scientific Instruments:
1. solar wind collector packages
2. imaging system

Results: Zond 8 was the last in the series of circumlunar spacecraft designed to rehearse a piloted circumlunar flight. The project was initiated in 1965 to compete with the Americans in the race to the Moon, but lost its importance once three astronauts circled the Moon on the Apollo 7 mission in December 1968. After a mid-course correction on 22 October at distance of 250,000 kilometers from Earth, Zond 8 reached the Moon without any apparent problems, circling its target on 24 October at a range of 1,110 kilometers. The spacecraft took black-and-white photographs of the lunar surface during two separate sessions. (Earlier, it took pictures of Earth during the outbound flight at a distance of 65,000 kilometers). After two mid-course corrections on the return leg, Zond 8 achieved a return trajectory over Earth's northern hemisphere instead of the standard southern approach profile, allowing Soviet ground control stations to maintain near-continuous contact with the craft. The guidance system, however, malfunctioned on the return leg, and the spacecraft performed a simple ballistic (instead of a guided) reentry into Earth's atmosphere. The vehicle's descent module splashed down safely in the Indian Ocean at 13:55 UT on 27 October about 730 kilometers southeast of the Chagos Islands, 24 kilometers from its original target point. Soviet recovery ships were on hand to collect it and bring it back to Moscow.

122

Luna 17 and Lunokhod 1

Nation: USSR (75)
Objective(s): lunar roving operations
Spacecraft: Ye-8 (no. 203)
Spacecraft Mass: 5,700 kg
Mission Design and Management: GSMZ imeni Lavochkina
Launch Vehicle: Proton-K + Blok D (8K82K no. 251-01 + 11S824 no. 406L)
Launch Date and Time: 10 November 1970 / 14:44:01 UT
Launch Site: NIIP-5 / Site 81/23
Scientific Instruments (on Lunokhod 1):

1. imaging system (two low resolution TV + four high resolution photometers)
2. RIF-MA x-ray spectrometer
3. PrOP penetrometer
4. TL-1 laser reflector
5. RV-2N radiation detector
6. RT-1 x-ray telescope
7. odometer/speedometer

Results: Luna 17 continued the spate of successes in Soviet lunar exploration begun by Luna 16 and Zond 8. Luna 17 carried Lunokhod 1, the first in a series of robot lunar roving vehicles, whose conception had begun in the early 1960s, originally as part of the piloted lunar landing operations. The undercarriage was designed and built by VNII Transmash although Lavochkin retained overall design conception of the vehicle. This was the second attempt to land such a vehicle on the Moon after a failure in February 1969. The descent stage was equipped with two landing ramps for the "ascent stage," i.e., the rover, to disembark on to the Moon's surface. The 756-kilogram rover stood about 1.35 meters high and was 2.15 meters across. Each of its eight wheels could be controlled independently for two forward and two reverse speeds. Top speed was about 100 meters/hour, with commands issued by a five-man team of "drivers" on Earth who had to deal with a minimum 4.1 second delay (which included the 2.6 second roundtrip of the signal plus time to exert pressure on levers on the control panel). These men were carefully selected from a pool of hundreds in a process that began as early as May 1968. Two crews, for two shifts, were selected, each comprising five men (commander, driver, flight-engineer, navigator, and narrow-beam antenna guidance operator). The commanders of these "sedentary cosmonauts," as they were called, were Yu. F. Vasil'yev and I. L. Fedorov, while the drivers were N. M. Yeremenko and V. G. Dovgan', respectively. The set of scientific instruments was powered by solar cells (installed on the inside of the hinged top lid of the rover) and chemical batteries. After two mid-course corrections en route to the Moon, Luna 17 entered an 85 × 141- kilometer lunar orbit inclined at 141°. Repeating the same dynamic descent activities as its predecessor, Luna 17 landed on the lunar surface at 03:46:50 UT on 17 November 1970 at 38° 24′ N / 34° 47′ W (as known at the time), about 2,500 kilometers from the Luna 16 site in the Sea of Rains. The lander settled in a crater-like depression 150–200 meters in diameter and 7 meters deep. As a result of Yeremenko's command, driver G. G. Latypov pushed a lever and then pressed a button to move the vehicle off its platform with a "First—Forward!" exclamation. It was 06:27:07 UT on 17 November. It took 20 seconds to roll down to the surface. During its 322 Earth days of operation, the rover traveled 10.47 kilometers (later, in 2013, revised down to 9.93 kilometers) and returned more than 20,000 TV images and 206 high resolution panoramas. In addition, Lunokhod 1 performed 25 soil analyses with its RIF-MA x-ray fluorescence spectrometer and used its penetrometer at 537 different locations over a 10.5-kilometer route (averaging one use every c. 20 meters). Lunokhod 1 also carried a 3.7-kilogram French-supplied instrument above the forward cameras, the TL-1, consisting of 14 10-centimeter silica glass prisms to bounce back pulses of ruby laser light fired from observatories in Crimea and France. Scientists first used this reflector on 5 and 6 December, allowing the Earth–Moon distance to be measured down to an accuracy of 30 centimeters. However, dust apparently covered the reflector and few further echoes were obtained. Controllers finished the last communications session with Lunokhod 1 at 13:05 UT on 14 September 1971. Attempts to reestablish contact were discontinued on 4 October, thus culminating one of the most successful robotic missions of the early space age. Lunokhod 1 clearly outperformed its expectations—its planned design life was only 3 lunar days (about 21 Earth days) but it operated for 11. Many years later, in March 2010, a team of scientists based at several U.S. academic institutions resumed laser ranging with the laser reflector on Lunokhod 1, based on data from NASA's Lunar Reconnaissance Orbiter (LRO) which allowed a precise determination (to 5 meters accuracy) of the location of the (former) Soviet rover. The new data provided a more precise location for Lunokhod 1 as 38.333° N / 35.037° W. The landing site of Luna 17 was also refined to 38.238° N / 34.997° W.

1971

123

Mariner 8

Nation: USA (48)
Objective(s): Mars orbit
Spacecraft: Mariner-71H / Mariner-H
Spacecraft Mass: 997.9 kg
Mission Design and Management: NASA / JPL
Launch Vehicle: Atlas Centaur (AC-24 / Atlas 3C no. 5405C / Centaur D-1A)
Launch Date and Time: 9 May 1971 / 01:11:01 UT
Launch Site: Cape Kennedy / Launch Complex 36A
Scientific Instruments:

1. imaging system
2. ultraviolet spectrometer
3. infrared spectrometer
4. infrared radiometer

Results: Mariner-71H (also called Mariner-H) was the first of a pair of American spacecraft intended to explore the physical and dynamic characteristics of Mars from Martian orbit. The overall goals of the series were: to search for an environment that could support life; to collect data on the origins and evolution of the planet; to gather information on planetary physics, geology, planetology, and cosmology; and to provide data that could aid future spacecraft such as the Viking landers. Launch of Mariner-71H was nominal until just after separation of the Centaur upper stage when a malfunction occurred in the stage's flight control system leading to loss of pitch control at an altitude of 148 kilometers at T+4.7 minutes. As a result, the stack began to tumble and the Centaur engines shut down. The stage and its payload reentered Earth's atmosphere approximately 1,500 kilometers downrange from the launch site, about 400 kilometers north of Puerto Rico. The problem was traced to a failed integrated circuit in the pitch guidance module.

124

Kosmos 419 [Mars]

Nation: USSR (76)
Objective(s): Mars orbit
Spacecraft: M-71S (3MS no. 170)
Spacecraft Mass: 4,549 kg
Mission Design and Management: GSMZ imeni Lavochkina
Launch Vehicle: Proton-K + Blok D (8K82K no. 253-01 + 11S824 no. 1101L)
Launch Date and Time: 10 May 1971 / 16:58:42 UT
Launch Site: NIIP-5 / Site 81/23
Scientific Instruments:

1. fluxgate magnetometer
2. infrared radiometer
3. infrared photometer
4. RIEP-2801 multi-channel plasma spectrometer
5. visible photometer
6. radiometer
7. ultraviolet photometer
8. cosmic ray detector
9. D-127 charged particle traps
10. imaging system (with Vega and Zufar cameras)
11. Stéréo-1 radio-astronomy experiment

Results: Kosmos 419 was the first "fifth generation" Soviet Mars probe (after those launched in 1960, 1962, 1963-64, and 1969). The original plan was to launch two orbiter-lander combinations known as M-71 during the 1971 Mars launch period, but in order to preempt the American Mariner H/I vehicles, Soviet planners added a third mission, the M-71S, a simple orbiter that could become the first spacecraft to go into orbit around Mars. The orbiter could also collect data important for aiming the two landers at precise locations on the Martian surface. These new vehicles were the first Soviet robotic spacecraft to have digital computers (the S-530), which was a simplified version of the Argon-11 carried on the 7K-L1 ("Zond") circumlunar spacecraft. The first of these Mars spacecraft entered Earth orbit successfully (145 × 159 kilometers at 51.5° inclination), but the Blok D upper stage failed to fire the second time to send the spacecraft to Mars. Later investigation showed that there had been human error in programming the firing time for the Blok. Apparently, the timer that would ignite the Blok D was incorrectly programmed to fire after 150 hours instead of 1.5 hours. (One report claimed that a programmer incorrectly programmed the time in years instead of hours). The stranded spacecraft, which was named Kosmos 419 by the Soviet press, reentered Earth's atmosphere within two days of launch. The Soviets had promised the French that two of their Stéréo-1 instruments would be sent to Mars during the 1971 window, but since one was lost on Kosmos 419 which had officially nothing to do with a Mars mission, Soviet officials were forced to keep silent about its fate.

125

Mars 2

Nation: USSR (77)
Objective(s): Mars orbit and landing
Spacecraft: M-71 (4M no. 171)
Spacecraft Mass: 4,650 kg
Mission Design and Management: GSMZ imeni Lavochkina
Launch Vehicle: Proton-K + Blok D (8K82K no. 255-01 + 11S824 no. 1201L)
Launch Date and Time: 19 May 1971 / 16:22:49 UT

A ground model of the PrOP-M mobile device that was installed on the Soviet Mars landers for the 1971 missions. These were capable of moving 15 meters away from the lander while hooked to a tether. *Credit: T. Varfolomeyev*

Launch Site: NIIP-5 / Site 81/24
Scientific Instruments:
Orbiter:
  1. infrared bolometer (radiometer)
  2. microwave radiometer (radiotelescope)
  3. infrared photometer (CO_2 gas absorption strips)
  4. IV-2 interference-polarized photometer
  5. photometer to measure brightness distribution
  6. 4-channel UV photometer
  7. imaging system (two cameras, each capable of 480 images)
  8. ferrozoid tricomponent magnetometer
  9. ion trap
  10. RIEP-2801 spectrometer for charged particles
  11. cosmic ray detector
  12. radiotransmitter (for determination of structure of atmosphere through refraction)
  13. D-127 charged particle traps [unconfirmed]
Lander:
  1. gamma-ray spectrometer
  2. x-ray spectrometer
  3. thermometer
  4. anemometer

5. barometer

6. imaging system (2 cameras)

7. mass spectrometer

8. penetrometer (on PrOP-M)

9. gamma-ray densitometer (on PrOP-M)

Results: Mars 2 was the first of two orbiter-lander combination spacecraft sent to Mars by the Soviet Union during the 1971 launch period. The orbiters were roughly cylindrical structures fixed to a large propellant tank base. The landers were egg-shaped modules with petals that would open up on the Martian surface. The 1,000 kilogram landers (of which 355 kilograms was the actual capsule on the surface) were fastened to the top of the bus and protected by a braking shell for entry into the Martian atmosphere. After jettisoning the shell, the landers would deploy parachutes to descend to the Martian surface. Each lander also carried 4-kilogram mini-rovers called PrOP-M (for *Pribor otsenki prokhodimosti-Mars*, or Device to Evaluate Mobility—Mars), a box-shaped robot that was equipped with skids; the device would move by moving the body forward (at a speed of 1 meter/minute) with the ski resting on the soil. The range of the device, connected by an umbilical cord to supply power and transmit data, was 15 meters. PrOP-M carried a penetrometer and a radiation densimeter. If the device ran into an obstacle, sensors at the front would send a signal to an automated control system on the lander, which would then send a command to back up PrOP-M, turn it "right" (if the obstacle was towards the left) or "left" (if the obstacle was towards the right), and then take one step forward again. As Mars 2 made its way to the Red Planet, controllers performed two successful mid-course corrections, on 17 June and 20 November 1971. On 27 November 1971, Mars 2 implemented its final mid-course correction after which the lander probe separated to initiate atmospheric entry. At this point, the onboard computer was designed to implement final corrections to the trajectory and spin up the lander around its longitudinal axis and then fire a solid propellant engine to initiate reentry in a specific direction. As

it happened, after the final course correction, the trajectory of the spacecraft had been so accurate that there was no need for further corrective measures. Because of pre-programmed algorithms that simply assumed a deviated trajectory, the lander was put into an incorrect attitude after separation to compensate for the "error." When the reentry engine fired, the angle of entry proved to be far too steep. The parachute system never deployed and the lander eventually crashed onto the Martian surface at 4° N / 47° W. It was the first human-made object to make contact with Mars. The Mars 2 orbiter meanwhile successfully entered orbit around Mars at 20:19 UT on 27 November 1971. Parameters were 1,380 × 25,000 kilometers at 48.9° inclination, with an orbital period of 1,080 minutes, slightly less than the expected 1,440 minutes. In a clear obfuscation of the truth, the Soviets claimed that one of the two French Stéréo-1 instruments was lost with the other instruments on Mars 2, when in fact, Stéréo-1 was not carried on Mars 2 but the earlier failed Kosmos 419. [See Mars 3 for Mars 2 orbiter program.]

126

Mars 3

Nation: USSR (78)

Objective(s): Mars orbit and landing

Spacecraft: M-71 (4M no. 172)

Spacecraft Mass: 4,650 kg

Mission Design and Management: GSMZ imeni Lavochkina

Launch Vehicle: Proton-K + Blok D (8K82K no. 249-01 + 11S824 no. 1301L)

Launch Date and Time: 28 May 1971 / 15:26:30 UT

Launch Site: NIIP-5 / Site 81/23

Scientific Instruments:

Orbiter:

1. infrared bolometer (radiometer)

2. microwave radiometer (radiotelescope)

3. infrared photometer (CO_2 gas absorption strips)
4. IV-2 interference-polarized photometer
5. photometer to measure brightness distribution
6. 4-channel UV photometer
7. imaging system (two cameras, each capable of 480 images)
8. ferrozoid tricomponent magnetometer
9. ion trap
10. RIEP-2801 spectrometer for charged particles
11. cosmic ray detector
12. radiotransmitter (for determination of structure of atmosphere through refraction)
13. D-127 charged particle traps [unconfirmed]
14. Stéréo-1 radio-astronomy experiment
15. modulation-type ion trap

Lander:

1. gamma-ray spectrometer
2. x-ray spectrometer
3. thermometer
4. anemometer
5. barometer
6. imaging system (two cameras)
7. mass spectrometer
8. penetrometer (on PrOP-M)
9. gamma-ray densitometer (on PrOP-M)

Results: Like its predecessor, Mars 3 was successfully sent on a trajectory to the Red Planet. The spacecraft completed three mid-course corrections on 8 June, 14 November, and 2 December 1971. At 09:14 UT on 2 December 1971, the lander separated from the orbiter and 4.5 hours later began entry into the Martian atmosphere. Finally, at 13:47 UT, the probe successfully set down intact on the Martian surface becoming the first human-made object to perform a survivable landing on the planet. Landing coordinates were 44.90° S / 160.08° W. The bus meanwhile entered orbit around Mars with parameters of 1,530 × 214,500 kilometers at 60.0° inclination, significantly more eccentric than originally planned—at least 11 times

higher than nominal. Immediately after landing, at 13:50:35 UT, the lander probe began transmitting a TV image of the Martian surface although transmissions abruptly ceased after 14.5 seconds (or 20 seconds according to some sources). Because of a violent dust storm that raged across the planet, controllers surmised that coronal discharge may have shorted all electric instrumentation on the lander. The received image showed only a gray background with no detail, probably because the two imaging "heads" had still not deployed in 20 seconds to their full height to see the surface. After the initial contact, the ground lost all contact with the lander probe. The Mars 3 orbiter, like the Mars 2 orbiter, had problems with its imaging mission. Because the orbiters had to perform their imaging mission soon after entering orbit—mainly because the chemicals on board for developing film had a finite lifetime—they could not wait until the dust storms subsided on the surface. As a result, the photographs from both orbiters showed few details of the surface. On 23 January 1972, *Pravda* noted that "the dust storm is still making photography and scientific measurement of the planet difficult" but added on 19 February that "information obtained [more recently] shows that the dust storm has ended." Later analysis showed that the dust storm began during the first 10 days of October and lasted three months although the atmosphere contained residual dust until late January. Additionally, controllers had set the cameras at the wrong exposure setting, making the photos far too light to show much detail. Despite the failure of the imaging mission, both orbiters carried out a full cycle of scientific experiments returning data on properties of the surface and atmosphere—including on the nature and dynamics of the dust storm and water content in the Martian atmosphere—until contact was lost almost simultaneously, according to Lavochkin, in July 1972. TASS announced completion of both orbital missions on 23 August 1972. The French Stéréo-1 instrument on Mars 3 operated successfully for 185 hours over nearly seven months returning a megabyte of data on solar radiation. In April 2013, NASA announced that its Mars Reconnaissance Orbiter (MRO) may have imaged hardware from Mars 3, including its parachute, heat shield, braking engine, and the lander itself. These were found in 2008 by a community of Russian space enthusiasts who were following the mission of Curiosity.

127

Mariner 9

Nation: USA (49)

Objective(s): Mars orbit

Spacecraft: Mariner-71I / Mariner-I

Spacecraft Mass: 997.9 kg

Mission Design and Management: NASA / JPL

Launch Vehicle: Atlas Centaur (AC-23 / Atlas 3C no. 5404C / Centaur D-1A)

Launch Date and Time: 30 May 1971 / 22:23:04 UT

Launch Site: Cape Kennedy / Launch Complex 36B

Scientific Instruments:

1. imaging system
2. ultraviolet spectrometer
3. infrared spectrometer
4. infrared radiometer

Results: Mariner 9 was the second in the pair of identical spacecraft launched in 1971 to orbit Mars. The first spacecraft, Mariner 8, failed to reach Earth orbit. Based on a wide octagonal structure, these vehicles used a bi-propellant propulsion system with a fixed thrust of 136 kgf for orbital insertion around Mars. All scientific instrumentation on the spacecraft were mounted on a movable scan platform "underneath" the main bodies. The span of the spacecraft over its extended solar panels was 6.9 meters. After an en route mid-course correction on 5 June 1971, at 00:18 UT on 14 November 1971, Mariner 9 ignited its main engine for 915.6 seconds to become the first human-made object to enter orbit around a planet. Initial orbital parameters were 1,398 × 17,916 kilometers at 64.3° inclination. (Another firing on the fourth revolution

around Mars refined the orbit to 1,394 × 17,144 kilometers at 64.34° inclination). The primary goal of the mission was to map about 70% of the surface during the first three months of operation. The dedicated imaging mission began in late November, but because of the major dust storm at the planet during this time, photos of the planet taken prior to about mid-January 1972 did not show great detail. Once the dust storm had subsided, from 2 January 1972 on, Mariner 9 began to return spectacular photos of the deeply pitted Martian landscape, for the first time showing such features as the great system of parallel rilles stretching more than 1,700 kilometers across Mare Sirenum. The vast amount of incoming data countered the notion that Mars was geologically inert. There was some speculation on the possibility of water having existed on the surface during an earlier period, but the spacecraft data could not provide any conclusive proof. By February 1972, the spacecraft had identified about 20 volcanoes, one of which, later named Olympus Mons, dwarfed any similar feature on Earth. Based on data from Mariner 9's spectrometers, it was determined that Olympus Mons, part of Nix Olympica—a "great volcanic pile" possibly formed by the eruption of hot magma from the planet's interior—is about 15–30 kilometers tall and has a base with a diameter of 600 kilometers. Another major surface feature identified was Valles Marineris, a system of canyons east of the Tharsis region that is more than 4,000 kilometers long, 200 kilometers wide, and in some areas, up to 7 kilometers deep. On 11 February 1972, NASA announced that Mariner 9 had achieved all its goals although the spacecraft continued sending back useful data well into the summer. By the time of last contact at 22:32 UT on 27 October 1972 when it exhausted gaseous nitrogen for attitude control, the spacecraft had mapped 85% of the planet at a resolution of 1–2 kilometers, returning 7,329 photos (including at least 80 of Phobos and Deimos). Thus ended one of the great early robotic missions of the space age and undoubtedly one of the most influential. The spacecraft is expected to crash onto the Martian surface sometime around 2020.

128

Apollo 15 Particle and Fields Subsatellite

Nation: USA (50)

Objective(s): lunar orbit

Spacecraft: Apollo 15 P&FS

Spacecraft Mass: 35.6 kg

Mission Design and Management: NASA / MSC

Launch Vehicle: Apollo 15 CSM-112 (itself launched by Saturn V SA-510)

Launch Date and Time: 26 July 1971 / 13:34:00 UT (subsatellite ejection on 4 August 1971 / 20:13:19 UT)

Launch Site: Kennedy Space Center / Launch Complex 39A

Scientific Instruments:

1. magnetometer
2. S-band transponder
3. charged particle detectors

Results: This small satellite was deployed by the Apollo 15 crew—David R. Scott, Alfred M. Worden, and James B. Irwin—shortly before leaving lunar orbit. The probe was designed around a hexagonal structure with a diameter of 35.6 centimeters that was equipped with three instrument booms. Power supply came from solar panels and chemical batteries. The instruments measured the strength and direction of interplanetary and terrestrial magnetic fields, detected variations in the lunar gravity field, and measured proton and electron flux. The satellite confirmed Explorer XXXV's finding that while Earth's magnetic field deflects the incoming solar wind into a tail, the Moon acts as a physical barrier due to its weak field and creates a "hole" in the wind. An electronic failure on 3 February 1972 formally ended the mission. Although it originally had a one-year design life, all mission objectives were fulfilled.

129

Luna 18

Nation: USSR (79)

Objective(s): lunar sample return

Spacecraft: Ye-8-5 (no. 407)

Spacecraft Mass: 5,725 kg

Mission Design and Management: GSMZ imeni Lavochkina

Launch Vehicle: Proton-K + Blok D (8K82K no. 256-01 + 11S824 no. 0601L)

A ground model of the Apollo 15 Particles and Fields Satellite. *Credit: NASA*

Launch Date and Time: 2 September 1971 / 13:40:40 UT

Launch Site: NIIP-5 / Site 81/24

Scientific Instruments:

1. stereo imaging system
2. remote arm for sample collection
3. radiation detector
4. radio altimeter

Results: This was the seventh Soviet attempt to recover soil samples from the surface of the Moon and the first after the success of Luna 16. After two mid-course corrections on 4 and 6 September 1971, Luna 18 entered a circular orbit around the Moon on 7 September at 100 kilometers altitude with an inclination of 35°, an orbit that was off-nominal since the orbit insertion engine cut off 15 seconds earlier than planned. To save propellant, mission planners decided to conduct one (instead of two) orbital corrections to bring the spacecraft into the proper orbit for descent. This firing, held outside direct radio contact, was also not nominal, leaving the vehicle in a 93.4 × 180.3-kilometer orbit (instead of 16.9 × 123.9 kilometers). Telemetry data showed that the pitch program was two orders less than expected due to the low effectiveness of one of the orientation engines, which was working only on fuel and not oxidizer. After a subsequent orbital correction, on 11 September, the vehicle began its descent to the lunar surface. However, due to its off-nominal orbit, the same orientation engine completely failed on all three axes (pitch, yaw, roll). As a result, the roll angle was 10° less than the computed value (since the other orientation engines were not able to fully compensate). Contact with the spacecraft was abruptly lost at 07:47:16.5 UT at the previously determined point of lunar landing. Impact coordinates were 3° 34′ N / 56° 30′ E near the edge of the Sea of Fertility. Officially, the Soviets announced that "the lunar landing in the complex mountainous conditions proved to be unfavorable." Later, in 1975, the Soviets published data from Luna 18's continuous-wave radio altimeter which determined the mean density of the lunar topsoil.

130

Luna 19

Nation: USSR (80)

Objective(s): lunar orbit

Spacecraft: Ye-8LS (no. 202)

Spacecraft Mass: 5,330 kg

Mission Design and Management: GSMZ imeni Lavochkina

Launch Vehicle: Proton-K + Blok D (8K82K no. 257-01 + 11S824 no. 4001L)

Launch Date and Time: 28 September 1971 / 10:00:22 UT

Launch Site: NIIP-5 / Site 81/24

Scientific Instruments:

1. 2 TV cameras
2. gamma-ray spectrometer (ARL)
3. RV-2NLS radiation detector
4. SIM-RMCh meteoroid detector
5. SG-59M magnetometer
6. radio altimeter

Results: Luna 19 was the first "advanced" lunar orbiter whose design was based upon the same Ye-8-class bus used for the lunar rovers and the sample collectors. For these orbiters, designated Ye-8LS, the basic "lander stage" was topped off by a wheel-less Lunokhod-like frame that housed all scientific instrumentation in a pressurized container. Luna 19 entered orbit around the Moon on 2 October 1972 with parameters of 141.2 × 133.9 kilometers at 40.5° inclination. After two mid-course corrections on 29 September and 1 October, a final correction on 6 October was to put the spacecraft into proper altitude to begin its imaging mission. However, a failure in a gyro-platform in the orientation system pointed the vehicle incorrectly and Luna 19 ended up in a much higher orbit than planned. As a result, the original high-resolution imaging mission as well as use of the radio altimeter was canceled. Instead, the mission was reoriented to provide panoramic images of the mountainous areas of the Moon between 30° and 60° S latitude and between

20° and 80° E longitude. Because the vehicle was in a contingency spin-stabilized mode, the images unfortunately came out blurred (41 pictures were returned). In an article in *Pravda* on 17 February 1973, Lavochkin Chief Designer Sergey Kryukov (1918–2005) noted that "prominence in the scientific mission of the Luna 19 station was given to the study of the gravitational field of the Moon" and the location of mascons. Occultation experiments in May and June 1972 allowed scientists to determine the concentration of charged particles at an altitude of 10 kilometers. Additional studies of the solar wind were coordinated with those performed by the Mars 2 and 3 orbiters and Veneras 7 and 8. The gamma-ray spectrometer apparently failed to provide any data. Communications with Luna 19 was lost on 1 November 1972 after 13 months of continuous operation (over 4,000 orbits of the Moon), far exceeding the planned lifetime of three months.

1972

Luna 20

Nation: USSR (81)
Objective(s): lunar sample return
Spacecraft: Ye-8-5 (no. 408)
Spacecraft Mass: 5,725 kg
Mission Design and Management: GSMZ imeni Lavochkina
Launch Vehicle: Proton-K + Blok D (8K82K no. 258-01 + 11S824 no. 0801L)
Launch Date and Time: 14 February 1972 / 03:27:58 UT
Launch Site: NIIP-5 / Site 81/24
Scientific Instruments:

1. stereo imaging system
2. remote arm for sample collection
3. radiation detector
4. radio altimeter

Results: This, the eighth Soviet spacecraft launched to return lunar soil to Earth, was sent to complete the mission that Luna 18 had failed to accomplish. After a four-and-a-half day flight to the Moon that included a single mid-course correction on 15 February, Luna 20 entered orbit around the Moon on February 18. Initial orbital parameters were 100 × 100 kilometers at 65° inclination. Three days later at 19:13 UT, the spacecraft fired its main engine for 267 seconds to begin descent to the lunar surface. A second firing at an altitude of 760 meters further reduced velocity before Luna 20 set down safely on the Moon at 19:19:09 UT on 21 February 1972 at coordinates 3° 32′ N / 56°

33′ E (as announced at the time), only 1.8 kilometers from the crash site of Luna 18, in a highland region between Mare Fecunditatis and Mare Crisium. The spacecraft settled on a slope of about 8° to 10°. After landing, the imaging system was used to locate a scientifically promising location to collect a sample, which was done on 23 February, but in stages due to increased soil resistance. After the sample was safely collected, the spacecraft's ascent stage lifted off at 22:58 UT on 22 February and quickly accelerated to 2.7 kilometers/second velocity, sufficient to return to Earth. The small spherical capsule parachuted down safely on an island in the Karkingir river, 40 kilometers northwest of the town of Dzhezkazgan in Kazakhstan at 19:12 UT on 25 February 1972. The 55-gram soil sample differed from that collected by Luna 16 in that the majority (50–60%) of the rock particles in the newer sample were ancient anorthosite (which consist largely of feldspar) rather than the basalt of the earlier one (which contained about 1–2% of anorthosite). They were, in fact, quite similar to samples collected by the Apollo 16 astronauts in April 1972. Like the Luna 16 soil, samples of the Luna 20 collection were exchanged with NASA officials who delivered a sample from the Apollo 15 mission, the exchange taking place on 13 April 1972. Samples were also shared with scientists from France, Czechoslovakia, Great Britain (both Luna 16 and 20), and India. Also, as with Luna 16, in 2009 and 2010, images from NASA's Lunar Reconnaissance Orbiter (LRO) were used to more precisely identify Luna 20's landing site as 3.7866° N / 56.6242° E.

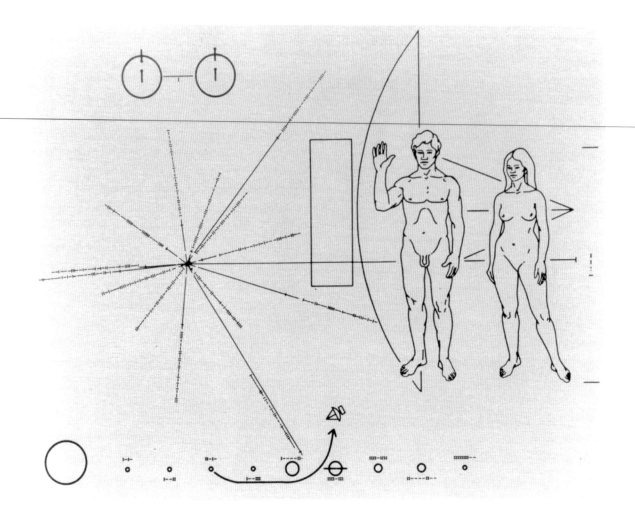

The plaque carried on Pioneers 10 and 11 showed a human male and female standing next to a Pioneer spacecraft. At the top left are two hydrogen atoms, each in a different energy state. Note the planets of the solar system at the bottom, with a line tracing Pioneer to the third planet from the Sun. *Credit: NASA*

132

Pioneer 10

Nation: USA (51)

Objective(s): Jupiter flyby

Spacecraft: Pioneer-F

Spacecraft Mass: 258 kg

Mission Design and Management: NASA / ARC

Launch Vehicle: Atlas Centaur (AC-27 / Atlas 3C no. 5007C / Centaur D-1A)

Launch Date and Time: 2 March 1972 / 01:49:04 UT

Launch Site: Cape Kennedy / Launch Complex 36A

Scientific Instruments:

1. imaging photopolarimeter
2. helium vector magnetometer (HVM)
3. infrared adiometer
4. quadrispherical plasma analyzer
5. ultraviolet photometer
6. charged particle instrument (CPI)
7. cosmic ray telescope (CRT)
8. Geiger tube telescope (GTT)
9. Sisyphus asteroid/meteoroid detector (AMD)
10. meteoroid detectors
11. trapped radiation detector (TRD)

Results: Pioneer 10, the first NASA mission to the outer planets, garnered a series of firsts perhaps unmatched by any other robotic spacecraft in the space era: the first vehicle placed on a trajectory to escape the solar system into interstellar space; the first spacecraft to fly beyond Mars; the first to fly through the asteroid belt; the first to fly past Jupiter; and the first to use all-nuclear electrical power (two SNAP-19 radioisotope thermal generators [RTGs] capable of delivering about 140 W during the Jupiter encounter). After launch by a three-stage version of the Atlas Centaur (with a TE-M-364-4 solid propellant engine modified from the Surveyor lander), Pioneer 10 reached a maximum escape velocity of 51,682 kilometers/hour, faster than any previous human-made object at that point in time. Controllers carried out two course corrections, on 7 March and 26 March, the latter to ensure an occultation experiment with Jupiter's moon Io. There were some initial problems during the outbound voyage when direct sunlight caused heating problems, but nothing to endanger the mission. On 15 July 1972, the spacecraft entered the asteroid belt, emerging in February 1973 after a 435 million-kilometer voyage. During this period, the spacecraft encountered some asteroid hits (although much less than expected) and also measured the intensity of Zodiacal light in interplanetary space. In conjunction with Pioneer IX (in solar orbit), on 7 August, Pioneer 10 recorded details of one of the most violent solar storms in recent record. At 20:30 UT on 26 November, the spacecraft reported a decrease in the solar wind from 420 to 250 kilometers/second and a 100-fold increase in temperature, indicating that it was passing through the front of Jupiter's bow shock where the solar wind clashed with the planet's magnetosphere. In other words, it had entered Jupiter's magnetosphere. By 1 December, Pioneer 10 was returning better images of the planet than possible from Earth. (It had already begun imaging as early as 6 November 1973). Command-and-return time was up to 92 minutes by this time. Pioneer 10's closest approach to Jupiter was at 02:26 UT on 4 December 1973 when the spacecraft raced by the planet at a range of 130,354 kilometers at a velocity

of approximately 126,000 kilometers/hour. Of the spacecraft's 11 scientific instruments, 6 operated continuously through the encounter. The spacecraft passed by a series of Jovian moons, obtaining photos of Callisto, Ganymede, and Europa (but not of Io, as the photopolarimeter succumbed to radiation by that time). Approximately 78 minutes after the closest approach, Pioneer 10 passed behind Jupiter's limb for a radio occultation experiment. In addition, the infrared radiometer provided further information on the planet's atmosphere. Between 6 November and 31 December, the vehicle took about 500 pictures of Jupiter's atmosphere with a highest resolution of 320 kilometers, clearly showing such landmarks as the Great Red Spot. The encounter itself was declared over on 2 January 1974. Pioneer 10 fulfilled all objectives except one due to false commands triggered by Jupiter's intense radiation. Based on incoming data, scientists identified plasma in Jupiter's magnetic field. The spacecraft crossed Saturn's orbit in February 1976, recording data that indicated that Jupiter's enormous magnetic tail, almost 800 million kilometers long, covered the whole distance between the two planets. Still operating nominally, Pioneer 10 crossed the orbit of Neptune (then the outermost planet) on 13 June 1983, thus becoming the first human-made object to go beyond the furthest planet. NASA maintained routine contact with Pioneer 10 for over two decades until 19:35 UT on 31 March 1997 (when the spacecraft was 67 AU from Earth) when routine contact was terminated due to budgetary reasons. Intermittent contact, however, continued, but only as permitted by the onboard power source, with data collections from the Geiger tube telescope and the charged particle instrument. Until 17 February 1998, Pioneer 10 was the farthest human-made object in existence (69.4 AU) when it was passed by Voyager 1. A NASA ground team received a signal on the state of spacecraft systems (still nominal) on 5 August 2000. The spacecraft returned its last telemetry data on 27 April 2002 and less than a year later, on 23 January 2003, it sent its last signal back to Earth when it was 12.23 billion kilometers from Earth. That signal took 11 hours and 20 minutes to reach Earth. By that time, it was clear that the spacecraft's RTG power source had decayed, thus delivering insufficient power to the radio transmitter. A final attempt to contact Pioneer 10 on 4 March 2006 failed. Originally designed for a 21-month mission, the mission's lifetime far exceeded expectations. By 5 November 2017, the inert Pioneer 10 spacecraft was roughly 118.824 AUs (or 17.776 billion kilometers) from Earth, a range second only to Voyager 1. The spacecraft is generally heading in the direction of the red star Aldebaran which forms the eye of the Taurus constellation. It is expected to pass by Aldebaran in about two million years. Pioneer 10 is heading out of the solar system in a direction very different from the two Voyager probes and Pioneer 11, i.e., towards the nose of the heliosphere in an upstream direction relative to the inflowing interstellar gas. In case of an intercept by intelligent life, Pioneer 10 carries an aluminum plaque with diagrams of a man and a woman, the solar system, and its location relative to 14 pulsars. The expectation is that such intelligent life would be able to interpret the diagram to determine the position of the Sun (and thus, Earth) at the time of launch relative to the Pulsars.

133

Venera 8

Nation: USSR (82)
Objective(s): Venus landing
Spacecraft: V-72 (3V no. 670)
Spacecraft Mass: 1,184 kg
Mission Design and Management: GSMZ imeni Lavochkina
Launch Vehicle: Molniya-M + Blok NVL (8K78M no. S1500-63)
Launch Date and Time: 27 March 1972 / 04:15:06 UT
Launch Site: NIIP-5 / Site 31/6

Scientific Instruments:
Spacecraft Bus:
1. KS-18-4M cosmic ray detector
Lander:
1. thermometers and barometers (ITD)
2. IOV-72 photometers
3. GS-4 gamma-ray spectrometer
4. IAV-72 gas (ammonia) analyzer
5. DOU-1M accelerometer
6. radar altimeter

Results: Venera 8 was the first in new pair of Soviet spacecraft designed to explore Venus. Although similar in design to its predecessors, the 495-kilogram lander was substantially modified, based on the results from Venera 7. Lavochkin Chief Designer Sergey Kryukov noted in an interview in *Pravda* on 5 August 1972 that "Venera 8 was a logical continuation of the previous Venera-7" but "the construction of the descent vehicle was almost completely new." The new capsule was designed to withstand pressures of "only" 105 atmospheres (versus 150 atmospheres on Venera 7) and 493°C (instead of 540°C), had an upgraded parachute system, and carried extra scientific instrumentation. After one mid-course correction on 6 April 1972, Venera 8's lander separated from the flyby bus at 07:40 UT on 22 July 1972 and entered the Venusian atmosphere 57 minutes later at a velocity of 11.6 kilometers/second. During the aerodynamic breaking segment, the descent module reduced its velocity from 11.6 kilometers/second to 250 meters/second, thus surviving a maximum g-force of 335; the gas temperature in the shock wave at the "front" of the vehicle was more than 12,000°C. Successful landing took place at 09:29 UT about 500 kilometers from the morning terminator on the sunlit side of Venus, the first such landing. Landing coordinates were a 10-mile radius of 10.7° S / 335.25° E. The probe transmitted data for another 50 minutes, 11 seconds from the hostile surface before succumbing to ground conditions. The transmitted information indicated that temperature and pressure at the landing site were 470±8°C and 90±1.5 atmospheres respectively, very close to values obtained on the planet's night side by Venera 7. Wind velocity was less than 1 kilometer/second below 10 kilometers altitude. Data from the gamma-ray spectrometer made it possible to make some determination of naturally occurring radioactive elements in the soil. Preliminary data suggested that the surface material contained 4% potassium, 0.002% uranium, and 0.00065% thorium. The lander answered one of the key questions about the surface of Venus, namely the degree of illumination on the ground. Based on data from the photometer, scientists concluded that "a certain portion of solar rays in the visible region of the spectrum penetrates to the surface of the planet and that there are significant differences in illumination between day and night." The data indicated that visibility on the ground was about one kilometer at the time Venera 8 landed. The spacecraft also recorded a sharp change in illumination between 30 and 35 kilometers altitude.

134

Kosmos 482 [Venera]

Nation: USSR (83)
Objective(s): Venus landing
Spacecraft: V-72 (3V no. 671)
Spacecraft Mass: c. 1,180 kg
Mission Design and Management: GSMZ imeni Lavochkina
Launch Vehicle: Molniya-M + Blok NVL (8K78M no. S1500-64)
Launch Date and Time: 31 March 1972 / 04:02:33 UT
Launch Site: NIIP-5 / Site 31/6
Scientific Instruments:
Spacecraft Bus:
1. KS-18-4M cosmic ray detector
Lander:
1. thermometers and barometers (ITD)
2. IOV-72 photometers
3. GS-4 gamma-ray spectrometer

4. IAV-72 gas (ammonia) analyzer
5. DOU-1M accelerometer
6. radar altimeter

Results: This was the sister craft to Venera 8, which was launched four days prior. Unfortunately, the spacecraft never left Earth orbit. The Blok NVL escape stage's main engine prematurely cut off after only 125 seconds of firing due to a failure in the onboard timer. As a result, the spacecraft entered an elliptical orbit around Earth. Officially, the Soviets named the probe Kosmos 482 to disguise its true mission. The main spacecraft reentered on 5 May 1981.

135

Apollo 16 Particles and Fields Subsatellite

Nation: USA (52)
Objective(s): lunar orbit
Spacecraft: Apollo 16 P&FS
Spacecraft Mass: 42 kg
Mission Design and Management: NASA / MSC
Launch Vehicle: Apollo 16 CSM-113 (itself launched by Saturn V SA-511)
Launch Date and Time: 16 April 1972 / 17:54:00 UT (subsatellite ejection on 24 April 1972 / 09:56:09 UT)
Launch Site: Kennedy Space Center / Launch Complex 39A
Scientific Instruments:
1. magnetometer
2. S-band transponder
3. charged particle detectors

Results: Nearly identical to its predecessor, the Apollo 16 Particles and Fields Subsatellite was ejected from the Apollo 16 Command and Service Module about 4 hours prior to the crew's trans-Earth injection burn which sent them home from the Moon. Because of problems with the Apollo CSM main engine, the crew were forced to release the subsatellite in a low lunar orbit of 100 × 100 kilometers at 10° inclination. Thus, the probe eventually crashed onto the lunar surface after 34 days in orbit rather than the planned one year. Impact point was at 10.2° N / 112° E at 21:00 UT on 29 May 1972. Because of its low orbit, the spacecraft did, however, return some valuable low-altitude data.

136

[N1 launch test, 7K-LOK no. 6A]

Nation: USSR (84)
Objective(s): lunar orbit
Spacecraft: 7K-LOK (no. 6A)
Spacecraft Mass: c. 9,500 kg
Mission Design and Management: TsKBEM
Launch Vehicle: N1 (no. 15007)
Launch Date and Time: 23 November 1972 / 06:11:55 UT
Launch Site: NIIP-5 / Site 110/37
Scientific Instruments: [unknown]

Results: This was fourth test launch of the giant Soviet N1 booster. The first two, launched in 1969, had attempted to send rigged up 7K-L1 spacecraft to lunar orbit. The third booster, launched in June 1971, had carried a payload mockup for tests in Earth orbit. All three had failed. This fourth launch was intended to send a fully equipped 7K-LOK spacecraft (similar to a beefed-up Soyuz) on a robotic lunar orbiting mission during which the spacecraft would spend 3.7 days circling the Moon (over 42 orbits) taking photographs of future landing sites for piloted missions. The booster lifted off without problems, but a few seconds prior to first stage cutoff, at T+107 seconds, a powerful explosion ripped apart the bottom of the first stage, destroying unequivocally Soviet hopes of sending cosmonauts to the Moon. There was never a conclusive reason for the failure, with some suggesting that there had been an engine failure and others convinced that the scheduled shutdown of six central engines just prior to the explosion had caused a structural shockwave that eventually caused the explosion.

1973

137

Luna 21 and Lunokhod 2

Nation: USSR (85)
Objective(s): lunar roving operations
Spacecraft: Ye-8 (no. 204)
Spacecraft Mass: 5,700 kg
Mission Design and Management: GSMZ imeni Lavochkina
Launch Vehicle: Proton-K + Blok D (8K82K no. 259-01 + 11S824 no. 205L)
Launch Date and Time: 8 January 1973 / 06:55:38 UT
Launch Site: NIIP-5 / Site 81/23
Scientific Instruments:

1. imaging system (three low resolution TV + four high resolution photometers)
2. RIFMA-M x-ray fluorescence spectrometer
3. PROP penetrometer
4. TL-2 laser reflector (with Rubin-1 photo receiver)
5. RV-2N radiation detector
6. x-ray telescope
7. odometer/speedometer
8. AF-3L visible/ultraviolet photometer
9. SG-70A tricomponent magnetometer
10. photodetector

Results: Luna 21 carried the second successful Soviet "8YeL" lunar rover, Lunokhod 2. Launched less than a month after the last Apollo lunar landing, Luna 21 entered orbit around the Moon on 12 January 1973 (after a single mid-course correction en route). Parameters were 110 × 90 kilometers at 60° inclination. On 15 January, the spacecraft deorbited, and after multiple engine firings, landed on the Moon at 22:35 UT the same day inside the 55-kilometer wide LeMonnier crater at 25° 51′ N /

30° 27′ E (as announced at the time) between Mare Serenitatis and the Taurus Mountains, about 180 kilometers north of the Apollo 17 landing site. Less than 3 hours later, at 01:14 UT on 16 January, the rover disembarked onto the lunar surface. The 840 kilogram Lunokhod 2 was an improved version of its predecessor, and was equipped with two types of television cameras. The first consisted of two vidicon cameras ("small frame television" cameras according to the Soviet media) for transmitting information to help basic navigation control. A second set consisted of four pair-mounted, side-carried panoramic opto-mechanical cycloramic cameras. The rover also included an improved 8-wheel traction system and additional scientific instrumentation, including significantly, a magnetometer. By the end of its first lunar day, 23 January 1973, Lunokhod 2 had already traveled further than Lunokhod 1 in its entire operational life. The main focus of investigations during the second lunar day, which ended on 22 February, was a study of the transitional mare highlands in the southern part of the LeMonnier Crater, including stereoscopic panoramic television imagery of the surface, measurement of the lunar soil's chemical properties, and taking magnetic readings. By this time, the rover had travelled a total of 11.067 kilometers. For the ground "crew" navigating Lunokhod, there were times of high stress, compounded by the time lag between Earth and Moon. *Izvestiya* reported on 13 March 1973 that at times "crew" members pulses reached 130–135 beats per minute with one individual holding their breath in nervousness for 15–20 seconds. During the fourth lunar day, which began on 9 April, Lunokhod 2 traveled right to the edge of a large tectonic fault in the eastern area of the littoral zone of the LeMonnier Crater, an area that was very difficult to traverse given the proliferation of rocks up to 2–3 meters in size. By the end of the fourth lunar day on 23 April it had travelled 36.2 kilometers. On 9 May the rover was commanded to leave the area of the fault but inadvertently rolled into a crater, with dust covering its solar panels, disrupting temperatures in the vehicle. Attempts to save the rover failed, and on 3 June the Soviet news agency announced that its mission was over. An official internal Soviet report on Lunokhod 2 noted that the rover had "ended its operations" earlier, at 12:25 UT on 10 May 1973, after temperatures had reached up to 43–47°C and on board systems had shut down. All subsequent attempts at contact apparently failed. Before last contact, the rover took 80,000 TV pictures and 86 panoramic photos and had performed hundreds of mechanical and chemical surveys of the soil, including 25 soil analyses with the RIF-MA (*Rentgenskiy izotopnyy fluorestsentnyy metod analyiza* or x-ray Isotopic Fluorescence Analysis Method) instrument. Despite the formal end of the mission, experiments with the French TL-2 laser reflector continued for decades, and were much more successful than the rangings carried out with the similar instrument on Lunokhod 1. At the time of the Lunokhod 2 mission, scientists calculated a total travel distance of 37.5 kilometers, about three-and-a-half times more than its predecessor. An extended summary of the scientific results from the mission was published in *Pravda* on 20 November 1973. The Soviets later revealed that during a conference on planetary exploration in Moscow held from 29 January to 2 February 1973 (i.e., after the landing of Luna 21), an American scientist had given photos of the lunar surface around the Luna 21 landing site to a Soviet engineer in charge of the Lunokhod 2 mission. These photos, taken prior to the Apollo 17 landing, were later used by the "driver team" to navigate the new rover on its mission on the Moon. Later, in 2013, based on imagery from the Lunar Reconnaissance Orbiter (LRO), Russian researchers led by Irina Karachevtseva at the Moscow State University's Institute of Geodesy and Cartography (MIIGAiK), recalculated the

total distance travelled and came to a more precise number of between 42.1 and 42.2 kilometers. The original landing site location was also sharpened to 25.99° N / 30.41° E.

138

Pioneer 11

Nation: USA (53)
Objective(s): Jupiter flyby, Saturn flyby
Spacecraft: Pioneer-G
Spacecraft Mass: 258.5 kg
Mission Design and Management: NASA / ARC
Launch Vehicle: Atlas Centaur (AC-30 / Atlas 3D no. 5011D / Centaur D-1A)
Launch Date and Time: 6 April 1973 / 02:11 UT
Launch Site: Cape Kennedy / Launch Complex 36B
Scientific Instruments:

1. imaging photopolarimeter
2. helium vector magnetometer (HVM)
3. infrared radiometer
4. quadrispherical plasma analyzer
5. ultraviolet photometer
6. charged particle instrument (CPI)
7. cosmic ray telescope (CRT)
8. Geiger tube telescope (GTT)
9. Sisyphus asteroid/meteoroid detector (AMD)
10. meteoroid detectors
11. trapped radiation detector (TRD)
12. triaxial fluxgate magnetometer

Results: Pioneer 11, the sister spacecraft to Pioneer 10, was the first human-made object to fly past Saturn and also returned the first pictures of the polar regions of Jupiter. After boost by the TE-M-364-4 engine, the spacecraft sped away from Earth at a velocity of 51,800 kilometers/hour, thus equaling the speed of its predecessor, Pioneer 10. During the outbound journey, there were a number of malfunctions on the spacecraft—including the momentary failure of one of the RTG booms to deploy, a problem with an attitude control thruster, and the partial failure of the asteroidal dust detector—but

Completed in 1973, Deep Space Station 63 (DSS-63) was the third 64-meter antenna of NASA's Deep Space Network. Located in Robledo de Chevala near Madrid, DSS-63 received its first signals from Pioneer 10 and Mariner 10. *Credit: NASA*

none of these jeopardized the mission. Pioneer 11 passed through the asteroid belt without damage by mid-March 1974. Soon, on 26 April 1974, it performed a mid-course correction (after an earlier one on 11 April 1973) to guide it much closer to Jupiter than Pioneer 10 and ensure a polar flyby. Pioneer 11 penetrated the Jovian bow shock on 25 November 1974 at 03:39 UT. The spacecraft's closest approach to Jupiter occurred at 05:22 UT on 3 December 1974 at a range of 42,500 kilometers from the planet's cloud tops, three times closer than Pioneer 10. By this time, it was travelling faster than any human-made object at the time, 171,000 kilometers/hour. Because of its high speed during the encounter, the spacecraft's exposure to Jupiter's radiation belts spanned a shorter time than its predecessor although it was actually closer to the planet. Pioneer 11 repeatedly crossed Jupiter's bow shock, indicating that the Jovian magnetosphere changes its boundaries as it is buffeted by the solar wind. Besides the many images of the planet (and better pictures of the Great Red Spot), Pioneer 11 took about 200 images of the moons of Jupiter. The vehicle then used Jupiter's massive gravitational field to swing back across the solar system to set it on a course to Saturn. After its Jupiter encounter, on 16 April 1975, the micrometeoroid detector was turned off since it was issuing spurious commands which were interfering with other instruments. Mid-course corrections on 26 May 1976 and 13 July 1978 sharpened its trajectory towards Saturn. Pioneer 11 detected Saturn's bow shock on 31 August 1979, about a million-and-a-half kilometers out from the planet, thus providing the first conclusive evidence of the existence of Saturn's magnetic field. The spacecraft crossed the planet's ring plane beyond the outer ring at 14:36 UT on 1 September 1979 and then passed by the planet at 16:31 UT for a close encounter at 20,900-kilometer range. It was moving at a relative velocity of 114,100 kilometers/hour at the point of closest approach. During the encounter, the spacecraft took 440 images of the planetary system, with about 20 at a resolution of 90 kilometers. Those of Saturn's moon Titan (at a resolution of 180 kilometers) showed a featureless orange fuzzy satellite. A brief burst of data on Titan indicated that the average global temperature of Titan was –193°C. Among Pioneer 11's many discoveries were a narrow ring outside the A ring named the "F" ring and a new satellite 200 kilometers in diameter. The spacecraft recorded the planet's overall temperature at –180°C and photographs indicated a more featureless atmosphere than that of Jupiter. Analysis of data suggested that the planet was primarily made of liquid hydrogen. After leaving Saturn, Pioneer 11 headed out of the solar system in a direction opposite to that of Pioneer 10, i.e., to the center of galaxy in the general direction of Sagittarius. Pioneer 11 crossed the orbit of Neptune on 23 February 1990 becoming the fourth spacecraft (after Pioneer 10, Voyager 1 and 2) to do so. Scientists expected that during their outbound journeys, both Pioneer 10 and 11 would find the boundary of the heliosphere where the solar wind slows down and forms a "termination shock," beyond which there would be the heliopause and finally the bow shock of the interstellar medium, i.e., space beyond the solar system. By 1995, 22 years after launch, two instruments were still operational on the vehicle. NASA Ames Research Center made last contact with the spacecraft on 30 September 1995 when Pioneer 11 was 44.1 AU from Earth. Scientists later received a few minutes of good engineering data on 24 November 1995 but lost contact again once Earth moved out of view of the spacecraft's antenna. Like Pioneer 10, Pioneer 11 also carries a plaque with a message for any intelligent beings. By 5 November 2017, it was estimated to be about 97.590 AU (or 14.599 billion kilometers) from Earth.

139

Explorer 49

Nation: USA (54)

Objective(s): lunar orbit

Spacecraft: RAE-B

Spacecraft Mass: 330.2 kg

Mission Design and Management: NASA / GSFC

Launch Vehicle: Delta 1913 (DSV-3P-11 no. 95 or "Delta-95" / Thor no. 581)

Launch Date and Time: 10 June 1973 / 14:13:00 UT

Launch Site: Cape Canaveral / Launch Complex 17B

Scientific Instruments:

1. galactic studies experiment
2. sporadic low-frequency solar radio bursts experiment
3. sporadic Jovian bursts experiment
4. radio emission from terrestrial magnetosphere experiment
5. cosmic source observation experiment

Results: Explorer 49 was the final U.S. lunar mission for 21 years (until Clementine in 1994). The spacecraft, part of a duo of Radio Astronomy Explorer (RAE) missions (other being Explorer 48), was designed to conduct comprehensive studies of low frequency radio emissions from the Sun, Moon,

Artist's impression of fully deployed Explorer 49 in orbit around the Moon. *Credit: NASA*

the planets, and other galactic and extra-galactic sources, while in a circular orbit around the Moon. Its location was driven by the need to avoid terrestrial radio interference. After launch on a direct ascent trajectory to the Moon and one mid-course correction on 11 June, Explorer 49 fired its insertion motor on 07:21 UT on 15 June to enter orbit around the Moon. Initial orbital parameters were 1,334 × 1,123 kilometers at 61.3° inclination. On 18 June the spacecraft jettisoned its main engine and, using its Velocity Control Propulsion System, circularized its orbit to 1,063 × 1,052 kilometers at 38.7° inclination. The spacecraft was the largest human-made object to orbit the Moon with its deployed antennas measuring 457.2 meters (nearly half a kilometer!) tip-to-tip. These antennas, as well as a 192-meter long damper boom and a 36.6-meter dipole antenna were all stored away on motor-driven reels which allowed them to unfurl in lunar orbit. Once in lunar orbit, the spacecraft deployed its various antennae in stages, assuming its full form by November 1974. During its mission, Explorer 49 studied low-frequency radio emissions from the solar system (including the Sun and Jupiter) and other galactic and extra-galactic sources. NASA announced completion of the mission in June 1975 although contact was maintained until August 1977.

140

Mars 4

Nation: USSR (86)
Objective(s): Mars orbit
Spacecraft: M-73S (3MS no. 52S)
Spacecraft Mass: 4,000 kg
Mission Design and Management: GSMZ imeni Lavochkina
Launch Vehicle: Proton-K + Blok D (8K82K no. 261-01 + 11S824 1701L)

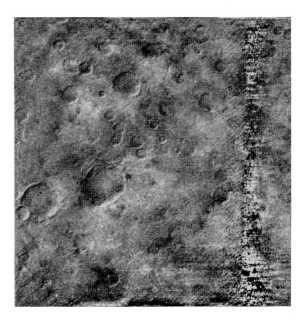

One of 12 images taken by the Vega imaging system on board the Soviet Mars 4 spacecraft. The vehicle failed to enter orbit around Mars but took these photos during its flyby on 10 February 1974. *Credit: Don Mitchell*

Launch Date and Time: 21 July 1973 / 19:30:59 UT
Launch Site: NIIP-5 / Site 81/23
Scientific Instruments:

1. atmospheric radio-probing instrument
2. radiotelescope
3. infrared radiometer
4. spectrophotometer
5. narrow-band photometer
6. narrow-band interference-polarization photometer
7. imaging system (OMS scanner + 2 TV cameras)
8. photometers
9. two polarimeters
10. ultraviolet photometer
11. scattered solar radiation photometer
12. gamma spectrometer
13. magnetometer
14. plasma traps
15. multi-channel electrostatic analyzer

Results: Mars 4 was one of four Soviet spacecraft of the 3MS (or M-73) series launched in 1973. Soviet planners were eager to preempt the American Viking missions planned for 1976 but were limited by the less advantageous positions of the planets which allowed the Proton-K/Blok D boosters to launch only limited payloads towards Mars. The Soviets thus separated the standard pair of orbiter-lander payload combinations into two orbiters and two landers. Less than four months prior to launch, ground testing detected a major problem with the 2T312 transistors (developed by the Pulsar Scientific-Research Institute) used on all four vehicles, apparently because the factory that manufactured it used aluminum contacts instead of gold-plated contacts. An analysis showed that the transistors' failure rate began to increase after 1.5 to 2 years operation, i.e., just about when the spacecraft would reach Mars. Despite the roughly 50% odds of success, the government decided to proceed with the missions. The first spacecraft, Mars 4, successfully left Earth orbit and headed towards Mars and accomplished a single mid-course correction on 30 July 1973, but soon two of three channels of the onboard computer failed due to the faulty transistors. As a result, the second mid-course correction by its main 11D425A engine could not be implemented. With no possibility for Mars orbit insertion, Mars 4 flew by the Red Planet at 15:34 UT on 10 February 1974 at a range of 1,844 kilometers. Ground control was able to command the vehicle to turn on its TV imaging system (Vega-3MSA) 2 minutes prior to this point (at 15:32:41) to begin a short photography session of the Martian surface during the flyby. (The other TV camera system known as Zufar-2SA was never turned on due to a failure). The TV camera took 12 standard images from ranges of 1,900 to 2,100 kilometers distance over a period of 6 minutes. The other OMS scanner also provided two panoramas of the surface. The spacecraft eventually entered heliocentric orbit.

141

Mars 5

Nation: USSR (87)

Objective(s): Mars orbit

Spacecraft: M-73S (3MS no. 53S)

Spacecraft Mass: 4,000 kg

Mission Design and Management: GSMZ imeni Lavochkina

Launch Vehicle: 8K82K + Blok D (Proton-K no. 262-01 + 11S824 no. 1801L)

Launch Date and Time: 25 July 1973 / 18:55:48 UT

Launch Site: NIIP-5 / Site 81/24

Scientific Instruments:

1. atmospheric radio-probing instrument
2. radiotelescope
3. infrared radiometer
4. spectrophotometer
5. narrow-band photometer
6. narrow-band interference-polarization photometer
7. imaging system (OMS scanner + 2 TV cameras)
8. photometers
9. VPM-73 polarimeter unit
10. ultraviolet photometer
11. scattered solar radiation photometer
12. gamma spectrometer
13. magnetometer
14. plasma traps
15. multi-channel electrostatic analyzer

Results: Mars 5 was the sister Mars orbiter to Mars 4. After two mid-course corrections on 3 August 1973 and 2 February 1974, Mars 5 successfully fired its main engine at 15:44:25 UT to enter orbit around the planet. Initial orbital parameters were 1,760 × 32,586 kilometers at 35° 19' 17" inclination. Soon after orbital insertion, ground controllers detected the slow depressurization of the main instrument compartment on the orbiter, probably as a result of an impact with a particle during or after orbital insertion. Calculations showed that at the current rate

Composite of images taken by the Soviet Mars 5 spacecraft from Martian orbit on 23 February 1974. *Credit: Don Mitchell*

of air loss, the spacecraft would be operational for approximately three more weeks. Scientists drew up a special accelerated science program that included imaging of the surface at 100-meter resolution. Five imaging sessions on 17, 21, 23, 25 and 26 February 1974 produced a total of 108 frames of comprising only 43 usable photographs. Both the high-resolution Vega-3MSA and the survey Zufar-2SA TV cameras were used. Additionally, Mars 5 used the OMS scanner to take five panoramas of the surface. The last communication with Mars 5, when the final panorama was transmitted back to Earth, took place on 28 February 1974, after which pressure in the spacecraft reduced below working levels. Mars 5's photos, some of which were of comparable quality to those of Mariner 9, clearly showed surface features which indicated erosion caused by free-flowing water. The first of these images taken by both the television cameras were published in the Academy of Sciences' journal in the fall of 1974. Among significant achievements claimed for Mars 5 was "receipt of mean data on the chemical composition of rocks on Mars for the first time." The vehicle was supposed to act as a data relay for the Mars 6 and Mars 7 landers which arrived in March 1974 but was obviously unable to do so.

142

Mars 6

Nation: USSR (88)
Objective(s): Mars flyby and Mars landing
Spacecraft: M-73P (3MP no. 50P)
Spacecraft Mass: 3,880 kg
Mission Design and Management: GSMZ imeni Lavochkina
Launch Vehicle: Proton-K + Blok D (8K82K no. 281-01 + 11S824 no. 1901L)
Launch Date and Time: 5 August 1973 / 17:45:48 UT
Launch Site: NIIP-5 / Site 81/23
Scientific Instruments:

Spacecraft Bus:
1. magnetometer
2. plasma traps
3. cosmic ray sensors
4. micrometeoroid detectors
5. Gémeaux-S1 and Gémeaux-T instruments for study of solar proton and electron fluxes
6. Stereo-5 antenna

Lander:
1. thermometer
2. barometer
3. accelerometer
4. radio-altimeter
5. mass spectrometer
6. soil analyzer

The Soviet Mars 6 spacecraft. The lander aeroshell is visible on top. *Credit: Don Mitchell*

Results: Mars 6 was one of two combination fly-by-lander launched by the Soviet Union during the 1973 launch period. The landers were very similar in design to the Mars 2 and Mars 3 landers dispatched by the Soviets in 1971, except the spacecraft was now comprised of a flyby vehicle (instead of an orbiter) and a 1,000-kilogram lander. Mars 6 completed its first mid-course correction en route to Mars at 23:45 UT on 12 August 1973, but immediately a tape recorder on board failed, forcing controllers to use a backup. Then on 3 September, there was a major failure in the telemetry system that transmitted scientific and operations data from the spacecraft. Only two channels remained operational, neither of which provided the ground with any direct data on the status of the flyby vehicle's systems. Controllers could only use a time-consuming "playback" mode for the reception of data. Ultimately, the flyby spacecraft automatically performed all its functions and at 05:01:56 UT (signal reception time) on 12 March 1974, the lander successfully separated from its mother ship at a distance of 46,000 kilometers from Mars. About 4 hours later, at 09:05:53 UT, it entered the Martian atmosphere at a velocity of 5,600 meters/second. The parachute system deployed correctly at an altitude of 20 kilometers (at 09:08:32) when speed had been reduced to about 600 meters/second, and scientific instruments began to collect and transmit data (to the flyby vehicle) as the probe descended. The only useful data was, however, directly from the lander to Earth, and its information was rather "weak" and difficult to decode. It appeared that the lander was rocking back and forth under its parachute far more vigorously than expected. Nevertheless, Mars 6 returned the first direct measurements of the temperature and pressure of the Martian atmosphere as well as its chemical composition (using the radio-frequency mass spectrometer) to Earth. The data indicated that argon made up about one-third of the Martian atmosphere. Moments before the expected landing, the ground lost contact with the probe. The last confirmed data was information on ignition of the soft-landing engines received about 2 seconds before impact, the probe landing at 09:11:05 UT at 23.9° S / 19.5° W. Later investigation never conclusively identified a single cause of loss of contact. Probable reasons included failure of the radio system or landing in a geographically rough area. The Mars 6 flyby bus, meanwhile, collected some scientific information during its short flyby (at a minimum range of 1,600 kilometers to the surface) before heading into heliocentric orbit.

143

Mars 7

Nation: USSR (89)
Objective(s): Mars flyby and Mars landing
Spacecraft: M-73P (3MP no. 51P)
Spacecraft Mass: 3,880 kg
Mission Design and Management: GSMZ imeni Lavochkina
Launch Vehicle: Proton-K + Blok D (8K82K no. 281-02 + 11S824 no. 2001L)
Launch Date and Time: 9 August 1973 / 17:00:17 UT
Launch Site: NIIP-5 / Site 81/24
Scientific Instruments:

Spacecraft Bus:
1. magnetometer
2. plasma traps
3. KM-73 cosmic ray detector
4. micrometeoroid detectors
5. Gémeaux-S1 and Gémeaux-T instruments for study of solar proton and electron fluxes
6. Stereo-5 antenna

Lander:
1. thermometer
2. barometer
3. accelerometer
4. radio-altimeter
5. mass spectrometer
6. soil analyzer

Results: Mars 7 was the last of the four Soviet spacecraft sent to Mars in the 1973 launch period (although it arrived at Mars prior to Mars 6). On its way to Mars, the spacecraft performed a single mid-course correction at 20:00 UT on 16 August 1973. En route to Mars, there were failures in the communications systems, and controllers were forced to maintain contact via the only remaining radio-communications complex. On 9 March 1974, the flyby spacecraft ordered the lander capsule to separate for its entry into the Martian atmosphere. Although the lander initially refused to "accept" the command to separate, it eventually did. Ultimately, the lander's main retro-rocket engine failed to fire to initiate entry into the Martian atmosphere. As a result, the lander flew by the planet at a range of 1,300 kilometers and eventually entered heliocentric orbit. The flyby probe did, however, manage to collect data during its encounter with the Red Planet while contact was maintained until 25 March 1974. Both the failures on Mars 4 (computer failure) and 7 (retro-rocket ignition failure) were probably due to the faulty transistors installed in the circuits of the onboard computer which were detected prior to launch. Data from Mars 7 was being analyzed as late as 2003 when researchers published results based on data collected by the KM-73 cosmic ray detector in September 1973 en route to Mars.

144

Mariner 10

Nation: USA (55)
Objective(s): Mercury flyby, Venus flyby
Spacecraft: Mariner-73J / Mariner-J
Spacecraft Mass: 502.9 kg
Mission Design and Management: NASA / JPL
Launch Vehicle: Atlas Centaur (AC-34 / Atlas 3D no. 5014D / Centaur D-1A)
Launch Date and Time: 3 November 1973 / 05:45:00 UT

Mariner 10 took this image of Venus on 5 February 1974. The original photo was color-enhanced to provide more contrast in Venus's cloudy atmosphere. *Credit: NASA*

Launch Site: Cape Canaveral / Launch Complex 36B
Scientific Instruments:

1. 2 telescopes/cameras
2. infrared radiometer
3. ultraviolet airglow spectrometer
4. ultraviolet occultation spectrometer
5. magnetometer
6. charged particle telescope
7. plasma analyzer

Results: Mariner 10 was the first spacecraft sent to the planet Mercury; the first mission to explore two planets (Mercury and Venus) during a single mission; the first to use gravity-assist to change its flight-path; the first to return to its target after an initial encounter; and the first to use the solar wind as a major means of spacecraft orientation during flight. The primary goal of the Mariner 10 was to study the atmosphere (if any), surface, and physical characteristics of Mercury. Soon after leaving Earth orbit, the spacecraft returned striking photos of both Earth and the Moon as it sped to its first destination, Venus. During the coast, there were numerous technical problems, including malfunctions in the high-gain antenna and the attitude control system. In January 1974, Mariner 10 successfully returned data (sans photographs) on Comet C/1973 E1 Kohoutek, the first time a spacecraft returned data on a long-period comet. After mid-course corrections on 13 November 1973 and 21 January 1974, Mariner 10 approached Venus for a gravity-assist maneuver to send it towards Mercury. On 5 February 1974, the spacecraft began returning images of Venus, the first picture showing the day-night terminator of the planet as a thin bright line. Overall, Mariner 10 returned a total of 4,165 photos of the Venus and collected important scientific data during its encounter. Closest flyby range was 5,768 kilometers at 17:01 UT on 5 February. Assisted by Venusian gravity, the spacecraft now headed to the innermost planet, which it reached after another mid-course correction on 16 March 1974. As Mariner 10 approached Mercury, photos began to show a very Moon-like surface with craters, ridges, and chaotic terrain. The spacecraft's magnetometers revealed a weak magnetic field. Radiometer readings suggested nighttime temperatures of −183°C and maximum daytime temperatures of 187°C. Closest encounter came at 20:47 UT on 29 March 1974 at a range of 703 kilometers. An occultation experiment as the vehicle crossed behind the nightside of the planet indicated a lack of an atmosphere or ionosphere. Leaving Mercury behind, the spacecraft looped around the Sun and headed back to its target, helped along by

subsequent course corrections on 9 May, 10 May, and 2 July 1974. Mariner 10 flew by Mercury once more on at 20:59 UT on 21 September 1974 at a more distant 48,069 kilometers range, adding imagery of the southern polar region. The spacecraft used solar pressure on its solar panels and high-gain antenna for attitude control. Mariner 10 once again sped away from Mercury before a final and third encounter with Mercury, enabled by three maneuvers (on 30 October 1974, 13 February 1975, and 7 March 1975), the last one actually to avoid impact with the planet. The third flyby, at 22:39 UT on 16 March 1975, was the closest to Mercury, at a range of 327 kilometers. Because of the failure of a tape recorder and restrictions in the rate of data reception, only the central quarter of each of 300 high resolution images was received during this encounter. Last contact with the spacecraft was at 12:21 UT on 24 March 1975 after the spacecraft exhausted its supply of gas for attitude control. Overall, Mariner 10 returned over 2,700 pictures during its three Mercury flybys that covered nearly half of the planet's surface. Some of the images showed detail as small as 100 meters wide. Perhaps the most impressive surface feature was the Caloris basin, characterized by a set of concentric rings and ridges and about 2,500 kilometers in diameter. The mission was the last visit to Mercury by a robotic probe for more than 30 years.

1974

Luna 22

Nation: USSR (90)
Objective(s): lunar orbit
Spacecraft: Ye-8LS (no. 206)
Spacecraft Mass: 5,700 kg
Mission Design and Management: GSMZ imeni Lavochkina
Launch Vehicle: Proton-K + Blok D (8K82K no. 282-02 + 11S824 no. 0701L)
Launch Date and Time: 29 May 1974 / 08:56:51 UT
Launch Site: NIIP-5 / Site 81/24
Scientific Instruments:

1. 2 TV cameras
2. ARL-M gamma-ray spectrometer
3. RV-2N-1 radiation detector
4. SIM-RMCh meteoroid detector
5. SG-70 magnetometer
6. AKR-1 low-frequency space radio wave detector
7. 8 pairs of friction materials with different lubricants
8. 12 kinds of coatings with different reflective properties
9. Vega radio-altimeter

Results: Luna 22 was the second of two "advanced" lunar orbiters (the first being Luna 19) designed to conduct extensive scientific surveys from orbit. Launched about a year after termination of Lunokhod 2 operations on the lunar surface, Luna 22 performed a single mid-course correction en route the Moon on 30 May before entering lunar orbit on 2 June 1974. Initial orbital parameters were 219 × 222 kilometers at 19° 35′ inclination. An orbital correction on 9 June put the spacecraft in its nominal orbit of 244 × 25 kilometers for its primary imaging mission. The spacecraft carried out four mapping sessions; a fifth one was canceled due to a significant decrease in the perilune from 24.5 kilometers (on 9 June) to 15.4 kilometers (on 12 June). Nevertheless, Luna 22 provided the best Soviet imagery of the Moon. In addition to its primary mission of surface photography, Luna 22 also performed investigations to determine the chemical composition of the lunar surface, recorded meteoroid activity, searched for a lunar magnetic field, measured solar and cosmic radiation flux, and continued studies of the irregular magnetic field. Through various orbital changes—including a burn on 11 November 1974 to put the vehicle into a high 1,437 × 171-kilometer orbit to conduct gravitational experiments—Luna 22 performed without any problems, continuing to return photos 15 months into the mission, although its primary mission ended by 2 April 1975. The spacecraft's maneuvering propellant was finally depleted on 2 September and the highly successful mission was formally terminated in early November 1975. Luna 22 remains the final Soviet or Russian dedicated lunar orbiter.

Luna 23

Nation: USSR (91)
Objective(s): lunar sample return
Spacecraft: Ye-8-5M (no. 410)
Spacecraft Mass: 5,795 kg
Mission Design and Management: GSMZ imeni Lavochkina

Launch Vehicle: Proton-K + Blok D (8K82K no. 285-01 + 11S824 no. 0901L)

Launch Date and Time: 28 October 1974 / 14:30:32 UT

Launch Site: NIIP-5 / Site 81/24

Scientific Instruments:

1. stereo imaging system
2. LB09 drill for sample collection
3. radiation detector
4. radio altimeter

Results: Luna 23 was the first modified lunar sample return spacecraft, designed to return a deep core sample of the Moon's surface (hence the change in index from Ye-8-5 to Ye-8-5M). The main differences were the use of a new drilling and sampling instrument, the LB09, the removal of the low-altitude Kvant altimeter, and lightening of the torroidal instrument compartment. The diameter of the container containing soil in the return capsule had also been increased from 68 to 100 mm. While Luna 16 and 20 had returned samples from a depth of 0.3 meters, the new spacecraft was designed to dig to 2.3 meters. After a mid-course correction on 31 October, Luna 23 entered orbit around the moon on 2 November 1974. Parameters were 104 × 94 kilometers at 138° inclination. Following several more changes to the orbit, the spacecraft descended to the lunar surface on November 6. The first part of the descent occurred without anomalies until the vehicle was at 2.28-kilometer altitude. At that point, the DA-018 Doppler radar was switched on to provide data on the final stage of descent. At a height of 130 meters, however, all altitude measurements stopped. Nevertheless, the vehicle managed to land in one piece despite a landing velocity of 11 meters/second (instead of 5 meters/second). Landing was in the southernmost portion of Mare Crisium at 12° 41′ N / 62° 17′ E. As a result of the hard landing, equipment on the lander was damaged—there was a depressurization of the instrument compartment and failure of a transmitter. Subsequent attempts to activate the drill all failed, preventing fulfillment of the primary mission, the return of lunar soil to Earth.

Controllers devised a makeshift plan to conduct a limited science exploration program with the stationary lander and maintained contact with the spacecraft until 9 November 1974. Images from NASA's Lunar Reconnaissance Orbiter (LRO) in 2012 showed that Luna 23 was actually laying on its side on the lunar surface.

147

Helios 1

Nation: Federal Republic of Germany (1)

Objective(s): heliocentric orbit

Spacecraft: Helios-A

Spacecraft Mass: 370 kg

Mission Design and Management: DFVLR / NASA / GSFC

Launch Vehicle: Titan IIIE-Centaur (TC-2 / Titan no. 23E-2 / Centaur D-1T)

Launch Date and Time: 10 November 1974 / 07:11:02 UT

Launch Site: Cape Canaveral / Launch Complex 41

Scientific Instruments:

1. plasma detector
2. flux gate magnetometer
3. 2nd flux gate magnetometer
4. plasma and radio wave experiment
5. cosmic ray detector
6. low energy electron and ion spectrometer
7. zodiacal light photometer
8. micrometeoroid analyzer
9. search coil magnetometer
10. Faraday effect experiment

Results: Helios 1 was a joint German-U.S. deep space mission to study the main solar processes and solar-terrestrial relationships. Specifically, the spacecraft's instruments were designed to investigate phenomena such as solar wind, magnetic and electric fields, cosmic rays, and cosmic dust in regions between Earth's orbit and approximately 0.3 AU from the Sun. It was the largest bilateral project to date for NASA with Germany paying about $180 million of the total $260 million cost. The Federal Republic of Germany (West Germany) provided the spacecraft, named after the Greek god of the Sun, and NASA the launch vehicles. After launch, Helios 1 entered into a parking orbit around Earth and then sent into an elliptical orbit around the Sun at 0.985 × 0.3095 AU at 0.02° inclination to the ecliptic. The Centaur boost stage on this mission, after separation from the first stage, conducted a set of maneuvers to provide data for the (then-called) Mariner Jupiter-Saturn missions planned for launches in 1977. By January 1975, control of the mission had been transferred from the U.S. to West Germany, which faced a few minor communications problems, especially with antennas for both the American and German plasma wave experiments. On 15 March 1975, Helios 1 passed within 46 million kilometers of the Sun (at perihelion) at a speed of 238,000 kilometers/hour, a distance that was the closest any human-made object had been to our nearest star. The spacecraft achieved a second close flyby of the Sun on 21 September 1975 when temperatures on its solar cells reached 132°C. During its mission, the spacecraft spun once every second to evenly distribute the heat coming from the Sun, 90% of which was reflected by optical surface mirrors. Its data indicated the presence of 15 times more micrometeoroids close to the Sun than there are near Earth. Helios 1's data was correlated with the Interplanetary Monitoring Platform (IMP) Explorers 47 and 50 in Earth orbit, the Pioneer solar orbiters, and Pioneer 10 and 11 leaving the solar system. Data was received through the late 1970s and early 1980s but after 1984, both primary and backup receivers failed and the high-gain antenna lost tracking of Earth. Last telemetry from the spacecraft was received on 10 February 1986 after which the spacecraft automatically shut down its transmitter due to lack of power.

Venera 9

Nation: USSR (92)

Objective(s): Venus orbit and landing

Spacecraft: 4V-1 (no. 660)

Spacecraft Mass: 4,936 kg (in Earth orbit)

Mission Design and Management: NPO imeni Lavochkina

Launch Vehicle: Proton-K + Blok D-1 (8K82K no. 286-01 + 11S824M no. 1L)

Launch Date and Time: 8 June 1975 / 02:38:00 UT

Launch Site: NIIP-5 / Site 81/24

Scientific Instruments:

Orbiter:

1. imaging system
2. infrared radiometer
3. infrared radiometer
4. photometer
5. photopolarimeter
6. ultraviolet imaging spectrometer
7. radiophysics experiment
8. magnetometer
9. plasma electrostatic spectrometer
10. charged particle traps

Lander:

1. panoramic imaging system
2. 5 thermometers
3. 6 barometers
4. mass spectrometer
5. anemometer (ISV)
6. IOV-75 photometers
7. MNV-75 nephelometers
8. gamma-ray spectrometer
9. radiation densitometer
10. accelerometers

Results: Venera 9 was the first of a new generation of Soviet space probes ("4V") designed to explore Venus, and designed on the basis of the M-71 and M-73 Mars platforms. Launched by the more powerful Proton-K launch booster, the new spacecraft were nearly five times heavier than their predecessors. Each spacecraft comprised of both a bus and a lander, the former equipped with a powerful 11D425A engine capable of 1,928 kgf thrust (throttleable down to 1,005 kgf). For this series of missions, the 2,300-kilogram (mass at Venus orbit insertion) buses would serve as orbiters photographing the planet in ultraviolet light and conducting other scientific investigations. The 660 kilogram landers, of a completely new design, employed aerodynamic braking during Venusian atmospheric entry and contained a panoramic photometer to take images of the surface. During

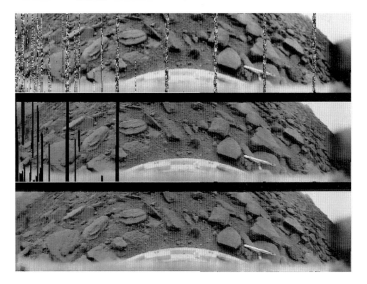

Three versions of the first surface panoramic image taken by Venera 9 on Venus. In the upper image, made up of the raw 6-bit data, the vertical lines represent telemetry bursts that interrupted the image data. The second and third images have been processed by American researcher Don Mitchell. These were the first clear images taken on the surface of a planet. *Credit: Don Mitchell*

the coast to Venus, they would be packed inside a 1,560-kilogram spherical reentry pod with a diameter of 2.4 meters. Without any apparent problems and two trajectory corrections (on 16 June and 15 October), Venera 9's lander separated from its parent on 20 October 1975 and two days later hit Venus' turbulent atmosphere at a velocity of 10.7 kilometers/hour. After aerodynamic deceleration, the cover of the parachute compartments was jettisoned at about 65 kilometers altitude, with two parachutes (one a drogue and the second to remove the upper portions of the heat shield casing) successively deployed. Descent velocity reduced, as a result, from 250 meters/second to 150 meters/second. At that point, a long drag parachute deployed and data transmission began. The drag parachute decreased descent velocity down to 50 meters/second before, finally, at 62 kilometers attitude, three large canopy parachutes deployed (with a total area of 180 m²). Four seconds after deployment, the lower hemisphere of the heat-shield casing was discarded. The now fully deployed descent vehicle descended for approximately 20 minutes before the main parachutes were jettisoned. The rest of the descent was slowed only by the capsule's own disc-shaped aerobrakes. The lander impacted on the surface at a velocity of approximately 7 meters/second. *Pravda* noted on 21 February 1976 that "the landing units, which are thin-walled torroidal shells, were deformed [as planned] during landing, thereby absorbing the energy of the impact and assured an oriented position of the descent vehicle on the planet." Landing occurred on the planet's dayside at 05:13:07 UT on 22 October. (Times were only announced for reception of landing signal on Earth). Landing coordinates were within a 150-kilometer radius of 31.01° N / 290.64° E at the base of a hill near Beta Regio. During its 53 minutes of transmissions from the surface, Venera 9 took and transmitted the very first picture of the Venusian surface, taken from a height of 90 centimeters. These were, in fact, the very first photos received of the surface of another planet. The lander was supposed to transmit a full 360° panorama, but because one of the two covers on the camera failed to release, only a 180° panorama was received. Illumination was akin to a cloudy day on Earth. The image clearly showed flat rocks strewn around the lander. The Venera 9 orbiter meanwhile entered a 1,510 × 112,200-kilometer orbit around the planet at 34° 10′ inclination and acted as a communications relay for the lander. It became the first spacecraft to go into orbit around Venus. The Soviets announced on 22 March 1976 that the orbiter's primary mission, which included using French-made ultraviolet cameras to obtain photographs in 1,200 kilometer swaths, had been fulfilled.

149

Venera 10

Nation: USSR (93)
Objective(s): Venus orbit and landing
Spacecraft: 4V-1 (no. 661)
Spacecraft Mass: 5,033 kg (in Earth orbit)
Mission Design and Management: NPO imeni Lavochkina
Launch Vehicle: Proton-K + Blok D-1 (8K82K no. 285-02 + 11S824M no. 2L)
Launch Date and Time: 14 June 1975 / 03:00:31 UT
Launch Site: NIIP-5 / Site 81/24
Scientific Instruments:
  Orbiter:
    1. imaging system
    2. infrared radiometer
    3. infrared radiometer
    4. photometer
    5. photopolarimeter
    6. ultraviolet imaging spectrometer
    7. radiophysics experiment
    8. magnetometer
    9. plasma electrostatic spectrometer
    10. charged particle traps
  Lander:
    1. panoramic imaging system

2. 5 thermometers

3. 6 barometers

4. mass spectrometer

5. anemometer (ISV)

6. IOV-75 photometers

7. MNV-75 nephelometers

8. gamma-ray spectrometer

9. radiation densitometer

10. accelerometers

Results: Venera 10, like its sister craft Venera 9, fully accomplished its mission to soft-land on Venus and return data from the surface. The spacecraft followed an identical mission to its twin, arriving only a few days later after two trajectory corrections on 21 June and 18 October 1975. The 1,560-kilogram lander separated from its parent on 23 October and entered the atmosphere two days later at 01:02 UT. During reentry, the lander survived loads as high as 168 g's and temperatures as high 12,000°C. It performed its complex landing procedures without fault (see Venera 9 for details) and landed without incident at 05:17:06 UT approximately 2,200 kilometers from the Venera 9 landing site. (Times were only announced for reception of landing signal on Earth). Landing coordinates were a 150-kilometer radius of 15.42° N / 291.51° E. Venera 10 transmitted for a record 65 minutes from the surface, although it was designed to last only 30 minutes. A photo of the Venera 10 landing site showed a smoother surface than that of its twin. The small image size was part of the original plan, and was determined by the slow telemetry rates and an estimated 30-minute lifetime. Like Venera 9, the Venera 10 lander was supposed to take a 360° panorama but covered only 180° of the surroundings because of a stuck lens cover. Soviet officials later revealed that termination of data reception from both Venera 9 and 10 landers was not due to the adverse surface conditions but because the orbiter relays for both spacecraft flew out of view. Gamma-ray spectrometer and radiation densitometer (shaped a bit like a paint-roller deployed on the surface) data indicated that the surface layer was akin to basalt rather than granite as hinted by the information from Venera 8. The Venera 10 orbiter meanwhile entered a 1,620 × 113,000-kilometer orbit around Venus inclined at 29° 30′, transmitting data until at least June 1976. Unlike the Venera 9 orbiter, photographs taken by the Venera 10 orbiter were never released and it remains unclear whether it actually carried a camera.

150

Viking 1

Nation: USA (56)

Objective(s): Mars landing and orbit

Spacecraft: Viking-B

Spacecraft Mass: 3,527 kg

Mission Design and Management: NASA / LaRC (overall) / JPL (Orbiter)

Launch Vehicle: Titan IIIE-Centaur (TC-4 / Titan no. E-4 / Centaur D-1T)

Launch Date and Time: 20 August 1975 / 21:22:00 UT

Launch Site: Cape Canaveral / Launch Complex 41

Scientific Instruments:

Orbiter:

1. imaging system (2 vidicon cameras) (VIS)

2. infrared spectrometer for water vapor mapping (MAWD)

3. infrared radiometer for thermal mapping (IRTM)

Lander:

1. imaging system (2 facsimile cameras)

2. gas chromatograph mass spectrometer (GCMS)

3. seismometer

4. x-ray fluorescence spectrometer

5. biological laboratory

6. weather instrument package (temperature, pressure, wind velocity)

7. remote sampler arm

Aeroshell:

1. retarding potential analyzer

2. upper-atmosphere mass spectrometer

3. pressure, temperature, and density sensors

Carl Sagan, a member of the Viking science team, posing with a life-size model of the Viking Lander in Death Valley, California in the mid-1970s. At the time, Sagan was professor of astronomy at Cornell University. *Credit: NASA/JPL*

Results: Viking 1 was the first of a pair of complex deep space probes that were designed to reach Mars and collect evidence on the possibility (or lack of) for life on Mars. Each spacecraft was composed of two primary elements, an orbiter (2,339 kilograms) and a lander (978 kilograms). The Orbiter design heavily borrowed from the Mariner buses, while the Lander looked superficially like a much larger version of the Surveyor lunar lander. Prior to launch, the batteries of the first spacecraft were discharged, prompting NASA to replace the original first spacecraft with the second, which was launched as Viking 1. After three mid-course corrections (on 27 August 1975, and 10 and 15 June 1976), the spacecraft entered orbit around Mars on 19 June 1976. Initial orbital parameters were 1,500 × 50,300 kilometers. The following day, the orbiter moved into an operational orbit at 1,500 × 32,800 kilometers. The same day, when

the Orbiter began transmitting back photos of the primary landing site in the Chryse region, scientists discovered that the area was rougher than expected. Using the new photos, scientists targeted the lander to a different site on the western slopes of Chryse Planitia ("Golden Plain"). The Lander separated from the Orbiter at 08:32 UT on 20 July 1976, and after a complicated atmospheric entry sequence during which the probe took air samples, Viking Lander 1 set down safely at 22.483° N / 47.94° N at 11:53:06 UT on 20 July 1976 (about 28 kilometers from its planned target). Once down, the spacecraft began taking high-quality photographs (in three colors) of its surroundings. Besides high-resolution images, the lander also took a 300° panorama of its surroundings that showed not only parts of the spacecraft itself but also the gently rolling plains of the environs. Instruments recorded temperatures ranging

NASA scientists at the Langley Research Center in Virginia stand in front of the aeroshell that protected the Viking Lander I during its entry into the Martian atmosphere in 1976. The aeroshell was made of the heat shield and a "back-shell" which contained parachutes and other components used for entry. *Credit: NASA/JPL*

from −86°C (before dawn) to −33°C (in the afternoon). The seismometer on the lander was, however, inoperable. On 28 July, the lander's robot arm scooped up the first soil samples and deposited them into a special biological laboratory that included a gas chromatograph mass spectrometer. The cumulative data from the four samples collected could be construed as indicating the presence of life ("weak positive"), but the major test for organic compounds using the gas chromatograph experiment, capable of detecting organic compounds that comprised more than 10–100 parts per billion in the soil, gave negative results. Data

showed an abundance of sulfur, certainly different from any known material found on Earth or the Moon. While the primary mission for both Viking 1 and Viking 2 ended in November 1976, activities continued through the Extended Mission (November 1976 to May 1978) and the Continuation Mission (May 1978 to July 1979). Viking 1's orbiter then continued a "Survey Mission" from July 1979 to July 1980. The Lander continued to return daily (and then eventually weekly) weather reports as part of the Viking Monitor Mission. In January 1982, it was renamed the Thomas Mutch Memorial Station in honor of Thomas A. Mutch

Panoramic image taken by Viking Lander 2, created by combining standard low resolution color images with standard high resolution black and white images. *Credit: NASA/JPL*

(1931–1980) the leader of the Viking imaging team who had passed away on 6 October 1980. The lander operated until 11 November 1982 when a faulty command sent from Earth resulted in an interruption of communications. Further attempts to regain contact proved to be unsuccessful. The Orbiter, after taking many more high-resolution images of the planet and its two moons, far superior than those from Mariner 9, was shut down on 7 August 1980 after it ran out of attitude control propellant on its 1,489th orbit around Mars. Current projections are that the orbiter will enter the Martian atmosphere sometime around 2019.

151

Viking 2

Nation: USA (57)
Objective(s): Mars landing and orbit
Spacecraft: Viking-A
Spacecraft Mass: 3,527 kg
Mission Design and Management: NASA / LaRC (overall) / JPL (Orbiter)
Launch Vehicle: Titan IIIE-Centaur (TC-3 / Titan no. E-3 / Centaur no. D-1T)

Viking 2 Lander image showing the spacecraft and part of Utopia Planitia, looking due south. The American flag, color grid, and bicentennial symbols (for the 200th anniversary of the Declaration of the Independence, celebrated in 1976) are visible in this image, and were used to calibrate color images as they were received on Earth. The image was taken on 2 November 1976. *Credit: NASA/JPL*

Launch Date and Time: 9 September 1975 / 18:39:00 UT

Launch Site: Cape Canaveral / Launch Complex 41

Scientific Instruments:

Orbiter:

1. imaging system (2 vidicon cameras) (VIS)
2. infrared spectrometer for water vapor mapping (MAWD)
3. infrared radiometer for thermal mapping (IRTM)

Lander:

1. imaging system (2 facsimile cameras)
2. gas chromatograph mass spectrometer (GCMS)
3. seismometer
4. x-ray fluorescence spectrometer
5. biological laboratory
6. weather instrument package (temperature, pressure, wind velocity)
7. remote sampler arm

Aeroshell:

1. retarding potential analyzer
2. upper-atmosphere mass spectrometer
3. pressure, temperature, and density sensors

Results: Viking-A was scheduled to be launched first, but had to be launched second due to a problem with its batteries that had to be repaired. After a successful launch and a mid-course correction on 19 September 1975, Viking 2 entered orbit around Mars nearly a year after launch on 7 August 1976. Initial orbital parameters were 1,502 × 35,728 kilometers inclined at 55.6°. As with Viking 1, photographs of the original landing site indicated rough terrain, prompting mission planners to select a different site at Utopia Planitia near the edge of the polar ice cap where water was located, i.e., where there was a better chance of finding signs of life. The Lander separated from the Orbiter without incident at 20:19 UT on 3 September 1976 and after atmospheric entry, landed safely at 22:37:50 UT, about 6,460 kilometers from the Viking 1 landing site. Touchdown coordinates were 47.968° N / 225.71° W. Photographs of the area showed a rockier, flatter site than that of Viking 1. The lander was in fact tilted 8.5° to the west. Panoramic views of the landscape showed a terrain different from that of Viking 1, with much less definition and very little in the way of horizon features. Because of the lack of general topographical references on the ground, imagery from the orbiters was unable to precisely locate the lander. The biology experiments with scooped up soil collected on three occasions (beginning on 12 September) produced similar results to its twin, i.e., inconclusive on the question of whether life exists or ever has existed on the surface of Mars. Scientists believed that Martian soil contained reactants created by ultraviolet bombardment of the soil that could produce characteristics of living organisms in Earth soil. On 16 November 1976, NASA announced that both Viking 1 and Viking 2 missions had successfully accomplished their mission goals and announced an Extended Mission that continued until May 1978 followed by a Continuation Mission until July 1979. The Orbiter continued its successful imaging mission, approaching as close as 28 kilometers to the Martian Moon Deimos in May 1977. A series of leaks prompted termination of Orbiter 2 operations on 24 July 1978 while the Lander 2 continued to transmit data until 12 April 1980. In July 2001, the Viking 2 lander was renamed the Gerald Soffen Memorial Station after Gerald Soffen (1926–2000), the NASA Project Scientist for Viking who had died recently. In total, the two Viking Orbiters returned 52,663 images of Mars and mapped about 97% of the surface at a resolution of 300 meters resolution. The Landers returned 4,500 photos of the two landing sites.

152

[Luna]

Nation: USSR (94)

Objective(s): lunar sample return

Spacecraft: Ye-8-5M (no. 412)

Spacecraft Mass: 5,795 kg

Mission Design and Management: NPO imeni Lavochkina

Launch Vehicle: Proton-K + Blok D (8K82K no. 287-02 + 11S824 no. 1401L)

Launch Date and Time: 16 October 1975 / 14:04:56 UT

Launch Site: NIIP-5 / Site 81/23

Scientific Instruments:

1. stereo imaging system
2. LB09 drill for sample collection
3. radiation detector
4. radio altimeter

Results: This was the second attempt by the Soviet Union to send an "advanced" lunar sample return craft to the Moon, equipped with the capability to dig for a deeper core. The first spacecraft (Luna 23) was damaged during landing on the Moon in October 1974. On this mission, the first three stages of the Proton-K launch vehicle worked without fault, but the Blok D stage, during its first burn for insertion into Earth orbit, failed. The expensive payload burned up in Earth's atmosphere without ever reaching Earth orbit.

1976

Helios 2

Nation: Federal Republic of Germany (2)
Objective(s): solar orbit
Spacecraft: Helios-B
Spacecraft Mass: 370 kg
Mission Design and Management: DFVLR / NASA / GSFC
Launch Vehicle: Titan IIIE-Centaur (TC-5 / Titan no. E-5 / Centaur D-1T)
Launch Date and Time: 15 January 1976 / 05:34:00 UT
Launch Site: Cape Canaveral / Launch Complex 41
Scientific Instruments:

1. plasma detector
2. flux gate magnetometer
3. 2nd flux gate magnetometer
4. plasma and radio wave experiment
5. cosmic ray detector
6. low energy electron and ion spectrometer
7. zodiacal light photometer
8. micrometeoroid analyzer
9. search coil magnetometer
10. Faraday effect experiment

A technician stands next to one of the twin Helios spacecraft. *Credit: NASA/ Max Planck Institute for Solar System Research*

Results: Helios 2 was the second spacecraft launched to investigate solar processes as part of cooperative project between the Federal Republic of Germany and the United States in which the former provided the spacecraft and the latter the launch vehicle. Although similar to Helios 1, the second spacecraft had improved systems designed to help it survive longer. Like its twin, the spacecraft was put into heliocentric orbit; all communications with the spacecraft was directed from the German Space Operation Center near Munich. In contrast to Helios 1, Helios 2, flew three million kilometers closer to the Sun, achieving perihelion on 17 April 1976 at a distance of 0.29 AU (or 43.432 million kilometers), a distance that makes Helios 2 the record holder for the closest ever flyby of the Sun. As a result, the spacecraft was exposed to 10% more heat (or 20°C more) than its predecessor. The spacecraft provided important information on solar plasma, the solar wind, cosmic rays, and cosmic dust, and also performed magnetic field and electrical field experiments. Besides its investigation of the Sun and solar environment, both Helios 1 and Helios 2 observed the dust and ion tails of at least three comets, C/1975V1 West, C/1978H1 Meier, and C/1979Y1 Bradfield. Helios 2's downlink transmitter, however, failed on 3 March 1980 and no further usable data was received from the spacecraft. Ground controllers shut down the spacecraft on 7 January 1981 to preclude any possible radio interference with other spacecraft in the future.

154

Luna 24

Nation: USSR (95)

Objective(s): lunar sample return

Spacecraft: Ye-8-5M (no. 413)

Spacecraft Mass: c. 5,800 kg

Mission Design and Management: NPO imeni Lavochkina

Launch Vehicle: Proton-K + Blok D (8K82K no. 288-02 + 11S824 no. 1501L)

Launch Date and Time: 9 August 1976 / 15:04:12 UT

Launch Site: NIIP-5 / Site 81/23

Scientific Instruments:

1. stereo imaging system
2. LB09 drill for sample collection
3. radiation detector
4. radio altimeter

Results: Luna 24 was the third attempt to recover a sample from the unexplored Mare Crisium (after Luna 23 and a subsequent launch failure in October 1975), the location of a large lunar mascon. After a trajectory correction on 11 August 1976, Luna 24 entered orbit around the Moon three days later. Initial orbital parameters were 115 × 115 kilometers at 120° inclination. After further changes to its orbit on 16 and 17 August, which brought the orbit down to an elliptical 120 × 12 kilometers, Luna 24 began its descent to the surface at perilune by firing its descent engine. Just 6 minutes later, the spacecraft set down safely on the lunar surface at 06:36 UT on 18 August 1976 at 12° 45′ N / 62° 12′ E (as announced at the time), not far from where Luna 23 had landed. Later analysis showed that the spacecraft landed just 10 meters from the rim of a 65-meter diameter impact crater.

After appropriate commands from ground control, within 15 minutes of landing, the lander deployed its sample arm and pushed its drilling head (using a rotary drilling mode) about 2.25 meters into the nearby soil. Because the drill entered at an angle, the probable surface depth of the sample was about 2 meters. The sample was safely stowed in the small return capsule, and after nearly a day on the Moon, Luna 24 lifted off successfully from the Moon at 05:25 UT on 19 August 1976. After an uneventful return trip lasting 84 hours, Luna 24's capsule entered Earth's atmosphere and parachuted down to Earth safely at 05:55 UT on 23 August 1976 about 200 kilometers southeast of Surgut in western Siberia. Study of the recovered 170.1 grams of soil indicated a laminated type structure, as if laid down in successive deposits. Tiny portions of the sample were exchanged with NASA in December 1976. At the time, the Luna 24 sample puzzled investigators because its titanium content and "maturity" (amount of time the sample was exposed to the space environment) were very different than expected at Mare Crisium. Images from NASA's Lunar Reconnaissance Orbiter (LRO) in 2012 showed that lander sampled impact ejecta from a nearby 64-meter diameter crater that brought up material from deeper lava flows that had not been previously exposed to the space environment. Thus, the Luna 24 sample probably represented subsurface materials only exposed to the space environment for a relatively short time. LRO images helped refine the landing target area as 12.7146° N / 62.2129° E, which was only 2.32 kilometers from the crashed Luna 23. Luna 24 remains the last Soviet or Russian probe to the Moon. A Japanese spacecraft (Hiten, page 179) returned to the Moon nearly 14 years later.

155

Voyager 2

Nation: USA (58)

Objective(s): Jupiter flyby, Saturn flyby, Uranus flyby, Neptune flyby

Spacecraft: Voyager-2

Spacecraft Mass: 721.9 kg

Mission Design and Management: NASA / JPL

Launch Vehicle: Titan IIIE-Centaur (TC-7 / Titan no. 23E-7 / Centaur D-1T)

Launch Date and Time: 20 August 1977 / 14:29:44 UT

Launch Site: Cape Canaveral / Launch Complex 41

Scientific Instruments:

1. imaging science system (ISS)
2. ultraviolet spectrometer (UVS)
3. infrared interferometer spectrometer (IRIS)
4. planetary radio astronomy experiment (PRA)
5. photopolarimeter (PPS)
6. triaxial fluxgate magnetometer (MAG)
7. plasma spectrometer (PLS)
8. low-energy charged particles experiment (LECP)
9. plasma waves experiment (PWS)
10. cosmic ray telescope (CRS)
11. radio science system (RSS)

One of the two Voyager Golden Records displayed with a Voyager spacecraft. *Credit: NASA/JPL*

Results: The two-spacecraft Voyager missions were designed to replace original plans for a "Grand Tour" of the planets that would have used four highly complex spacecraft to explore the five outer planets during the late 1970s. NASA canceled the plan in January 1972 largely due to anticipated costs (projected at $1 billion) and instead proposed to launch only two spacecraft in 1977 to Jupiter and Saturn. The two spacecraft were designed to explore the two gas giants in more detail than the two Pioneers (Pioneers 10 and 11) that preceded them. In 1974, mission planners proposed a mission in which, if the first Voyager was successful, the second one could be redirected to Uranus and then Neptune using gravity assist maneuvers. Each of the two spacecraft were equipped with slow-scan color TV to take images of the planets and their moons and also carried an extensive suite of instruments to record magnetic, atmospheric, lunar, and other data about the planetary systems. The design of the two spacecraft was based on the older Mariners, and they were known as Mariner 11 and Mariner 12 until 7 March 1977 when NASA Administrator James C. Fletcher (1919–1991) announced that they would be renamed Voyager.

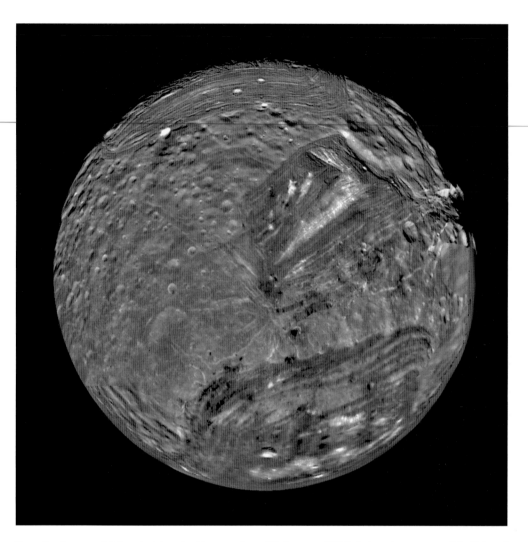

Stunning image of Miranda taken by Voyager 2 on 24 January 1986 during the spacecraft's flyby of Uranus. This particular image was made from nine separate photos combined to obtain a full disc. *Credit: NASA/JPL*

Power was provided by three plutonium dioxide radioisotope thermoelectric generators (RTGs) mounted at the end of a boom. Although launched after Voyager 1, Voyager 2 exited the asteroid belt after its twin and then followed it to Jupiter and Saturn. Its primary radio transmitter failed on 5 April 1978 and the spacecraft used its backup past that point. Voyager 2 began transmitting images of Jupiter on 24 April 1979 for time-lapse movies of atmospheric circulation. Unlike Voyager 1, Voyager 2 made close passes to the Jovian moons on its way *into* the system, with scientists especially interested in more information from Europa and Io (which necessitated a 10-hour long "volcano watch"). During its encounter, it relayed back spectacular photos of the entire Jovian system, including its moons Callisto, Ganymede, Europa (at 205,720-kilometer range, much closer than Voyager 1), Io, and Amalthea, all of which had already been surveyed by Voyager 1. Voyager 2's closest encounter to Jupiter was at 22:29 UT on 9 July 1979 at a range of 645,000 kilometers. It transmitted new data on the planet's clouds, its newly discovered four moons, and ring system as well as 17,000 new pictures. When the earlier Pioneers had flown by Jupiter, they noticed few atmospheric changes

High-altitude cloud streaks are visible in Neptune's atmosphere in a picture taken during Voyager 2's flyby of the gas giant in 1989. *Credit: NASA/JPL*

from one encounter to the second; in this case, Voyager 2 detected many significant changes, particular a drift in the Great Red Spot as well as changes in its shape and color. With the combined cameras of the two Voyagers, at least 80% of the surfaces of Ganymede and Callisto were mapped out to a resolution of 5 kilometers. Following a mid-course correction 2 hours after its closest approach to Jupiter, Voyager 2 sped to Saturn, its

trajectory determined to a large degree by the decision, taken in January 1981, to try and send the spacecraft to Uranus and Neptune later in the decade. Its encounter with the sixth planet began on 22 August 1981, two years after leaving the Jovian system, with imaging of the moon Iapetus. Once again, Voyager 2 repeated the photographic mission of its predecessor, although it actually flew 23,000 kilometers closer to Saturn. Closest

The 34 meter High Efficiency Antenna in the foreground was nicknamed the "Uranus Antenna" because it was built in the 1980s to receive signals during Voyager 2's Uranus encounter. The antenna is located at the Goldstone Deep Space Communications Complex in California. *Credit: NASA*

encounter was at 01:21 UT on 26 August 1981 at 101,000 kilometer range. The spacecraft provided more detailed images of the ring "spokes" and kinks, and also the F-ring and its shepherding moons, all found by Voyager 1. Voyager 2's data suggested that Saturn's A-ring was perhaps only 300 meters thick. As it flew behind and up past Saturn, the probe passed through the plane of Saturn's rings at a speed of 13 kilometers/second; for several minutes during this phase, the spacecraft was hit by thousands of micron-sized dust grains that created "puff" plasma as they were vaporized. Because the vehicle's attitude was repeatedly shifted by the particles, attitude control jets automatically fired many times to stabilize the vehicle. During the encounter, Voyager 2 also

photographed the Saturn moons Hyperion (the "hamburger moon"), Enceladus, Tethys, and Phoebe as well as the more recently discovered Helene, Telesto, and Calypso. Although Voyager 2 had fulfilled its primary mission goals with the two planetary encounters, mission planners directed the veteran spacecraft to Uranus on a four-and-a-half-year-long journey during which it covered 33 AU's. In fact, its encounter with Jupiter was optimized in part to ensure that future planetary flybys would be possible. The Uranus encounter's geometry was also defined by the possibility of a future encounter with Neptune: Voyager 2 had only 5.5 hours of close study during its flyby, the first of any human-made spacecraft past the planet Uranus. Long-range observations of the planet began on

4 November 1985 when signals took approximately 2.5 hours to reach Earth. Light conditions were 400 times less than terrestrial conditions. Closest approach to Uranus took place at 17:59 UT on 24 January 1986 at a range of 81,500 kilometers. During its flyby, Voyager 2 discovered 10 new moons (given such names as Puck, Portia, Juliet, Cressida, Rosalind, Belinda, Desdemona, Cordelia, Ophelia, and Bianca—obvious allusions to Shakespeare), two new rings in addition to the "older" nine rings, and a magnetic field tilted at 55° off-axis and off-center. The spacecraft found wind speeds in Uranus' atmosphere as high as 724 kilometers/hour and found evidence of a boiling ocean of water some 800 kilometers below the top cloud surface. Its rings were found to be extremely variable in thickness and opacity. Voyager 2 also returned spectacular photos of Miranda, Oberon, Ariel, Umbriel, and Titania, five of Uranus' larger moons. In flying by Miranda at a range of only 28,260 kilometers, the spacecraft came closest to any object so far in its nearly decade-long travels. Images of the moon showed a strange object whose surface was a mishmash of peculiar features that seemed to have no rhyme or reason. Uranus itself appeared generally featureless in the photographs taken. The spectacular news of the Uranus encounter was interrupted the same week by the tragic Challenger accident that killed seven astronauts during their Space Shuttle launch on 28 January. Following the Uranus encounter, the spacecraft performed a single mid-course correction on 14 February 1986—the largest ever made by Voyager 2—to set it on a precise course to Neptune. Voyager 2's encounter with Neptune capped a 7 billion-kilometer journey when on 25 August 1989 at 03:56 UT, it flew 4,800 kilometers over the cloud tops of the giant planet, the closest of its four flybys. It was the first human-made object to fly by the planet. Its 10 instruments were still in working order at the time. During the encounter, the spacecraft discovered six new moons (Proteus, Larissa, Despina, Galatea, Thalassa, and Naiad) and four new rings. The planet itself was found to be more active than previously believed, with 1,100 kilometer winds. Hydrogen was found to be the most common atmospheric element, although the abundant methane gave the planet its blue appearance. Images revealed details of the three major features in the planetary clouds—the Lesser Dark Spot, the Great Dark Spot, and Scooter. Voyager photographed two-thirds of Neptune's largest moon Triton, revealing the coldest known planetary body in the solar system and a nitrogen ice "volcano" on its surface. Spectacular images of its southern hemisphere showed a strange, pitted "cantaloupe"-type terrain. The flyby of Neptune concluded Voyager 2's planetary encounters, which spanned an amazing 12 years in deep space, virtually accomplishing the originally planned Grand Tour of the solar system, at least in terms of targets reached if not in science accomplished. Once past the Neptune system, Voyager 2 followed a course below the ecliptic plane and out of the solar system. Approximately 56 million kilometers past the encounter, Voyager 2's instruments were put in low power mode to conserve energy. After the Neptune encounter, NASA formally renamed the entire project the Voyager Interstellar Mission (VIM). Of the four spacecraft sent out to beyond the environs of the solar system in the 1970s, three of them—Voyagers 1 and 2 and Pioneer 11—were all heading in the direction of the solar apex, i.e., the apparent direction of the Sun's travel in the Milky Way galaxy, and thus would be expected to reach the heliopause earlier than Pioneer 10 which was headed in the direction of the heliospheric tail. In November 1998, 21 years after launch, non-essential instruments were permanently turned off, leaving seven instruments still operating. Through the turn of the century, JPL continued to receive ultraviolet and particle/fields data. For example, on 12 January 2001, an immense shock wave that had blasted out of the outer heliosphere on 14 July 2000, finally reached Voyager 2. During the six-month journey, the shock wave had ploughed through the solar wind, sweeping up and accelerating charged particles. The spacecraft

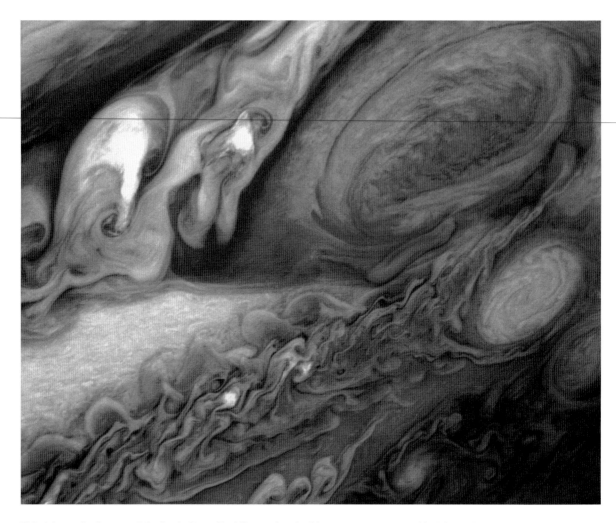

This false color image of Jupiter's Great Red Spot taken by Voyager 1 was assembled from three black-and-white negatives. *Credit: NASA/JPL*

provided important information on high-energy shock-energized ions. On 30 August 2007, Voyager 2 passed the termination shock and then entered the heliosheath. By November 5, 2017, the spacecraft was 116.167 AU (17.378 billion kilometers) from Earth, moving at a velocity of 15.4 kilometers/second relative to the Sun, heading in the direction of the constellation Telescopium. At this velocity, it would take about 19,390 years to traverse a single light-year. Data from the remaining five operating instruments—the cosmic ray telescope, the low-energy charged particles experiment, the magnetometer, the plasma waves experiment, and the plasma spectrometer—could be received as late as 2025.

156

Voyager 1

Nation: USA (57)

Objective(s): Jupiter flyby, Saturn flyby

Spacecraft: Voyager-1

Spacecraft Mass: 721.9 kg

Mission Design and Management: NASA / JPL

Launch Vehicle: Titan IIIE-Centaur (TC-6 / Titan no. 23E-6 / Centaur D-1T)

Launch Date and Time: 5 September 1977 / 12:56:01 UT

Launch Site: Cape Canaveral / Launch Complex 41

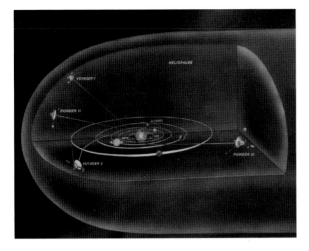

Image showing the general trajectories of the four NASA probes, Pioneers 10 and 11 and Voyagers 1 and 2, sent out of the solar system. As of February 2017, Voyager 1 was at a distance of 20.6 billion kilometers from the Sun while Voyager 2 was at 17 billion kilometers. Both Voyagers are headed towards the outer boundary of the solar system in search of the heliopause, the region where the Sun's influence wanes and the beginning of interstellar space can be sensed. *Credit: NASA/JPL*

Scientific Instruments:

1. imaging science system (ISS)
2. ultraviolet spectrometer (UVS)
3. infrared interferometer spectrometer (IRIS)
4. planetary radio astronomy experiment (PRA)
5. photopolarimeter (PPS)
6. triaxial fluxgate magnetometer (MAG)
7. plasma spectrometer (PLS)
8. low-energy charged particles experiment (LECP)
9. plasma waves experiment (PWS)
10. cosmic ray telescope (CRS)
11. radio science system (RSS)

Results: Voyager 1 was launched after Voyager 2, but because of a faster route, it exited the asteroid belt earlier than its twin, having overtaken Voyager 2 on 15 December 1977. It began its Jovian imaging mission in April 1978 at a range of 265 million kilometers from the planet; images sent back by January the following year indicated that Jupiter's atmosphere was more turbulent than during the Pioneer flybys in 1973–1974. Beginning on 30 January, Voyager 1 took a picture every 96

seconds for a span of 100 hours to generate a color timelapse movie to depict 10 rotations of Jupiter. On 10 February 1979, the spacecraft crossed into the Jovian moon system and in early March, it had already discovered a thin (less than 30 kilometers thick) ring circling Jupiter. Voyager 1's closest encounter with Jupiter was at 12:05 UT on 5 March 1979 at a range of 280,000 kilometers, following which it encountered several of Jupiter's Moons, including Amalthea (at 420,200-kilometer range), Io (21,000 kilometers), Europa (733,760 kilometers), Ganymede (114,710 kilometers), and Callisto (126,400 kilometers), in that order, returning spectacular photos of their terrain, opening up a completely new world for planetary scientists. The most interesting find was on Io, where images showed a bizarre yellow, orange, and brown world with at least eight active volcanoes spewing material into space, making it one of the most (if not the most) geologically active planetary body in the solar system. The presence of active volcanoes suggested that the sulfur and oxygen in Jovian space may be a result of the volcanic plumes from Io which are rich in sulfur dioxide. The spacecraft also discovered two new moons, Thebe and Metis. Following the Jupiter encounter, Voyager 1 completed an initial course correction on 9 April 1979 in preparation

Voyager 1 acquired this image of Io on 4 March 1979 about 11 hours before closest approach to the Jupiter Moon, at a range of 490,000 kilometers from the target. Visible is an enormous volcanic explosion silhouetted against dark space. *Credit: NASA/JPL*

for its meeting with Saturn. A second correction on 10 October 1979 ensured that the spacecraft would not hit Saturn's moon Titan. Its flyby of the Saturn system in November 1979 was as spectacular as its previous encounter. Voyager 1 found five new moons, a ring system consisting of thousands of bands, wedge-shaped transient clouds of tiny particles in the B ring that scientists called "spokes," a new ring (the "G-ring"), and "shepherding" satellites on either side of the F-ring—satellites that keep the rings well-defined. During its flyby, the spacecraft photographed Saturn's moons Titan, Mimas, Enceladus, Tethys, Dione, and Rhea. Based on incoming data, all the moons appeared to be composed largely of water ice. Perhaps the most interesting target was Titan, which Voyager 1 passed at 05:41 UT on 12 November at a range of 4,000 kilometers. Images showed a thick atmosphere that completely hid the surface. The spacecraft found that the moon's atmosphere was composed of 90% nitrogen. Pressure and temperature at the surface was 1.6 atmospheres and –180°C, respectively. Atmospheric data suggested that Titan might be the first body in the solar system (apart from Earth) where liquid might exist on the surface. In addition, the presence of nitrogen, methane, and more complex hydrocarbons indicated that prebiotic chemical reactions might be possible on Titan. Voyager 1's closest approach to Saturn was at 23:46 UT on 12 November 1980 at a range of 126,000 kilometers. Following the encounter with Saturn, Voyager 1 headed on a trajectory escaping the solar system at a speed of about 3.5 AU per year, 35° out of the ecliptic plane to the north, in the general direction of the Sun's motion relative to nearby stars. Because of the specific requirements for the Titan flyby, the spacecraft was not directed to Uranus and Neptune. The final images taken by the Voyagers comprised a mosaic of 64 images taken by Voyager 1 on February 14, 1990 at a distance of 40 AU of the Sun and all the planets of the solar system (although Mercury and Mars did not appear, the former because it was too close to the Sun and the latter because Mars was on the same

side of the Sun as Voyager 1 so only its dark side faced the cameras). This was the so-called "pale blue dot" image made famous by Cornell University professor and Voyager science team member Carl Sagan (1934-1996). These were the last of a total of 67,000 images taken by the two spacecraft. All the planetary encounters finally over in 1989, the missions of Voyager 1 and 2 were declared part of the Voyager Interstellar Mission (VIM), which officially began on 1 January 1990. The goal was to extend NASA's exploration of the solar system beyond the neighborhood of the outer planets to the outer limits of the Sun's sphere of influence, and "possibly beyond." Specific goals include collecting data on the transition between the heliosphere, the region of space dominated by the Sun's magnetic field and solar field, and the interstellar medium. On 17 February 1998, Voyager 1 became the most distant human-made object in existence when, at a distance of 69.4 AU from the Sun when it "overtook" Pioneer 10. On 16 December 2004, Voyager scientists announced that Voyager 1 had reported high values for the intensity for the magnetic field at a distance of 94 AU, indicating that it had reached the termination shock and had now entered the heliosheath. The spacecraft finally exited the heliosphere and began measuring the interstellar environment on 25 August 2012, the first spacecraft to do so. On 5 September 2017, NASA marked the 40th anniversary of its launch, as it continues to communicate with NASA's Deep Space Network and send data back from four still-functioning instruments—the cosmic ray telescope, the low-energy charged particles experiment, the magnetometer, and the plasma waves experiment. Each of the Voyagers contain a "message," prepared by a team headed by Carl Sagan, in the form of a 30-centimeter diameter gold-plated copper disc for potential extraterrestrials who might find the spacecraft. Like the plaques on Pioneers 10 and 11, the record has inscribed symbols to show the location of Earth relative to several pulsars. The records also contain instructions to play them using a cartridge and a needle, much like a vinyl record player. The audio on the disc includes greetings in 55 languages, 35 sounds from life on Earth (such as whale songs, laughter, etc.), 90 minutes of generally Western music including everything from Mozart and Bach to Chuck Berry and Blind Willie Johnson. It also includes 115 images of life on Earth and recorded greetings from then U.S. President Jimmy Carter (1924–) and then-UN Secretary-General Kurt Waldheim (1918–2007). By 5 November 2017, Voyager 1 was 140.931 AU (21.083 billion kilometers) from Earth, the farthest object created by humans, and moving at a velocity of 17.0 kilometers/second relative to the Sun.

157

Pioneer Venus 1

Nation: USA (60)
Objective(s): Venus orbit
Spacecraft: Pioneer Venus Orbiter
Spacecraft Mass: 582 kg
Mission Design and Management: NASA / ARC
Launch Vehicle: Atlas Centaur (AC-50 / Atlas no. 5030D)
Launch Date and Time: 20 May 1978 / 13:13:00 UT
Launch Site: Cape Canaveral / Launch Complex 36A
Scientific Instruments:

1. charged particle retarding potential analyzer (ORPA)
2. ion mass spectrometer (OIMS)
3. thermal electron temperature Langmuir probe (OETP)
4. neutral particle mass spectrometer (ONMS)
5. cloud photopolarimeter (OCPP)
6. temperature sounding infrared radiometer (OIR)
7. magnetic field fluxgate magnetometer (OMAG)
8. solar wind plasma analyzer (OPA)
9. surface radar mapper (ORAD)
10. electric field detector (OEFD)
11. transient gamma ray burst detector (OGBD)
12. radio occultation experiment
13. atmospheric and solar corona turbulence experiment
14. drag measurements experiment
15. 2 radio science experiments to determine gravity field
16. ultraviolet spectrometer (OUVS)

Results: Formally approved by NASA in August 1974, the Pioneer Venus project comprised two spacecraft to explore the atmosphere and surface of Venus. Both spacecraft used a basic cylindrical bus. Pioneer Venus 1, the orbiter, was designed to spend an extended period in orbit around Venus mapping the surface using a radar package. After a six-and-a-half-month-long journey, the spacecraft entered an elliptical orbit around Venus at 15:58 UT on 4 December 1978. It was the first American spacecraft to enter orbit around Venus, about three years after the Soviets accomplished the same feat. The initial orbital period was 23 hours, 11 minutes, which was altered within two orbits to the desired 24 hours—a maneuver that would allow the orbit's high and low points (about 160 kilometers) to occur at the same time each Earth day. Data from the radar mapper allowed scientists to produce a topographical map of most of the Venusian surface between 73° N and 63° S at a resolution of 75 kilometers. The data indicated that Venus was much more smooth and spherical than Earth. The orbiter identified the highest point on Venus as Maxwell Montes, which rises 10.8 kilometers above the mean surface. Infrared observations implied a clearing in the planet's atmosphere over the north pole. In addition, ultraviolet light photos showed dark markings that covered the clouds in the visible hemisphere. Cameras also detected almost continuous lightning activity in the atmosphere. The spacecraft confirmed that Venus has little, if any magnetic field. Because of the nature of its orbit, Pioneer Venus 1 passed through the planet's bow shock twice per revolution, and using its magnetometer, scientists were able to observe how the planet's ionosphere interacted with the solar wind. Although the mapping radar was switched off on 19 March 1981 (having mapped 93% of the band between 74° N and 63° S), it was reactivated again in 1991, 13 years after launch, to explore the previously inaccessible southern portions of the planet. In May 1992, Pioneer Venus 1 began the final phase of its mission, maintaining its periapsis between 150 and 250 kilometers until propellant depletion. The last transmission was received at 19:22 UT on 8 October 1992, as its decaying orbit no longer permitted communications. The spacecraft burned up the atmosphere soon after, ending a successful 14-year mission that was planned to last only eight months.

158

Pioneer Venus 2

Nation: USA (61)
Objective(s): Venus impact
Spacecraft: Pioneer Venus Multiprobe
Spacecraft Mass: 904 kg
Mission Design and Management: NASA / ARC
Launch Vehicle: Atlas Centaur (AC-51 / Atlas no. 5031D)
Launch Date and Time: 8 August 1978 / 07:33 UT
Launch Site: Cape Canaveral / Launch Complex 36A
Scientific Instruments:

Spacecraft Bus:
1. neutral mass spectrometer (BNMS)
2. ion mass spectrometer (BIMS)

Large Probe:
1. neutral mass spectrometer
2. solar flux radiometer
3. gas chromatograph
4. infrared radiometer
5. cloud particle size spectrometer
6. nephelometer

Small Probes (each):
1. neutral mass spectrometer
2. gas chromatograph
3. solar flux radiometer
4. infrared radiometer
5. cloud particle size spectrometer

Pioneer Project Manager Charlie Hall inspects the Pioneer Venus multiprobe at Hughes Aircraft Company in December 1976. *Credit: NASA/TRW*

6. temperature, pressure, acceleration sensors
7. nephelometer

Results: Pioneer Venus 2, the twin to Pioneer Venus 1, comprised a main bus, a Large Probe (316.5 kilograms), and three identical Small Probes, all of which were designed to collect data during independent atmospheric entry into Venus. The probes were each shaped like cones and not designed to survive past surface impact. After a course correction on 16 August 1978, Pioneer Venus 2 released the 1.5 diameter Large Probe on 16 November 1978, while about 11.1 million kilometers from the planet. Four days later, the bus released the three Small Probes (North, Day, and Night Probes) while 9.3 million kilometers from Venus. All five components reached the Venusian atmosphere on 9 December 1978, with the Large Probe entering first. Using a combination of air drag and a parachute, the Large Probe descended through the atmosphere, entering at a velocity of 11.6 kilometers/second, slowing down, until it impacted on the Venusian surface at 19:40 UT, landing at 4.4° N / 304.0° longitude at a velocity of 32 kilometers/hour. Transmissions ceased at impact as expected. The three 76-centimeter diameter Small Probes arrived in the atmosphere within minutes of the bigger one and descended rapidly through the atmosphere without the benefit of parachutes. They each opened their instrument doors at altitudes of about 70 kilometers and began to transmit information about the Venusian atmosphere immediately. Each took about 53–56 minutes to reach the surface. Amazingly, two of three probes survived the hard impact. The so-called Day Probe transmitted data from the surface for 67 minutes, 37 seconds, before succumbing to the high temperatures, pressures, and power depletion. Information from its nephelometer indicated that dust raised from its impact took several minutes to settle back to the ground. All three Small Probes suffered instrument failures, but not significant enough to jeopardize their main missions. Their landing coordinates were: 60° N / 4° E longitude (North Probe); 32° S / 318° E (Day Probe); and 27° S / 56° E (Night Probe). The main bus, meanwhile, burned up in the atmosphere at an altitude of 120 kilometers—about 1.5 hours after the other probes—and provided key data on higher regions. Data from the probes indicated that between 10 and 50 kilometers there is almost no convection in the atmosphere, while below a haze layer at 30 kilometers, the atmosphere is relatively clear. In addition, below an altitude of 50 kilometers, the temperatures reported from the four probes indicated very little differences even though their entry sites were separated by thousands of kilometers.

159

ISEE-3

Nation: USA (62)

Objective(s): Sun–Earth L1 Lagrange Point, Comet Giacobini-Zinner flyby

Spacecraft: ISEE-C

Spacecraft Mass: 479 kilograms

Mission Design and Management: NASA / GSFC

Launch Vehicle: Delta 2914 (no. 144 / Thor no. 633)

Launch Date and Time: 12 August 1978 / 15:12 UT

Launch Site: Cape Canaveral / Launch Complex 17B

Scientific Instruments:

1. solar wind plasma detector
2. vector helium magnetometer
3. low energy cosmic ray experiment
4. medium energy cosmic ray experiment
5. high energy cosmic ray experiment
6. plasma waves spectrum analyzer
7. energetic particle anisotropy spectrometer (EPAS)
8. interplanetary and solar electrons experiment
9. radio mapping of solar disturbances experiment
10. solar wind ion composition experiment
11. cosmic ray isotope spectrometer
12. x-rays and gamma-ray bursts experiment
13. gamma-ray bursts experiment
14. cosmic-ray energy spectrum charged-particle telescope

Results: ISEE-3 was the third of three International Sun-Earth Explorers (ISEE) designed and operated by NASA in cooperation with the European Space Agency (ESA). NASA built the first and third spacecraft while ESA built the second. The three spacecraft were to simultaneously investigate a wide range of phenomena in interplanetary space. After launch, on 20 November 1978, ISEE-3 was successfully placed in a halo orbit around the L1 Sun–Earth Lagrange Point on the sunward side of Earth, about 1.5 million kilometers from Earth where the gravitational forces of Earth and the Sun are exactly counterbalanced. ISEE-3 became not only the first spacecraft to be put into orbit around a libration point, but also the first spacecraft to monitor the solar wind approaching Earth. ISEE-3 completed its primary mission in 1981, but Goddard Space Flight Center scientists proposed sending the spacecraft first, through

Earth's magnetic tail, and second, into position to intercept a comet. By 10 June 1982, the spacecraft began to use its thrusters to move into the tail of Earth's magnetosphere. ISEE-3 completed the first deep survey of Earth's tail and detected a huge plasmoid of electrified gas that was ejected from Earth's magnetosphere. After a proposal by NASA scientist Robert Farquhar, NASA agreed in August 1982 to redirect the spacecraft for a rendezvous with Comet 21P/Giacobini-Zinner. Subsequently, after a series of five complex flybys of the Moon (the last on 22 December 1983 at a range of only 120 kilometers), ISEE-3 was sent on a trajectory to encounter the comet. At this point, the spacecraft was renamed the International Cometary Explorer (ICE). On 11 September 1985 at 11:02 UT, ICE passed within 7,862 kilometers of the comet's nucleus, becoming the first spacecraft to fly past a comet. The spacecraft returned excellent data on the comet's tail, confirming theories that comets are essentially "dirty snowballs" of ice, with surface material sleeting off during motion. ICE also flew to 40.2 million kilometers of the sunward side of Comet Halley on 28 March 1986 and provided upstream solar wind data. After daily data return was ended in December 1995, NASA eventually terminated ICE operations and support on 5 May 1997 although the spacecraft's transmitter was left on in order to facilitate tracking. In 2014, a team of independent scientists, engineers, and programmers organized an effort to "recapture" the spacecraft during its planned return to the vicinity of Earth in August 2014. The stated goal was to have ICE enter an orbit near Earth and "resume its original mission" with its 5 remaining operational instruments. With funding that came entirely from the public, the project achieved a new footing by signing an agreement with NASA in May 2014 that allowed the ISEE-3 Reboot Team to make use of a defunct NASA spacecraft. A first firing of thrusters on 2 July, the first time in 27 years, was successful but a longer firing on 8 July failed, probably due to a lack of nitrogen. In the event, the spacecraft passed by the Moon at 18:16 UT on 10 August 2014 at a range of 15,600 kilometers, just as had been expected years before. The spacecraft will continue in its existing trajectory (in heliocentric orbit) and return to the vicinity of Earth in 17 years.

160

Venera 11

Nation: USSR (96)

Objective(s): Venus flyby and landing

Spacecraft: 4V-1 (no. 360)

Spacecraft Mass: 4,447.3 kg

Mission Design and Management: NPO imeni Lavochkina

Launch Vehicle: Proton-K + Blok D-1 (8K82K no. 296-01 + 11S824M no. 3L)

Launch Date and Time: 9 September 1978 / 03:25:39 UT

Launch Site: NIIP-5 / Site 81/23

Scientific Instruments:

Spacecraft Bus:
1. plasma spectrometer
2. Konus gamma-ray detector
3. Signe-2MZ1 gamma and x-ray burst detector
4. DUMS-1 ultraviolet spectrometer
5. magnetometer
6. solar wind detectors
7. cosmic ray detectors

Lander:
1. imaging system
2. Sigma gas chromatograph
3. mass spectrometer
4. gamma-ray spectrometer
5. Groza lightning detector
6. temperature and pressure sensors
7. nephelometer
8. anemometer
9. optical spectrophotometer
10. remote soil collector

11. x-ray fluorescence cloud aerosol analyzer
12. Arakhis x-ray fluorescence spectrometer + drill
13. PrOP-V penetrometer

Results: Venera 11 was one of two identical probes (the other being Venera 12) that followed up on the highly successful Soviet missions to Venus in 1975. Venera 11 and 12 differed from their predecessors principally in the fact each carried a flyby bus + lander combination instead of the previous orbiter + lander combination. Engineers reverted to the flyby combination partly because of the weight limitations of the 1978 launch window, but also because flyby probes afforded better transmission time for landers. Several of the scientific instruments were also modified and new ones added. During the outbound trip to Venus, Venera 11 was beset with technical problems. In the very first communications session with the spacecraft, it became apparent that the solar orientation system failed (coinciding with the deployment of the VHF and UHF antennae). Proper orientation was restored soon enough, with Venera 11 in a nominal constant solar-stellar orientation mode, although this was maintained only intermittently. There were also failures of the primary and backup fans of the cooling system on the descent module detected immediately after launch. On 17 October, a "curtain" on the radiative heater on the lander was closed, a month earlier than planned, to accommodate the higher than expected heating. These measures appeared to partially work to bring temperatures down on the lander. Venera 11 arrived at Venus after two course corrections on 16 September and 17 December 1978. On 23 December 1978, the lander separated from the flyby probe and entered the Venusian atmosphere two days later at a velocity of 11.2 kilometers/second. The lander probe descended through the atmosphere using a system of three progressively larger parachutes (areas of 1.4 m², 6 m², and 24 m²). At an altitude of 46 kilometers, the final parachute was jettisoned and the lander effective dropped the rest of the way—for 52 minutes 58 seconds—guided only by a circular airbrake (with flaps). During this phase the lander carried out chemical analyses of the composition of the atmosphere and clouds, spectral analyses of the scattered solar radiation in the atmosphere, studied the electrical discharges in the atmosphere, and measured temperature, pressure, and windspeeds. The lander safely landed on Venus at 03:24 UT on 15 December 1978, impacting at a velocity of approximately 8 meters/second. Landing coordinates were 14° S / 299° longitude. Within 32 seconds, the imaging system and the soil sampling system simultaneously began operation. The latter collected soil for chemical and physical analysis, but soil analysis was unsuccessful because the soil was not properly deposited to an examination container for analysis (probably due to leaking air which disturbed the soil). The lander also failed to take color panoramas of the Venusian surface due to unopening of the lens covers of the camera system. The lander relayed information for a total of 95 minutes, the point of transmission cutoff being a function of the range of visibility of the flyby probe. The spectrophotometer on Venera 11 reported that only 3–6% of sunlight actually reached the Venusian surface. One of the odd findings was its recording of a loud noise some 32 minutes after landing. The Venera 11 flyby probe entered heliocentric orbit after flying past the planet at a range of 35,000 kilometers. A major course correction on 7 February 1979 (c. 350 meters/second delta-V) was done to ensure ideal conditions for the use of the Soviet-French Signe-2MZ1 experiment to study the localization of gamma-ray bursts, but it exhausted almost all of its propellants. As the distance from the Sun increased, the bus depended on its gyroscopes for orientation. Last contact was on 1 February 1980, probably due to improper orientation. The bus's orbit ranged from 1.715 × 1.116 AU.

161

Venera 12

Nation: USSR (97)

Objective(s): Venus flyby and landing

Spacecraft: 4V-1 (no. 361)

Spacecraft Mass: 4,457.9 kg

Mission Design and Management: NPO imeni
Lavochkina

Launch Vehicle: Proton-K + Blok D-1 (8K82K no.
296-02 / 11S824M no. 4L)

Launch Date and Time: 14 September 1978 / 02:25:13
UT

Launch Site: NIIP-5 / Site 81/24

Scientific Instruments:

Spacecraft Bus:

1. plasma spectrometer
2. Konus gamma-ray detector
3. Signe-2MZ1 gamma and x-ray burst
detector
4. DUMS-1 ultraviolet spectrometer
5. magnetometer
6. solar wind detectors
7. cosmic ray detectors

Lander:

1. imaging system
2. Sigma gas chromatograph
3. mass spectrometer
4. gamma-ray spectrometer
5. Groza lightning detector
6. temperature and pressure sensors
7. nephelometer
8. anemometer
9. optical spectrophotometer
10. remote soil collector
11. x-ray fluorescence cloud aerosol analyzer
12. Arakhis x-ray fluorescence spectrometer +
drill
13. PrOP-V penetrometer

Results: Venera 12 was the identical sister craft to Venera 11 with a bus-lander configuration, the latter weighing 1,645 kilograms (mass on the surface of Venus would be 750 kilograms). Launched successfully towards Venus, the spacecraft performed two mid-course corrections on 21 September and 14 December 1978. During the outbound trip one of the telemetry tape recorders failed, leading to more frequent use of the remaining tape recorder to relay information. As planned, on 17 November controllers closed the "curtain" of the radiation heater to begin cooling down the lander. Further such actions (on 9–10 and 13–14 December) brought down the temperature on the lander to a nominal −12.3°C although by the time of atmospheric entry the temperature was up to 5°C. Flying a shorter trajectory than its sister, Venera 12 moved "ahead" of Venera 11, and as with its twin, two days prior to the planetary encounter, the flyby probe released its lander. On 21 December the lander entered the Venusian atmosphere at a velocity of 11.2 kilometers/second and performed a descent profile almost identical to the earlier Veneras 9 and 10 in 1975. Release of the final chute occurred at 46 kilometers altitude and for the next 52 minutes and 11 seconds, the lander descended freely with only some airbraking flaps installed on its circular ring for resistance. A vast amount of information was collected and relayed to the flyby probe. The lander safely touched down at 03:30 UT on 21 December 1978. Landing coordinates were 7° S / 294° E longitude, about 800 kilometers from its twin. From the ground, the probe relayed data for a record 110 minutes, although like Venera 11, the spacecraft suffered two major failures: its soil sample delivery instrument failed to deposit the soil appropriately for scientific analysis; and the lens cover on the camera failed to release, effectively rendering the color imaging system useless. The flyby probe passed by the planet at a range of 35,000 kilometers after performing its data transmission mission and then entered heliocentric orbit with parameters of 0.715 × 1.156 AU. As with the Venera 11 flyby probe, a major course correction (c. 350 meters/second delta-V) on 5 February 1979 on Venera 12 was designed to ensure proper data collection from the Soviet-French Signe-2MZ1 experiment on the localization of gamma-ray bursts. About a year-and-a-half after launch, Venera 12 was fortuitously in position to study the newly discovered Comet C/1979 Y1 (Bradfield). On 13 February 1980, the spacecraft was properly oriented (with the use of its gyro-platform) to study the comet (which was about 50 kilometers closer to the Sun than the spacecraft). The DUMS-1 ultraviolet spectrometer was used to get a spectral signature of the comet both that day and later on 17 March 1980. In mid-April, the three-axis orientation system on the bus failed and the spacecraft was spin-stabilized, which prevented further communication after a last session on 18 April 1980.

1981

162

Venera 13

Nation: USSR (98)
Objective(s): Venus flyby and landing
Spacecraft: 4V-1M (no. 760)
Spacecraft Mass: 4,397.85 kg
Mission Design and Management: NPO imeni Lavochkina
Launch Vehicle: Proton-K + Blok D-1 (8K82K no. 311-01 + 11S824M no. 5L)
Launch Date and Time: 30 October 1981 / 06:04 UT
Launch Site: NIIP-5 / Site 200/40

Scientific Instruments:

Spacecraft Bus:
1. magnetometer
2. cosmic ray detector
3. solar wind detectors
4. Signe-2MZ2 gamma-ray burst detector

Lander:
1. x-ray fluorescence spectrometer + VB02 drill
2. x-ray fluorescence spectrometer for aerosols
3. imaging system
4. pressure and temperature sensors
5. mass spectrometer
6. Groza-2 lightning detector
7. Sigma-2 gas chromatograph
8. nephelometer

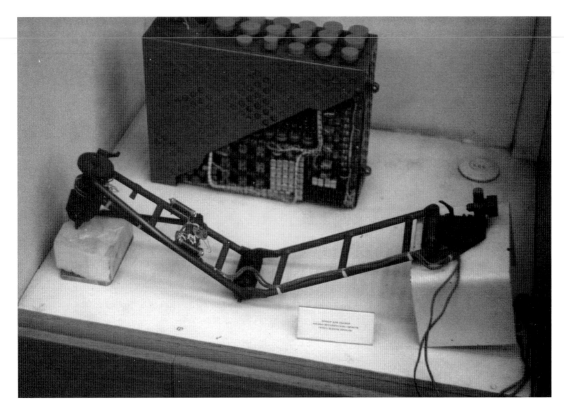

A model of the PrOP-V penetrometer installed on Veneras 13 and 14. *Credit: T. Varfolomeyev*

9. spectrophotometer

10. Bizon-M accelerometer

11. humidity sensor

12. PrOP-V soil mechanical/electrical probe

13. seismometer

Results: Venera 13 was part of the third pair of heavy Venus flyby/lander probes launched towards Venus by the Soviet Union in the 1970s (after Venera 9/10 and Venera 11/12). The Soviets picked the landing site for Venera 13 based on information passed on by NASA from the Pioneer Venus Orbiter vehicle. They were supposed to fly in 1980 but the failures of the soil sampler and cameras on Venera 11/12 forced a longer redesign process and these vehicles—especially the landers (each 1,643.72 kilograms) may have been the most rigorously tested deep space vehicles ever launched by the Soviets. The "new" spaceships had an improved set of instruments (such as the spectrophotometer, the gas chromatograph, and mass spectrometer) including a redesigned soil sampler as compared with its predecessors. Controllers implemented two mid-course corrections on 10 November 1981 and 21 February 1982. During the outbound phase, there were some anomalies: after the first mid-course correction, for example, there anomalies detected in two of the science instruments (Konus and RPP-01) and a data tape recorder failed. The Venera 13 lander separated from its parent on 27 February 1982. The capsule entered the Venusian atmosphere and began relaying atmospheric data back to the flyby probe which continued to fly past the planet after a 36,000-kilometer range encounter. After a roughly 1-hour-long descent, the lander set down on the Venusian surface at 03:57:21 UT on March 1, 1982. Landing velocity was about 7.5 meters/second and landing coordinates were 7.55° S / 303.69° E longitude. The probe continued to transmit for another additional 127 minutes, far beyond the planned lifetime of 32 minutes. The probe found temperature and pressure to be 462°C and 88.7 atmospheres, respectively. Venera 13 repeated the attempts at color surface photography (using red, green, and blue filters) that failed on Venera 11 and 12, and succeeded by relaying to Earth the first color photographs of the surface of Venus. Venera 13 returned eight successive panoramas showing a field of orange-brown angular rocks and loose soil. Successful soil analysis (which failed on Venera 11/12) showed soil similar to terrestrial leucitic basalt with a high potassium content. The flyby module entered a heliocentric orbit after relaying back to Earth all the data collected from the lander. Its engine was fired twice, on 10 June and 14 October 1982, as part of a test to provide engineering data for the anticipated Halley's Comet flyby by the Vega spacecraft. The bus continued to provide data until at least 25 April 1983.

163

Venera 14

Nation: USSR (99)

Objective(s): Venus flyby and landing

Spacecraft: 4V-1M (no. 761)

Spacecraft Mass: 4,394.5 kg

Mission Design and Management: NPO imeni Lavochkina

Launch Vehicle: Proton-K + Blok D-1 (8K82K no. 311-02 + 11S824M no. 6L)

Launch Date and Time: 4 November 1981 / 05:31 UT

Launch Site: NIIP-5 / Site 200/39

Scientific Instruments:

Spacecraft Bus:

1. magnetometer

2. cosmic ray detector

3. solar wind detectors

4. Signe-2MZ2 gamma-ray burst detector

Lander:

1. x-ray fluorescence spectrometer + VB02 drill

2. x-ray fluorescence spectrometer for aerosols

3. imaging system

4. pressure and temperature sensors

5. mass spectrometer

6. Groza-2 lightning detector

7. Sigma-2 gas chromatograph
8. nephelometer
9. spectrophotometer
10. Bizon-M accelerometer
11. humidity sensor
12. PrOP-V soil mechanical/electrical probe
13. seismometer

Results: Venera 14 was identical to its twin, Venera 13. The spacecraft carried out two mid-course corrections on the way to Venus, on 14 November 1981 and 25 February 1982. There were some anomalies en route: failures in the Konus device and erratic temperatures detected in the lander module (the outer shell holding the lander). Like Venera 13, one of the data tape recorders also failed. Also, the second mid-course correction imparted slightly lower velocity than expected (a shortfall of 1.9 meters/second). Later analysis indicated a possible failure in the turbopump assembly of the main engine. During flyby and release of the lander, the planned burn of the flyby vehicle was increased by 1 minute due to less than expected thrust, to ensure that the flyby bus did not burn up in the Venusian atmosphere and to give sufficient time to receive data from the lander. The lander probe separated from its flyby parent on 3 March 1982 before the entry cycle began but the flyby vehicle's main engine provided much less impulse than anticipated—a shortage of 56.4 meters/second imparted velocity. Fortunately, this was still enough to make sure that the flyby probe did not impact on Venus but it significantly reduced the relay time for data. The lander's main parachute opened at an altitude of 63.5 kilometers, thus activating the atmospheric instruments. The parachute was released at an altitude of 47 kilometers, and the 750-kilogram lander fell to the surface using only the atmosphere as a retarding medium. The probe made safe contact with the Venusian surface at 07:00:11 UT on 3 March 1982 and continued 57 minutes of transmissions. Landing coordinates were 13.055° S / 310.19° E longitude, about 1,000 kilometers from the Venera 13 landing site. As with its twin, Venera 14 returned color photographs of its surroundings and examined a soil sample (about one cubic centimeter taken from a 30-millimeter deep sample). Soil was deposited in a chamber sealed off from the outside environment and was then successively transferred through a series of chambers by blowing air until the sample was deposited in its final chamber with a temperature of only 30°C. Here it was examined by the x-ray fluorescence spectrometer. Temperature and pressure outside were considerably higher than at the Venera 13 site: 470°C and 93.5 atmospheres respectively. One minor failing of the mission was that one of the ejected caps from the cameras fell just where the PrOP-V penetrometers, designed to test soil properties, were about to operate; instead of testing the soil, PrOP-V simply tested the material of the lens cap. The flyby probe meanwhile passed Venus at a range of 36,000 kilometers—much closer than originally planned—and entered heliocentric orbit, continuing to provide data on solar x-ray flares. It performed one trajectory change on 14 November 1982 to provide engineering data for the upcoming Vega missions to encounter Halley's Comet. The bus continued to return data until 9 April 1983. A minor controversy surrounded the later Venera missions when a scientist from the Academy of Sciences' Institute of Space Research, Leonid Ksanformaliti, claimed that images from these missions revealed the possibility of "fauna" on Venus. Despite many Russians and Western analysts who debunked his claims, in a series of articles in 2012, 2013, and 2014 in the official journal of the Academy, Ksanformaliti insisted that the presence of "artificial" and repeating shapes ("stems") in the images from Venera 9, 10, 13, and 14 proved his hypothesis. All his claims, however, could be explained by the technical limitations of transmitting the images back to Earth.

1983

Venera 15

Nation: USSR (100)
Objective(s): Venus orbit
Spacecraft: 4V-2 (no. 860)
Spacecraft Mass: 5,250 kg
Mission Design and Management: NPO imeni
   Lavochkina
Launch Vehicle: Proton-K + Blok D-1 (8K82K no.
   321-01 / 11S824M no. 8L)
Launch Date and Time: 2 June 1983 / 02:38:39 UT
Launch Site: NIIP-5 / Site 200/39
Scientific Instruments:

1. Polyus-V side-looking radar
2. Omega-V 4-channel radiometric system
3. Radio occultation experiment
4. FS-1/4 infrared spectrometer
5. cosmic ray detectors
6. solar wind detectors
7. KS-18-6V to measure galactic and solar protons

Results: Venera 15/16 were a pair of two dedicated radar mappers designed to extend the studies began by the American Pioneer Venus Orbiter in constructing a detailed map of the surface down to about 1–2-kilometer resolution. For these missions, Soviet designers lengthened the central bus of the earlier Veneras (by 1 meter), installed much larger solar batteries, and attached a large side-looking radar antenna in place of the descent lander module on the earlier spacecraft. The infrared spectrometer was provided by the German Democratic Republic. Venera 15 carried out two mid-course corrections on 10 June 1983 and 1 October 1983 before successfully entering orbit around Venus at 03:05 UT on 10 October. Initial orbital parameters were 904.5 × 64,687 kilometers at 87.5° inclination, i.e., a near polar orbit. Two more orbital corrections were carried out on 17 October and 2 November leaving the spacecraft in an 873 × 65,000-kilometer orbit. The spacecraft's mapping operations began six days after entering orbit, over the north pole. About three months after entering Venusian orbit, it was discovered that the main omni-directional antennae on both Venera 15 and 16 were not naturally facing the direction of Earth (connected as they were by the need to have the solar panels constantly facing the direction of the Sun). Instead the solar panels were angled at 45° to direct sunlight. Using a special command to re-orient the antenna into proper orientation (using springs) in January 1984 proved successful on Venera 16 but not Venera 15. A subsequent engine firing on 9 April 1984 shook Venera 15 sufficiently that the antenna finally moved into proper orientation. Because of the nature of the spacecraft's orbit, the two orbiters mapped only the area from 30° N to the pole, about 115 million square kilometers. The official mission of both vehicles was to end on 10 March 1984 but was extended on account that both spacecraft were in excellent condition. The primary photography mission was completed on 10 July 1984. After that, an "optional program" of scientific work continued through the remainder of the year. On 30 December 1984, controllers found that the orientation system had exhausted its nitrogen. Last contact was made with the spacecraft on 5 January 1985.

165

Venera 16

Nation: USSR (101)
Objective(s): Venus orbit
Spacecraft: 4V-2 (no. 861)
Spacecraft Mass: 5,300 kg
Mission Design and Management: NPO imeni Lavochkina
Launch Vehicle: Proton-K + Blok D-1 (8K82K no. 321-02 + 11S824M no. 9L)
Launch Date and Time: 7 June 1983 / 02:32 UT
Launch Site: NIIP-5 / Site 200/40
Scientific Instruments:

1. Polyus-V side-looking radar
2. Omega-V 4-channel radiometric system
3. Radio occultation experiment
4. KS-18-6V infra-red spectrometer
5. cosmic ray detectors
6. solar wind detectors
7. KS-18-6V to measure galactic and solar protons

Results: Venera 16 arrived at Venus orbit at 06:22 UT on 14 October 1983 after en-route course corrections on 15 June and 5 October 1983. Its initial orbital parameters were 977.3 × 67,078 kilometers. It began its mapping operations six days later in its 24-hour, 34-minute period near-polar orbit. Its operational orbit, reached on 22 October was 944 × 65,336 kilometers, its orbital plane 4° 27′ inclined to that of its sister craft. Venera 16 typically followed Venera 15 over the same surface area after a three-day gap. Mapping resolution of both Venera 15 and 16 was comparable to that possible with the 300-meter dish at Arecibo in Puerto Rico although the Soviet orbiters provided coverage over latitudes higher than 30°, too far north for Earth-based observations. Both spacecraft also used an East German infrared spectrometer to map the planet in infrared wavelengths to provide a "heat atlas" of the atmosphere. The original missions of both spacecraft were to end by 10 March 1984 but due to the excellent condition of the vehicles, ground controllers extended the primary mission into the summer. On 21 June 1984, controllers altered Venera 16's orbit, and less than a month later, on 10 July, both vehicles finished their main missions, to photograph the northern hemisphere of Venus at a resolution of 1–2 kilometers. Venera 16 ended its extended mission rather strangely: on 13 June 1985, controllers were faced with an emergency situation on the Vega 1 spacecraft related to loss of orientation. A command was sent to Vega 1 to activate the recovery mode (constant solar-stellar mode orientation). Because the operating frequency of Vega 1 and Venera 16 were identical, the command activated Venera 16's search for an appropriate star to orient itself. The process unfortunately exhausted all the remaining propellant on Venera 16 and controllers lost contact. Because it had already fulfilled its primary mission, controllers did not consider this a great loss. In 1988, the Main Directorate of Geodesy and Cartography published a full album entitled "Atlas of Venus" (*Atlasa venery*) of annotated images generated by the Venera 14 and 15 missions. A 2013 article in a Russian journal claimed that these two orbital missions—Venera 15 and 16—were the "most successful [missions] in the history of automated spaceflight in the 1970s–1980s."

166

Vega 1

Nation: USSR (102)

Objective(s): Venus atmospheric entry and landing, Halley's Comet flyby

Spacecraft: 5VK (no. 901) = 5VS (Venus orbiter) + 5VP (Halley flyby + Venus lander)

Spacecraft Mass: c. 4,840 kg

Mission Design and Management: NPO imeni Lavochkina

Launch Vehicle: Proton-K + Blok D-1 (Proton-K no. 329-01 + 11S824M no. 11L)

Launch Date and Time: 15 December 1984 / 09:16:24 UT

Launch Site: NIIP-5 / Site 200/39

Scientific Instruments:

Lander:

1. meteorological complex to measure temperature and pressure
2. VM-4 instrument to measure moisture content
3. Sigma-3 gas chromatograph
4. IFP aerosol x-ray fluorescence spectrometer
5. ISAV-A optical aerosol analyzer
6. ISAV-S ultraviolet spectrometer
7. LSA laser aerosol meter
8. Malakhit-M aerosol mass spectrometer (MS 1S1)
9. BDRP-AM25 soil x-ray fluorescence spectrometer + drill
10. GS-15STsV gamma-ray spectrometer
11. PrOP-V penetrometer/soil ohmmeter

Balloon:

1. temperature and pressure sensors
2. vertical wind anemometer

3. nephelometer
4. light level/lighting detector

Bus:

1. magnetometer (MISHA)
2. PLAZMAG-1 plasma energy analyzer
3. Tyunde-M energetic particle analyzer
4. neutral gas mass spectrometer (ING)
5. APV-V high-frequency plasma wave analyzer
6. APV-N low-frequency plasma wave analyzer
7. dust mass spectrometer (PUMA)
8. SP-1 and SP-2 dust particle counters
9. dust counter and mass analyzer (DUCMA)
10. Foton dust particle recorder
11. imaging system (TVS)
12. infrared spectrometer (IKS)
13. ultraviolet, visible, infrared imaging spectrometer (TKS)

Results: The twin-spacecraft Vega project, named after the combination of Venera and Galley, the Russian words for "Venus" and "Halley," was perhaps the most ambitious deep space Soviet mission to date. The mission had three major goals: to place advanced lander modules on the surface of Venus, to deploy balloons (two each) in the Venusian atmosphere, and by using Venusian gravity, to fly the remaining buses past the Comet Halley. The entire mission was a cooperative effort between the Soviet Union (who provided the spacecraft and launch vehicle) with contributions from Austria, Bulgaria, Hungary, the German Democratic Republic (East Germany), Poland, Czechoslovakia, France (whose contribution was significant), and the Federal Republic of Germany (West Germany). In addition, one instrument, a comet dust (flux) analyzer, was provided by an American cosmic ray physicist, John A. Simpson (1916–2000), at the University of Chicago. This instrument turned out to be the only American instrument that directly studied Halley's comet during its 1986 encounter. While the landers were similar to ones used before for exploring Venus, the balloon gondolas were completely new Soviet-made vehicles that carried U.S.-French nephelometers to measure aerosol distribution in the atmosphere. The cometary flyby probes, which contained a 120-kilogram scientific package, were protected against high-velocity impacts from dust particles. After a successful flight to Venus, Vega 1 released its 1,500-kilogram descent module on 9 June 1985, two days before atmospheric entry. As the lander descended, at 61 kilometers altitude, it released the first helium-inflated plastic balloon with a hanging gondola underneath it. Mass was around 20.8 kilograms. As the balloon drifted through the Venusian atmosphere (controlled partly by ballast), it transmitted important data on the atmosphere back to a network of tracking antennas on Earth. Balloon 1 survived for 46.5 hours, eventually succumbing due to battery failure, having traversed about 11,600 kilometers. The lander set down safely on the ground at 03:02:54 UT on 11 June 1985 at 8.1° N / 176.7° longitude on the night side of Venus in the Mermaid Plain north of Aphrodite and transmitted from the surface for 56 minutes. Temperature and pressure at the landing site were 467°C and 93.88 atmospheres, respectively. The soil sample drill failed to complete its soil analysis, having spuriously deployed about 15 minutes before reaching the surface, but the mass spectrometer returned important data. The Vega 1 bus flew by Venus at a range of 39,000 kilometers and then headed for its encounter with Halley. After course corrections on 25 June 1985 and 10 February 1986, the spacecraft began its formal studies of the comet on 4 March when it was 14 million kilometers from its target. During the 3-hour encounter on 6 March 1986, the spacecraft approached to within 8,889 kilometers (at 07:20:06 UT) of Halley. Vega 1 took more than 500 pictures via different filters as it flew through the gas cloud around the coma. Although the spacecraft was battered by dust, none of the instruments were disabled during the encounter.

Vega 1 collected a wealth of information on Halley including data on its nucleus, its dust production rate, its chemical composition, and its rotational rate. After subsequent imaging sessions on 7 and 8 March 1986, Vega 1 headed out to heliocentric orbit where it continued to transmit data from at least some of its instruments until last contact on 30 January 1987.

167

Vega 2

Nation: USSR (103)

Objective(s): Venus atmospheric entry, Halley's Comet flyby

Spacecraft: 5VK (no. 902) = 5VS (Venus orbiter) + 5VP (Halley flyby + Venus lander)

Spacecraft Mass: 4,840 kg

Mission Design and Management: NPO imeni Lavochkina

Launch Vehicle: Proton-K + Blok D-1 (8K82K no. 325-02 + 11S824M no. 12L)

Launch Date and Time: 21 December 1984 / 09:13:52 UT

Launch Site: NIIP-5 / Site 200/40

Scientific Instruments:

Lander:

1. meteorological complex to measure temperature and pressure
2. VM-4 instrument to measure moisture content
3. Sigma-3 gas chromatograph
4. IFP aerosol x-ray fluorescence spectrometer
5. ISAV-A optical aerosol analyzer
6. ISAV-S ultraviolet spectrometer
7. LSA laser aerosol meter
8. Malakhit-M aerosol mass spectrometer (MS 1S1)
9. BDRP-AM25 soil x-ray fluorescence spectrometer + drill
10. GS-15STsV gamma-ray spectrometer
11. PrOP-V penetrometer/soil ohmmeter

Balloon:

1. temperature and pressure sensors
2. vertical wind anemometer
3. nephelometer
4. light level/lighting detector

Bus:

1. magnetometer (MISHA)
2. PLAZMAG-1 plasma energy analyzer
3. Tyunde-M energetic particle analyzer
4. neutral gas mass spectrometer (ING)
5. APV-V high-frequency plasma wave analyzer
6. APV-N low-frequency plasma wave analyzer
7. dust mass spectrometer (PUMA)
8. SP-1 and SP-2 dust particle counters
9. dust counter and mass analyzer (DUCMA)
10. Foton dust particle recorder
11. imaging system (TVS)
12. infrared spectrometer (IKS)
13. ultraviolet, visible, infrared imaging spectrometer (TKS)

Results: Vega 2 was the sister spacecraft to Vega 1, and essentially performed a near-identical mission to its twin. The main lander probe set down without problems at 03:00:50 UT on 15 June 1985 in the northern region of Aphrodite, about 1,500 kilometers south-east of Vega. Landing coordinates were 7.2° S / 179.4° longitude. Temperature and pressure were recorded as 462°C and 87.11 atmospheres, respectively. The spacecraft transmitted from the surface for 57 minutes. Unlike its twin, the Vega 2 lander was able to collect and investigate a soil sample; the experiment identified an anorthosite-troctolite rock, rarely found on Earth, but present in the lunar highlands. According to the lander's data, the area was probably the oldest explored by any Venera vehicle. The mass spectrometer did not return any data. The balloon, released upon entry into the atmosphere, flew through the Venusian atmosphere, collecting data like its twin and survived for 46.5 hours of data transmission, traveling a slightly longer distance than its compatriot from Vega 1. Neither balloon on Vega 1 nor Vega 2 detected any lightning in the Venusian atmosphere. After releasing its lander, the flyby probe continued on its flight to Comet Halley. The spacecraft initiated its encounter on 7 March 1986 by taking 100 photos of the comet from a distance of 14 million kilometers. Vega 2's closest approach (8,030 kilometers) to Halley was at 07:20 UT two days later when the spacecraft was traveling at a velocity of 76.8 kilometers/second (slightly lower than Vega 1's 79.2 kilometers/second). During the encounter, Vega 2 took 700 images of the comet of much better resolution than those from its twin, partly due to the presence of less dust outside of the coma during this transit, although many of the images were overexposed due to a failure in the primary pointing software. Ironically, Vega 2 sustained an 80% power loss during the encounter (as compared to Vega 1's 40%). Seven instruments between the two spacecraft were partially damaged, although no instrument on both were incapacitated. After further imaging sessions on 10 and 11 March 1986, Vega 2 finished its primary mission and headed out into heliocentric orbit. Like Vega 1, Vega 2 continued "a series of scientific investigations" until last contact on 24 March 1987.

1985

Sakigake

Nation: Japan (1)
Objective(s): Halley's Comet flyby
Spacecraft: MS-T5
Spacecraft Mass: 138.1 kg
Mission Design and Management: ISAS
Launch Vehicle: Mu-3S-II (no. 1)
Launch Date and Time: 7 January 1985 / 19:26 UT
Launch Site: Kagoshima / Launch Complex M1
Scientific Instruments:

1. solar wind ion detector
2. plasma wave probe
3. 2 magnetometers

Results: The MS-T5 spacecraft, named Sakigake ("pathfinder") after launch, was the first deep space spacecraft launched by any other country apart from the Soviet Union or the United States (Two German Helios probes had been launched by NASA). Japan's goal had been to launch a single modest probe to fly past Comet Halley as part of a test to prove out the technologies and mission operations of the actual mission. Japan's Institute of Space and Astronautical Sciences (ISAS) launched this test spacecraft, known as MS-T5, nearly identical to the "actual" spacecraft launched later. The spin-stabilized spacecraft was launched by a new Japanese launch vehicle, the Mu-3S-II. Following two course corrections on 10 January and 14 February 1985, Sakigake was sent on a long-range encounter with Halley. The original distance to the comet was planned to be 3 million kilometers but was altered to a planned 7.6 million kilometers when the launch had to be delayed due to bad weather and problems with the launch vehicle. The spacecraft served as a reference vehicle to permit scientists to eliminate Earth atmospheric and ionospheric contributions to the variations in Giotto's transmissions from within the coma. The spacecraft's closest approach to Halley was at 04:18 UT on 11 March 1986 when it was 6.99 million kilometers from the comet. Nearly six years after the Halley encounter, Sakigake performed a gravity assist by Earth on 8 January 1992 at 88,790-kilometer range. After two more distant flybys through Earth's tail (in June 1993 and July 1994), Sakigake maintained weekly contact with the ground until telemetry was lost on 15 November 1995. Earlier, Japanese scientists had hoped to send the spacecraft on a flyby past Comet 21P/Giacobini-Zinner in 1998 but these were abandoned due to lack of sufficient propellant. Although telemetry was lost, ground control continued to receive a beacon signal until 7 January 1999, 14 years after launch.

Giotto

Nation: European Space Agency (1)
Objective(s): Halley's Comet flyby
Spacecraft: Giotto
Spacecraft Mass: 960 kg
Mission Design and Management: ESA
Launch Vehicle: Ariane 1 (V14)
Launch Date and Time: 2 July 1985 / 11:23:16 UT
Launch Site: CSG / ELA-1
Scientific Instruments:

1. neutral mass spectrometer (NMS)
2. ion mass spectrometer (IMS)
3. Giotto radio experiment (GRE)
4. dust impact detection system (DID)

5. Rème plasma analyzer (RPA)

6. Johnstone plasma analyzer (JPA)

7. energetic particles analyzer (EPA)

8. magnetometer (MAG)

9. optical probe experiment (OPE)

10. Halley multicolor camera (HMC)

11. particulate impact analyzer (PIA)

Results: Giotto was the first deep space probe launched by the European Space Agency (ESA). It was named after famed Italian Renaissance painter Giotto di Bondone (c. 1267–1337) who had depicted Halley's Comet as the Star of Bethlehem in his painting Adoration of the Magi. Because the cylindrical spacecraft was designed to approach closer to Halley than any other probe, it was equipped with two dust shields separated by 23 centimeters, the first to bear the shock of impact and spread the impact energy over larger areas of the second thicker rear sheet. The design of the spacecraft was based on the spin-stabilized magnetospheric Geos satellites launched in 1977 and 1978. After launch, and further course corrections on 26 August 1985, 12 February 1986, and 12 March 1986, Giotto was put on a 500-kilometer-range flyby trajectory to the comet's core. Ballistics data on its precise voyage was based upon tracking information from the Soviet Vega 1 and 2 probes. The Giotto spacecraft eventually passed by Halley on 14 March 1986. Closest encounter was at a distance of 596 kilometers at 00:03:02 UT, the spacecraft being the only one among the large armada of spacecraft sent to investigate Halley that actually entered the ionosphere of the comet. At a range of 137.6 million kilometers from Earth, just 2 seconds before closest approach, telemetry stopped due to impact with a heavy concentration of dust that probably knocked the spacecraft's high gain antenna out of alignment with Earth. Fortunately, data transmission was restored within 21.75 seconds (with proper orientation of the antenna restored after 32 minutes). On average, Giotto had been hit 100 times a second by particles weighing up to 0.001 grams. By the end of its encounter with Halley, the spacecraft was covered in at least 26

kilograms of dust stemming from 12,000 impacts. Giotto returned 2,000 images of Halley. After the encounter, ESA decided to redirect the vehicle for a flyby of Earth. The spacecraft was officially put in hibernation mode on 2 April 1986. Course corrections on 10, 20, and 21 March 1986, however, set it on a 22,000-kilometer flyby of Earth on 2 July 1990 for a gravity assist (the first time that Earth had been used for such a purpose) to visit a new target: Comet 26P/Grigg-Skjellerup, which Giotto flew by at 15:30 UT on 10 July 1992 at range of approximately 200 kilometers. Eight experiments provided extensive data on a wide variety of cometary phenomena during this closest ever flyby of a comet. After formal conclusion of the encounter, Giotto was put in hibernation on 23 July 1992. Later, in September 1999, ESA scientists revealed that a second comet or cometary fragment may have been accompanying Grigg-Skjellerup during the encounter in 1992. The spacecraft repeated a flyby of Earth at 02:40 UT on 1 July 1999 at range of 219,000 kilometers but was not reactivated.

170

Suisei

Nation: Japan (2)

Objective(s): Halley's Comet flyby

Spacecraft: Planet-A

Spacecraft Mass: 139.5 kg

Mission Design and Management: ISAS

Launch Vehicle: Mu-3S-II (no. 2)

Launch Date and Time: 18 August 1985 / 23:33 UT

Launch Site: Kagoshima / Launch Complex M1

Scientific Instruments:

1. ultraviolet imaging system

2. electrostatic analyzer

Results: Planet-A, named Suisei ("comet") after launch, was the second of two Japanese probes launched towards Halley during the 1986 Earth encounter. The cylindrical spacecraft was launched directly on a deep space trajectory without entering intermediate Earth orbit. The main payload of the spacecraft was an ultraviolet-based imaging system that could study the huge hydrogen corona around the comet. After a course correction on 14 November 1985, Suisei flew within 152,400 kilometers of the comet's nucleus on 8 March 1986 at 13:06 UT, returning ultraviolet images of the 20 million-kilometer diameter hydrogen gas coma. Even at that relatively large distance from the comet, the spacecraft was hit by at least two dust particles, each 1 millimeter in diameter. After the Halley encounter, in 1987, ISAS decided to send the spacecraft through an elaborate trajectory for an encounter with Comet 21P/Giacobini-Zinner on 24 November 1998, 13 years after launch. Suisei performed a series of trajectory corrections on 5–10 April 1987 to send it on a gravity assist around Earth on 20 August 1992 at a range of 60,000 kilometers. Unfortunately, hydrazine for further corrections had been depleted by 22 February 1991. The planned encounter on 28 February 1998 with Giacobini-Zinner (as well as a far distance flyby of Comet 55P/Tempel-Tuttle) had to be cancelled, formally ending the mission.

1988

Fobos 1

Nation: USSR (104)
Objective(s): Mars flyby, Phobos encounter
Spacecraft: 1F (no. 101)
Spacecraft Mass: 6,220 kg
Mission Design and Management: NPO imeni Lavochkina
Launch Vehicle: Proton-K + Blok D-2 (8K82K no. 356-02 + 11S824F no. 2L)
Launch Date and Time: 7 July 1988 / 17:38:04 UT
Launch Site: NIIP-5 / Site 200/39
Scientific Instruments:

Orbiter:

1. Lima-D laser mass spectrometer analyzer
2. Dion secondary ion mass analyzer
3. radar system (RLK) (of which Plazma ionosphere study instrument only on Fobos 1)
4. videospectrometric system (VSK)
5. KRFM-ISM infrared spectrometer
6. Termoskan infrared spectrometer
7. IPNM-3 neutron detector (only on Fobos 1)
8. GS-14 STsF gamma-emission spectrometer
9. Ogyust optical radiation spectrometer (ISO)
10. scanning energy-mass spectrometer (ASPERA)
11. plasma spectrometer (MPK)
12. Ester electron spectrometer
13. plasma wave analyzer (APV-F / PWS)
14. flux gate magnetometer (FGMM)
15. magnetometer (MAGMA)
16. Terek solar telescope/coronograph (only on Fobos 1)
17. RF-15 x-ray photometer
18. ultrasound spectrometer (SUFR)
19. gamma-ray burst spectrometer (VGS)
20. Lilas gamma-ray burst spectrometer
21. solar photometer (IFIR)
22. Taus proton and alpha-particle spectrometer [not listed in all sources]
23. Harp ion and electron spectrometer [not listed in all sources]
24. Sovikoms energy, mass, and charge spectrometer [not listed in all sources]
25. Sled charged-particle spectrometer [not listed in all sources]

DAS:

1. Al'fa x-ray and alpha-particle backscattering spectrometer
2. Stenopee (Libratsiya) sun sensor to measure librations
3. 2 cameras
4. vibration measurement instrument (VIK) + temperature sensor
5. transponder

Results: Fobos 1 and 2 were part of an ambitious mission to Mars and its 27-kilometer diameter moon Phobos that culminated a decade-long program of development. A truly multinational project that was the last hurrah for Soviet planetary exploration, the missions involved contributions from 14 other nations including Austria, Bulgaria, Czechoslovakia, Finland, France (undoubtedly the most active partner), East Germany, West Germany, Hungary, Ireland, Poland, Switzerland, Sweden, and the European Space Agency. NASA provided some tracking support through its Deep Space Network. Each spacecraft, with a newly designed standardized bus known as the UMVL, comprised a Mars orbiter for long-term studies of the planet and a 67-kilogram Long-Term Autonomous Station (DAS) which would land on Phobos, anchored by a harpoon driven into the soil, to study its geological and climactic conditions. The core of the bus was an autonomous engine unit (using the S5.92 engine, later named Fregat) that essentially acted as the fifth stage of the Proton launch vehicle, boosting the spacecraft to Mars from a highly eccentric Earth orbit attained after a firing of the Blok D upper stage. After each spacecraft entered orbit around Mars, they were designed to make very close (c. 50 meters) flybys of Phobos (on 7 April 1989 for Fobos 1), and sample surface material using two innovative methods (using a laser and using a beam of krypton ions) that would actively disturb the soil of Phobos. Instruments would then measure and evaluate the response. After the Phobos flyby, it was planned for the spacecraft to continue science missions directed at Mars from Martian orbit. The spacecraft were loaded with an unprecedented array of experiments, making them probably the most highly instrumented deep space mission ever launched. Fobos 1 performed a course correction en route to Mars on 16 July 1988. On 29 August 1988, instead of a routine command to switch on the GS-14STsF gamma-emission spectrometer, an erroneous command was issued as a result of a programming error, to turn off the orientation and stabilization system. As a result, the spacecraft lost proper solar orientation, i.e., the solar panels faced away from the Sun and thus began to lose power. There was no word from Fobos 1 at the next scheduled communications session on 2 September. Continuing attempts to establish contact failed, and on 3 November 1988, the Soviets officially announced that there would be no further attempts at contact. The engineer who sent the false command was apparently barred from working on the shift teams for Fobos 2. Fobos 1 meanwhile flew by Mars without entering orbit (scheduled for 23 January 1989) and eventually entered heliocentric orbit. The most significant scientific data from the mission came from the Terek solar telescope, which returned important information on some of the then-least studied layers of the solar atmosphere, the chromosphere, the corona, and the transition layer.

172

Fobos 2

Nation: USSR (105)
Objective(s): Mars flyby, Phobos encounter
Spacecraft: 1F (no. 102)
Spacecraft Mass: 6,220 kg
Mission Design and Management: NPO imeni Lavochkina
Launch Vehicle: Proton-K + Blok D-2 (8K82K no. 356-01 + 11S824F no. 1L)
Launch Date and Time: 12 July 1988 / 17:01:43 UT
Launch Site: NIIP-5 / Site 200/40
Scientific Instruments:
Orbiter:
1. Lima-D laser mass spectrometer analyzer
2. Dion secondary ion mass analyzer
3. radar system (without Plazma ionosphere study instrument, only on Fobos 1) (RLK)
4. videospectrometric system (VSK)
5. KRFM-ISM infrared spectrometer
6. Termoskan infrared spectrometer
7. GS-14 STsF gamma-emission spectrometer
8. Ogyust optical radiation spectrometer (ISO)
9. scanning energy-mass spectrometer (ASPERA)
10. plasma spectrometer (MPK)
11. Ester electron spectrometer
12. plasma wave analyzer (APV-F / PWS)
13. flux gate magnetometer (FGMM)
14. magnetometer (MAGMA)
15. x-ray photometer (RF-15)
16. ultrasound spectrometer (SUFR)
17. gamma-ray burst spectrometer (VGS)
18. Lilas gamma-ray burst spectrometer
19. solar photometer (IFIR)
20. Termoskop
21. Taus proton and alpha-particle spectrometer [not listed in all sources]
22. Harp ion and electron spectrometer [not listed in all sources]

23. Sovikoms energy, mass, and charge spectrometer [not listed in all sources]
24. Sled charged-particle spectrometer [not listed in all sources]

DAS:
1. Alfa x-ray and alpha-particle backscattering spectrometer
2. Stenopee (Libratsiya) sun sensor to measure librations
3. 2 cameras
4. vibration measurement instrument (VIK) + temperature sensor
5. transponder

PrOP-FP:
1. penetrometer with ground sampler
2. accelerometers
3. x-ray fluorescence spectrometer
4. magnetometer (MFP)
5. kappameter
6. gravimeter
7. surface temperature sensors
8. instrument to measure surface electrical resistance
9. instrument to measure position tilt

Results: Fobos 2 had the same mission as its twin Fobos 1, to orbit Mars and fly past Phobos (on 13 June 1989) but had an additional payload, a small 43-kilogram instrumented "hopper" known as PrOP-FP that would make 20-meter jumps across the surface of Phobos for about 4 hours. The orbiter also had a slightly different instrument complement—it did not carry the IPNM, Plazma, and Terek instruments carried on Fobos 1. PrOP-FP (Device to Evaluate Mobility—Phobos) had its own power supply system, radio transmitter, and a suite of scientific instruments. This "rover" was designed to perform hops ranging from 10 to 40 meters and take data measurements after each hop. Despite some onboard anomalies, Fobos 2 carried out two en route course corrections on 21 July 1988 and 23 January 1989. One of the two radio transmitters failed, while its Buk computer was acting erratically due to faulty capacitors in the computer's power supply, a fact that was known before launch. At 12:55 UT on 29 January 1989, the spacecraft fired its engine to enter orbit around Mars. Initial orbital parameters were 819 × 81,214 kilometers at 1.5° inclination. In the initial months in orbit around Mars, the spacecraft conducted substantive investigations of the Red Planet and also photographed areas of its surface. During this period, controllers implemented four further orbital corrections in order to put its trajectory on an encounter course with Phobos. The spacecraft also jettisoned its Fregat upper stage (which had fired its engine to enter Mars orbit). Fobos 2 took high resolution photos of the moon on 23 February (at 860 kilometers range), 28 February (320 kilometers), and 25 March 1989 (191 kilometers), covering about 80% of its surface. Release of its lander was scheduled for 4–5 April 1989, but on 27 March during a regularly planned communications session at 15:58 UT, there was no word from the spacecraft. A weak signal was received between 17:51 and 18:03 UT, but there was no telemetry information. The nature of the signal indicated that the spacecraft had lost all orientation and was spinning. Future attempts to regain communication were unsuccessful and the mission was declared lost on 14 April 1989. The most probable cause was failure of the power supply for the Buk computer, something that had actually happened earlier in the mission (on 21 January 1989). On this occasion, controllers failed to revive the vehicle. Roald Sagdeev (1932–), the director of the Institute of Space Research noted in the journal *Priroda* (Nature) in mid-1989 that "I think we should seek the cause of the failure in the very organization of the project, in its planning" adding that "the relationships between the customer [the science community] and contractor [Lavochkin]…are clearly abnormal."

1989

Magellan

Nation: USA (63)

Objective(s): Venus orbit

Spacecraft: Magellan

Spacecraft Mass: 3,445 kg

Mission Design and Management: NASA / JPL

Launch Vehicle: STS-30R Atlantis

Launch Date and Time: 4 May 1989 / 18:47:00 UT

Launch Site: Kennedy Space Center / Launch Complex 39B

Scientific Instruments:

1. synthetic aperture radar (RDRS)

Results: Magellan, named after the Portuguese explorer Ferdinand Magellan (1480–1521), was the first deep space probe launched by the United States in almost 11 years, and also the first launched by the Space Shuttle. The Challenger disaster in January 1986 profoundly impacted the Shuttle launch manifest into the 1990s, which included a number of planetary missions. Magellan, for example, was delayed by at least a year. The spacecraft was designed to use a Synthetic Aperture Radar (SAR) to map 70% of the Venusian surface down to a resolution of 120–300 meters. The basic bus was assembled using spare parts left over from various prior missions including Voyager, Galileo, Ulysses, and Mariner 9. Magellan was deployed by the STS-30R crew and released at 01:01 UT on 5 May 1989 from Atlantis' payload bay. One hour later, a two-stage Inertial Upper Stage (IUS) fired to send the spacecraft on a trajectory to rendezvous with Venus. After three en route trajectory corrections (the first two on 21 May 1989, 13 March 1990, and 25 July 1990), Magellan arrived in Venus orbit on 10 August 1990. Orbital parameters were 297 × 8,463 kilometers at 85.5° inclination. Six days after entering orbit, Magellan suffered a communications outage lasting 15 hours. After a second 17 hour-interruption on 21 August, the ground sent up new preventative software to reset the system in case of such anomalies. Beginning 15 September 1990, the spacecraft began returning high-quality radar images of the Venusian terrain that showed evidence of volcanism, tectonic movement, turbulent surface winds, kilometers of lava channels, and pancake-shaped domes. Magellan completed its first 243-day cycle (i.e., the time it took for Venus to rotate once under Magellan's orbit) of radar mapping on 15 May 1991, providing the first clear views of 83.7% of the surface. The spacecraft returned 1,200 Gbits of data, far exceeding the 900 Gbits of data from all NASA planetary missions combined at the time. The spacecraft's second mapping cycle, already beyond the original goals of the mission, ended on 15 January 1992, raising coverage to 96%. A third cycle that focused on stereo imaging ended on 13 September 1992, finished coverage at 98%. Further cycles—a fourth (ending on 23 May 1993), a fifth (ending on 29 August 1994), and a sixth (ending on 13 October 1994)—focused on obtaining gravimetric data on the planet. In the summer of 1993, controllers commanded the spacecraft to drop into the outermost regions of the Venusian atmosphere and then successfully used an aerobraking method to circularize its orbit. Contact was lost after 10:05 UT on 13 October 1994 as the spacecraft was commanded to plunge into the atmosphere to gather aerodynamic data. The spacecraft burned up in the Venusian atmosphere about 10 hours later after one of the most successful deep space missions. Magellan found that at least 85% of the Venusian surface is covered

This simulated color global view of the surface of Venus taken by Magellan is centered at 180°E longitude. Data gaps in Magellan's data were filled in by information from Pioneer Venus. Most notably, the simulated hues are based on color images from the Soviet Venera 13 and Venera 14 spacecraft. *Credit: NASA/JPL*

with volcanic flows. The spacecraft's data suggested that despite the high surface temperatures (475°C) and high atmospheric pressures (92 atmospheres), the complete lack of water makes erosion an extremely slow process on the planet. As a result, surface features can persist for hundreds of millions of years. In addition, the spacecraft found that such phenomena as continental drift are not evident on the planet. Its imagery contributed to the best high-resolution radar maps of Venus' surface to date, improving on the images returned by the Soviet Venera 15 and 16 in the 1980s.

Galileo

Nation: USA (64)

Objective(s): Jupiter orbit and atmospheric entry

Spacecraft: Galileo (Orbiter and Entry Probe)

Spacecraft Mass: 2,380 kg

Mission Design and Management: NASA / JPL

Launch Vehicle: STS-34R Atlantis

Launch Date and Time: 18 October 1989 / 16:53:40 UT

Launch Site: Kennedy Space Center / Launch Complex 39B

Scientific Instruments:

Orbiter:

1. solid state imager (SSI)
2. near-infrared mapping spectrometer (NIMS)
3. ultraviolet spectrometer / extreme ultraviolet spectrometer (UVS/EUV)
4. photopolarimeter-radiometer (PPR)
5. magnetometer (MAG)
6. energetic particles detector (EPD)
7. plasma subsystem (PLS)
8. plasma wave subsystem (PWS)
9. heavy ion counter (HIC)
10. dust detector subsystem (DDS)

Atmospheric Entry Probe:

1. atmospheric structure instrument
2. neutral mass spectrometer
3. helium abundance interferometer
4. net-flux radiometer
5. nephelometer
6. lightning/radio-emission instrument

Results: Galileo, one of NASA's most ambitious deep space exploration projects, was the result of plans dating back to the early 1980s to deploy a Jupiter orbiter and probe. In its final configuration, the orbiter was a 4.6-meter tall spacecraft designed to operate for 22 months in Jovian orbit using 10 instruments/experiments to study the planet's atmosphere, satellites, and magnetosphere. Galileo, named after the Italian astronomer Galileo Galilei (1564–1642), carried a 337-kilogram probe designed to return data as it entered the Jovian atmosphere (by parachute) to identify atmospheric materials and conditions that cannot be detected from outside. Because of limitations of a Space Shuttle/IUS combination, NASA decided to use a complex multiple gravity assist scheme that required three flybys (two of Earth and one of Venus) on its way to Jupiter. The STS-34R crew released the spacecraft 6.5 hours after launch; an hour later, the two-stage IUS fired to send Galileo on its way. Galileo flew past Venus at 05:58:48 UT on 10 February 1990 at a range of 16,106 kilometers during which it conducted an extensive survey of the planet (including imaging). Having gained 8,030 kilometers/hour in velocity, the spacecraft flew by Earth twice, the first time at 960 kilometers range at 20:34:34 UT on 8 December 1990 when it detected chemical signatures associated with life-form activity in atmospheric trace elements on our home planet. The spacecraft also conducted lunar observations. A major problem occurred on 11 April 1991 when the high-gain antenna failed to fully deploy, thus eliminating the possibility of data transmission during its flyby of the asteroid 951 Gaspra. A low-gain antenna was instead used for the remainder of the mission, augmented by ingenious strategies including the use of data compression software that allowed a higher data throughput than was originally possible with the low-gain antenna. Becoming the first human-made object to fly past an asteroid, Galileo approached the minor planet Gaspra to a distance of 1,604 kilometers at 22:37 UT on 29 October 1991. The encounter provided much data including 150 images of the asteroid. Galileo then sped to its second encounter with the Earth–Moon system, with a flyby of Earth at 303.1 kilometers at 15:09:25 UT on 8 December 1992, adding 3.7 kilometers/second to its cumulative velocity. Galileo flew by a second asteroid, 243 Ida, at 16:51:59 UT on 28 August 1993 at a range of 2,410 kilometers, providing further data on minor planets. Later in July 1994, as it was speeding towards Jupiter, Galileo provided astronomers' only direct observations of Comet D/1993 F2

Shoemaker-Levy 9's impact into the Jovian atmosphere. Galileo's atmospheric entry probe, based on the design of the large probe of the Pioneer Venus multiprobe, was finally released on 13 July 1995 when the spacecraft was still 80 million kilometers from Jupiter. The probe hit the atmosphere at 6.5° N / 4.4° W at 22:04:44 UT on 7 December 1995, traveling at a relative velocity of 48 kilometers/hour, and returned valuable data for 58 minutes as it plunged into the Jovian cauldron. The entry probe endured a maximum deceleration of 228 g's about a minute after entry when temperatures scaled up to 16,000°C. The probe's transmitter failed 61.4 minutes after entry when the spacecraft was 180 kilometers below its entry ceiling, evidently due to the enormous pressure (22.7 atmospheres). Data, originally transmitted to its parent and then later transmitted back to Earth, indicated an intense radiation belt 50,000 kilometers above Jupiter's clouds, few organic compounds, and winds as high as 640 meters/second. The entry probe also found less lightning, less water vapor, and half the helium than had been expected in the upper atmosphere. The Galileo orbiter, meanwhile, fired its engine at 00:27 UT on 8 December, becoming Jupiter's first human-made satellite. Its orbital period was 198 days. Soon after, Galileo began its planned 11 tours over 22 months exploring the planet and its moons, including flybys of Ganymede (for the first time on 27 June 1996) and Europa (on 6 November 1997). Having fulfilled its original goals, NASA implemented a two-year extension to 31 December 1999 with the Galileo Europa Mission (GEM) during which the spacecraft conducted numerous flybys of Jupiter's moons, each encounter yielding a wealth of scientific data. These included flybys of Europa nine times (eight between December 1997 and February 1999 and once in January 2000), Callisto four times (between May 1999 and September 1999), and Io three times (in October 1999, November 1999, and February 2000). On the last flyby of Io, Galileo flew only 198 kilometers from the surface of Io sending back the highest resolution photos

yet of the volcanically active moon. On 8 March 2000, NASA announced plans to extend Galileo's mission again, with a new phase beginning October 2000 called the Galileo Millennium Mission. The idea was to coordinate investigations of Jupiter and its environs—particular the interaction of the solar wind with the planet's magnetosphere—with the Cassini spacecraft (on its way to Saturn) that was expected to fly past Jupiter in December 2000. This second extension was largely possible due to the extreme accuracy of navigation during the prior phases of the mission that had saved a significant amount of maneuvering propellant. The RTGs on board the spacecraft were still delivering about 450 W in 2000, although power capacity was declining at a rate of about 7 W per year. Under the Millennium Mission, Galileo flew by Ganymede, the largest moon in the solar system, on 20 May at a range of 809 kilometers and made its final flyby (also of Ganymede) on 28 December 2000 at a range of 2,337 kilometers. In January 2001, both Galileo and Cassini together encountered the magnetosphere bow shock within a half hour of each other. As the mission entered its final phase, mission scientists arranged for three final flybys of Io, primarily to obtain more data on the moon's magnetic field and heat generation that drives its volcanism. A final flyby of Callisto (in May 2001) resulted in some very high-resolution photographs and led to the Io encounters in August and October 2001 and in January 2002, the last one at a staggering 101.5 kilometers range, the closest it had gotten to any moon in its entire mission. This close encounter increased Galileo's orbit such that it could be easily commanded in the future to terminate its mission in Jupiter's atmosphere. A flyby of Amalthea in November 2002 provided key information on the moon's density and preceded its closest flyby of Jupiter itself. Beginning March 2003, Galileo was contacted once a week only to verify its status. Many years of operation in the Jovian system had exposed the spacecraft to intense radiation, taking a toll on many systems and instruments. Because Galileo had not been sterilized, to prevent contamination, it was decided to have the vehicle burn up in the Jovian atmosphere instead of risking impact on a moon such as Europa. Having completed its 35th orbit around Jupiter and after accompanying the planet for three-quarters of a circuit around the Sun, Galileo flew into the atmosphere at a velocity of 48.2 kilometers/second, just south of equator, on 21 September 2003 at 18:57 UT. In its nearly eight-year mission around Jupiter, Galileo had returned an unprecedented amount of data on the planet and its environs. For example, Galileo discovered far less lightning activity (about 10% of that found in an equal area on Earth) than anticipated, helium abundance in Jupiter very nearly the same as in the Sun (24% compared to 25%), extensive resurfacing of Io's surface due to continuing volcanic activity since the Voyagers flew by in 1979, and evidence for liquid water ocean under Europa's icy surface (with indications of similar liquid saltwater layers under the surfaces of Ganymede and Callisto). Jupiter's ring system was found to be made of dust from impacts on the four small inner moons. Galileo also discovered materials linked to organic compounds (clay-like minerals known as phyllosilicates) on the icy crust of Europa, perhaps produced by collisions with an asteroid or a comet. The spacecraft also identified the first internal magnetic field of a moon (Ganymede) that produces a "mini-magnetosphere" within Jupiter's larger magnetosphere. By March 2000, the spacecraft had returned about 14,000 images back to Earth.

1990

Hiten and Hagoromo

Nation: Japan (3)
Objective(s): lunar flyby and lunar orbit
Spacecraft: MUSES-A and Hagoromo subsatellite
Spacecraft Mass: 185 kg (MUSES-A), 12 kg (Hagoromo)
Mission Design and Management: ISAS
Launch Vehicle: Mu-3S-II (no. 5)
Launch Date and Time: 24 January 1990 / 11:46:00 UT
Launch Site: Kagoshima / Launch Complex M1
Scientific Instruments:

Hiten:

1. cosmic dust detector (MDC)

Results: This two-module Japanese spacecraft was designed to fly past the Moon and release an orbiter. It was the first Japanese lunar mission and also the first robotic lunar probe since the flight of the Soviet Luna 24 in 1976. MUSES-A (for Mu-launched Space Engineering Satellite), named Hiten ("musical angel") after launch, was put into a highly elliptical orbit around Earth that intersected with the Moon's orbit. Due to a problem with the orbital injection burn, the probe's orbital apogee was 290,000 kilometers, much less than the hoped for 476,000 kilometers, but after a number of subsequent maneuvers, Hiten reached its originally planned nominal orbit. At 19:37 UT on 18 March, during its first flyby of the Moon, Hiten released into lunar orbit, a small 12-kilogram "grandchild" satellite named Hagoromo. Hagoromo did not carry any science instruments and was designed to only transmit telemetry and diagnostic data back to Earth, but it made Japan the first nation besides the Soviet Union and the United States to put a spacecraft into lunar orbit. Hagoromo's initial orbital parameters were 20,000 × 7,400 kilometers. Although the maneuver successfully demonstrated the use of the swingby technique to enter lunar orbit, communications with Hagoromo had already been lost shortly before release on 21 February when its S-band transmitter failed. Hiten, meanwhile, passed by the Moon at 20:04:09 UT on 18 March at a distance of 16,472.4 kilometers and continued on its trajectory, simulating the orbital path of the proposed Geotail spacecraft. By 4 March 1991, Hiten had carried out seven more lunar flybys and began a phase of aerobraking into Earth's atmosphere—a feat it carried out for the first time by any spacecraft on 19 March (at 00:43 UT) at an altitude of 125.5 kilometers, which lowered Hiten's relative velocity by 1.712 meters/second and its orbital apogee by 8,665 kilometers. A second aerobraking over Earth occurred at 11:36 UT on 30 March at 120 kilometers altitude, reducing velocity by 2.8 kilometers/second and apogee by another 14,000 kilometers. With its primary mission now over, Hiten began an unexpected extended mission to experiment with a novel method to enter lunar orbit. After a ninth and tenth flyby of the Moon, the latter on 2 October 1991, Hiten was put into a looping orbit that passed the Earth–Moon L4 and L5 Lagrange Points in October 1991 and January 1991, respectively, where it activated its MDC cosmic dust detector, jointly built with Germany. Circling back to the Moon for its eleventh flyby of the Moon at 13:33 UT on 15 February 1992 at a range of 422 kilometers, Hiten used a portion of its last remaining propellant to fire its engine (for 10 minutes) and insert itself into lunar orbit. This profile to enter lunar orbit that required very little delta-V on the part of the spacecraft was developed by JPL

mathematician Edward Belbruno (1951–). Initial orbital parameters around the Moon were 422 × 49,400 kilometers. A final amount of propellant was then used two months later, on 10 April 1993 to deorbit Hiten, which crashed onto the lunar surface at 18:03:25.7 UT at 55.3° E / 34.0° S.

176

Ulysses

Nation: ESA and USA (1)
Objective(s): heliocentric orbit
Spacecraft: Ulysses
Spacecraft Mass: 371 kg
Mission Design and Management: ESA / NASA / JPL
Launch Vehicle: STS-41 Discovery
Launch Date and Time: 6 October 1990 / 11:47:16 UT
Launch Site: Kennedy Space Center / Launch Complex 39B
Scientific Instruments:
1. solar wind plasma experiment (BAM)
2. solar wind ion composition experiment (GLG)
3. magnetic fields experiment (HED)
4. energetic particle composition/neutral gas experiment (KEP)
5. low energy charged particle composition/anisotropy experiment (LAN)
6. cosmic rays and solar particles experiment (SIM)
7. radio/plasma waves experiment (STO)
8. solar x-rays and cosmic gamma-ray bursts experiment (HUS)
9. cosmic dust experiment (GRU)

Results: The Ulysses mission was an outgrowth of the abandoned International Solar Polar Mission (ISPM) that originally involved two spacecraft—one American and one European—flying over opposite solar poles to investigate the Sun in three dimensions. Eventually, NASA cancelled its spacecraft, significantly eroding the confidence of international partners in the reliability of NASA

as a partner, and the mission merged into a single spacecraft, provided by ESA. The scientific payload was shared by ESA and NASA, with the latter providing the RTG power source (similar to one used on Galileo), a Space Shuttle launch, and tracking from its Deep Space Network. Ground control operations were shared by both the Americans and Europeans. The vehicle was designed to fly a unique trajectory that would use a gravity assist from Jupiter to take it below the ecliptic plane and pass the solar south pole and then above the ecliptic to fly over the north pole. Eventually, 13 years after ESA's science council had approved the mission, and considerably delayed by the Challenger disaster, on 6 October 1990, about 7.5 hours after launch, Ulysses was sent on its way into heliocentric orbit via an Inertial Upper Stage/PAM-S motor combination. Escape velocity was 15.4 kilometers/second, higher than had been achieved by either of the Voyagers or Pioneers, and the fastest velocity ever achieved by a human-made object at the time. After a mid-course correction on 8 July 1991, Ulysses passed within 378,400 kilometers of Jupiter at 12:02 UT on 8 February 1992, becoming the fifth spacecraft to reach Jupiter. After a 17-day period passing through and studying the Jovian system, the spacecraft headed downwards and back to the Sun. From about mid-1993 on, Ulysses was constantly in the region of space dominated by the Sun's southern pole, as indicated by the constant negative polarity measured by the magnetometer. South polar observations extended from 26 June to 6 November 1994, when the vehicle was above 70° solar latitude. It reached a maximum of 80.2° in September 1994. Its instruments found that the solar wind blows faster at the south pole than at the equatorial regions. Flying up above the solar equator on 5 March 1995, Ulysses passed over the north polar regions between 19 June and 30 September 1995 (maximum latitude of 80.2°). Closest approach to the Sun was on 12 March 1995 at a range of 200 million kilometers. ESA officially extended Ulysses' mission on 1 October 1995, renaming this portion as the Second Solar

Orbit. Three times during its mission, the spacecraft unexpectedly passed through comet tails—the first time in May 1996 (Comet C/1996 B2 Hyakutake), the second time in 1999 (Comet C/1999 T1 McNaught-Hartley), and the third time in 2007 (Comet C/2006 P1 McNaught). The spacecraft made a second pass over the solar south pole between September 2000 and January 2001 and the northern pole between September and December 2001. At the time, the Sun was at the peak of its 11-year cycle; Ulysses found that the southern magnetic pole was much more dynamic than the north pole and lacked any fixed clear location. ESA's Science Program Committee, during a meeting on 5–6 June 2000, agreed to extend the Ulysses mission from the end of 2001 to 30 September 2004. In 2003–2004, Ulysses spun out towards its aphelion (furthest point in its orbit) and made distant observations of Jupiter. ESA's Science Program Committee approved a fourth extension of the Ulysses mission so that it could continue investigations over the Sun's poles in 2007 and 2008. In early 2008, ESA and NASA announced that the mission would finally terminate within the subsequent few months, having operated more than four times its design life. With communications systems failing as well as power depleting due to the decline of the RTGs (and thus allowing the hydrazine fuel in its attitude control system to freeze), the spacecraft was on its last breath at that point. Mission operations continued at reduced capacity until loss of contact on 30 June 2009, more than 18.5 years after launch. Ulysses' principal findings include data that showed that there is a weakening of the solar wind over time (which was at a 50-year low in 2008), that the solar magnetic field at the poles is much weaker than previously assumed, that the Sun's magnetic field "reverses" in direction every 11 years, and that small dust particles coming in from deep space into the solar system are 30 times more abundant than previously assumed.

177

Geotail

Nation: USA/Japan (1)

Objective(s): high elliptical Earth orbit

Spacecraft: Geotail

Spacecraft Mass: 1,009 kg

Mission Design and Management: ISAS / NASA

Launch Vehicle: Delta 6925 (no. D212)

Launch Date and Time: 24 July 1992 / 14:26 UT

Launch Site: Cape Canaveral Air Force Station / Launch Complex 17A

Scientific Instruments:

1. magnetic fields measurement monitor (MGF)
2. low energy particles experiment (LEP)
3. electric field monitor (EFD)
4. energetic particles and ion composition experiment (EPIC)
5. high-energy particle monitors (HEP)
6. plasma wave instrument (PWI)
7. comprehensive plasma instrument (CPI)

Results: The Geotail mission was a joint project of Japan's ISAS (and later, from 2003, JAXA), and NASA, executed as part of the International Solar Terrestrial Physics (ISTP) project, which also included the later Wind, Polar, SOHO, and Cluster missions. This particular mission's goal was to study the structure and dynamics of the long tail region of Earth's magnetosphere, which is created on the nightside of Earth by the solar wind. During active periods, the tail couples with the near-Earth magnetosphere, and often releases energy that is stored in the tail, thus activating auroras in the polar ionosphere. Although technically not a deep space or planetary mission, Geotail, in its extremely elliptical orbit, performed numerous lunar flybys, some closer than the distance at which the Soviet Luna 3 took the first pictures of the farside of the Moon. The spin-stabilized spacecraft (20 rpms) was designed with a pair of 100-meter tip-to-tip antennae and two 6-meter-long masts. On its fifth orbit around Earth, near apogee, on 8 September 1992, the spacecraft flew by the Moon at a range of 12,647 kilometers. The flyby raised apogee from 426,756 kilometers to 869,170 kilometers. Such flybys continued almost every month subsequent to that, and ultimately raised the spacecraft's apogee to 1.4 million kilometers. During these orbits, Geotail observed the magnetotail's far region (from 80 to 220 times the radius of Earth or "Re"). Geotail's 14th and last flyby of the Moon occurred on 25 October 1994 at a range of 22,445 kilometers and as a result, placed the spacecraft in orbits with progressively lower apogees. In November 1994, Geotail's apogee was 50 Re and by February 1995, it was down to 30 Re. The lower orbit was designed to allow the spacecraft to begin the second part of its mission, to study magnetotail sub-storms near Earth. During these orbits, perigee was about 10 Re, while the orbital inclination to the ecliptic plane was about −7° in order that the apogee would be located in the magnetotail's neutral plane during the winter solstice. Later in the decade, Geotail's orbit was adjusted so that it passed just inside Earth's magnetosphere's boundary plane on the dayside. In 2012, the spacecraft celebrated twenty years of continuous operation, and in 2014, despite having an original lifetime of only four years, was still sending back data on the formation of auroras, the nature of energy funneled from the Sun into near-Earth space, and the ways in which Earth's magnetic field lines move and rebound, thus producing explosive bursts that affect our magnetic environment.

178

Mars Observer

Nation: USA (65)

Objective(s): Mars orbit

Spacecraft: Mars Observer

Spacecraft Mass: 1,018 kg

Mission Design and Management: NASA / JPL

Launch Vehicle: Titan III (CT-4)

Launch Date and Time: 25 September 1992 / 17:05:01 UT

Launch Site: Cape Canaveral Air Force Station / Space Launch Complex-40

Scientific Instruments:

1. Mars Observer camera (MOC)
2. thermal emission spectrometer (TES)
3. pressure modulator infrared radiometer (PMIRR)
4. Mars Observer laser altimeter (MOLA)
5. magnetometer/electron reflectometer (MAG/ER)
6. gamma-ray spectrometer (GRS)
7. radio science experiment (RS)
8. Mars balloon relay receiver (MBR)

Results: Mars Observer was designed to carry out a high-resolution photography mission of the Red Planet over the course of a Martian year (687 days) from a 378 × 350-kilometer polar orbit. Building on the research done by the Viking missions, it carried a suite of instruments to investigate Martian geology, atmosphere, and climate in order to fill in gaps on our knowledge of planetary evolution. A mere 31 minutes after launch, the new Transfer Orbit Stage (TOS), using the Orbus 21 solid rocket motor, fired to boost the spacecraft on an encounter trajectory with Mars. After a 725 million-kilometer voyage lasting 11 months, at 00:40 UT on 22 August 1993, just two days prior to planned entry into Mars orbit, the spacecraft stopped sending telemetry (as planned), but then never resumed 14 minutes later. Despite vigorous efforts to regain contact, Mars Observer remained quiet. When the spacecraft did not reestablish command as a result of a stored program that was designed to do in case of five days of silence, mission planners finally gave up hope on salvaging the mission. The results of a five-month investigation proved to be inconclusive, but one likely cause of the catastrophic failure may have been a fuel line rupture that could have damaged the spacecraft's electronics, throwing the vehicle into a spin. In addition, the fact that the Mars Observer bus was a repurposed Earth science satellite bus may have also compromised the spacecraft's ability to adapt to the deep space environment. While none of the primary mission objectives were accomplished, the spacecraft did return data during its interplanetary cruise. Scientific instruments developed for Mars Observer were later used on several subsequent Mars probes, including Mars Global Surveyor (launched in 1996), Mars Climate Orbiter (1998), Mars Odyssey (2001), and Mars Reconnaissance Orbiter (2005).

1994

179

Clementine

Nation: USA (66)
Objective(s): lunar orbit
Spacecraft: Clementine
Spacecraft Mass: 424 kg
Mission Design and Management: BMDO / NASA
Launch Vehicle: Titan IIG (no. 23G-11)
Launch Date and Time: 25 January 1994 / 16:34 UT
Launch Site: Vandenberg AFB / SLC-4W
Scientific Instruments:

1. ultraviolet/visible camera (UV/Vis)
2. near-infrared camera (NIR)
3. laser image detection and ranging system (LIDAR)
4. long-wave infrared camera (LWIR)
5. high-resolution camera (HIRES)
6. 2 star tracker cameras
7. bistatic radar experiment
8. S-band transponder Doppler gravity experiment
9. charged particle telescope (CPT)

Results: Clementine was the first U.S. spacecraft launched to the Moon in over 20 years (since Explorer 49 in June 1973). The spacecraft, also known as the Deep Space Program Science Experiment (DSPSE), was designed and built to demonstrate a set of lightweight technologies such as small-imaging sensors and lightweight gallium arsenide solar panels for future low-cost missions flown by the Department of Defense. Specifically, Clementine was a technology proving mission for the DOD's Brilliant Pebbles program for the Strategic Defense Initiative (SDI), which required a large fleet of inexpensive spacecraft. Clementine carried 15 advanced flight-test components and 9 science instruments. After launch, the spacecraft remained in a temporary parking orbit until 3 February 1994, at which time a solid-propellant rocket ignited to send the vehicle to the Moon. After two subsequent Earth flybys on 5 and 15 February, on 19 February Clementine successfully entered an elliptical polar orbit (430 × 2,950 kilometers) around the Moon with a period of five days. In the following two months, it transmitted about 1.6 million digital images of the lunar surface, many of them with resolutions down to 100–200 meters, in the process, providing scientists with their first look at the total lunar landscape including polar regions. After completing its mission goals over 297 orbits around lunar orbit, controllers fired Clementine's thrusters on 3 May to inject it on a rendezvous trajectory (via an Earth flyby) with the asteroid 1620 Geographos in August 1994. However, due to a computer problem at 14:39 UT on 7 May that caused a thruster to fire and use up all propellant, the spacecraft was put in an uncontrollable tumble at about 80 rpms with no spin control. Controllers were forced to cancel the asteroid flyby and return the vehicle to the vicinity of Earth. A power supply problem further diminished the operating capacity of the vehicle. Eventually, on 20 July, lunar gravity took control of Clementine and propelled it into heliocentric orbit. The mission was terminated on 8 August when falling power supply levels no longer allowed clear telemetry exchange. Surprisingly, because the spacecraft was fortuitously in the correct attitude to power up again, ground controllers were able to briefly regain contact between 20 February and 10 May 1995. On 3 December 1996, the Department of Defense announced that Clementine data indicated that there was ice in the bottom of a permanently shadowed crater on the lunar south pole. Scientists estimated the deposit to be approximately 60,000 to 120,000 cubic meters in volume, i.e., comparable to a small lake that is four football fields in surface area and 5 meters deep. This estimate was very uncertain, however, due to the nature of the data. An accounting of Clementine's legacy should include the fact that methods developed for the project became the basis for NASA's "Faster, Better, Cheaper" initiative which ultimately paved the way for the Agency's Discovery program.

180

Wind

Nation: USA (67)
Objective(s): Sun–Earth L1 Lagrange Point
Spacecraft: Wind
Spacecraft Mass: 1,250 kg
Mission Design and Management: NASA / GSFC
Launch Vehicle: Delta 7925-10 (no. D227)
Launch Date and Time: 1 November 1994 / 09:31:00 UT
Launch Site: Cape Canaveral Air Force Station / Launch Complex 17B
Scientific Instruments:
1. radio and plasma wave experiment (WAVES)
2. energetic particle acceleration, composition, and transport experiment (EPACT)
3. solar wind and suprathermal ion composition experiment (SMS)
   - solar wind ion composition spectrometer (SWICS)
   - high mass resolution spectrometer (MASS)
   - suprathermal ion composition spectrometer (STICS)
4. solar wind experiment (SWE)
5. 2 triaxial fluxgate magnetometers (MFI magnetic field investigation)
6. three-dimensional plasma and energetic particle investigation (3DP)

7. transient gamma-ray spectrometer (TGRS)
8. Konus gamma-ray burst studies experiment

Results: Wind was part of the International Solar Terrestrial Physics (ISTP) program, a joint project between (principally) the United States, Japan, and the European Space Agency (although a few other countries such as Russia, the Czech Republic, and France made significant contributions) to study the behavior of the solar-terrestrial system. Participating spacecraft included Geotail, Wind, Polar, SOHO, and Cluster. The first of two NASA-sponsored Global Geospace Science (GGS) vehicles, the Wind spacecraft carried eight instruments (including the French WAVES and the Russian Konus) to investigate the solar wind's encounters with Earth's magnetosphere and ionosphere in order to determine the origins and three-dimensional characteristics of the solar wind. The spacecraft's original mission was to directly move to the L1 Lagrange point but because the SOHO and ACE spacecraft were directed to that location, the WIND mission was reformulated to operate for some time in a unique figure-eight-shaped elliptical orbit around Earth at 28,000 × 1.6 million kilometers, partially maintained by periodic "double flybys" of the Moon. In this orbit, Wind was positioned so as to make use of lunar gravity assists to maintain its apogee over the day hemisphere of Earth and conduct magnetospheric observations. The closest of its 19 flybys of the Moon between 1 December 1994 and 17 November 1998 took place on 27 December 1994 at a range of 11,834 kilometers. Finally, by November 1996, Wind was in a "halo orbit" around the Sun–Earth L1 libration point where solar and terrestrial gravity are approximately equal. Parameters varied between 235 and 265 Earth radii. In this orbit, Wind measured the incoming solar wind, and magnetic fields and particles on a continuous basis, providing about an hour warning to the other ISTP-related spacecraft on changes in the solar wind. On 17 November 1998, Wind began to move into a series of "petal" orbits, designed to take it out of the ecliptic plane. Wind's trips above and below the ecliptic (up to

60°) allowed the spacecraft to sample regions of interplanetary space and the magnetosphere that had not been previously studied. By 2004, it was back at L1 where it has remained. The original projected lifetime of the vehicle was anticipated to be three to five years but WIND continues to be largely operational in 2017, 22 years after its launch, with one instrument, the TGRS gamma-ray spectrometer turned off, and a couple of detectors on the EPACT instrument, non-functional. Current projections suggest that it will have enough fuel to remain at L1 for at least another 60 years. Despite the formal conclusion of the ISTP in December 2001, Wind continues to play a supporting role for a variety of other spacecraft supporting solar research, including Polar, Cluster, Geotail, Image, SOHO, and ACE.

1995

181

SOHO

Nation: ESA and USA (2)

Objective(s): Sun–Earth L1 Lagrange Point

Spacecraft: SOHO

Spacecraft Mass: 1,864 kg

Mission Design and Management: ESA / NASA

Launch Vehicle: Atlas Centaur IIAS (AC-121 / Atlas IIAS no. 8206 / Centaur II)

Launch Date and Time: 2 December 1995 / 08:08:01 UT

Launch Site: Cape Canaveral Air Force Station / Launch Complex 36B

Scientific Instruments:

1. solar-ultraviolet measurements of emitted radiation experiment (SUMER)
2. coronal diagnostic spectrometer (CDS)
3. extreme ultraviolet imaging telescope (EIT)
4. ultraviolet coronograph spectrometer (UVCS)
5. large angle and spectrometric coronograph (LASCO)
6. solar wind anisotropies experiment (SWAN)
7. charge, element, and isotope analysis experiment (CELIAS)
8. comprehensive suprathermal and energetic particle analyzer (COSTEP)
9. energetic and relativistic nuclei and electron experiment (ERNE)
10. global oscillations at low frequencies experiment (GOLF)
11. variability of solar irradiance and gravity oscillations experiment (VIRGO)
12. Michelson Doppler imager/solar oscillations investigation (MDI/SOI)

Results: The ESA-sponsored Solar and Heliospheric Observatory (SOHO) carries 12 scientific instruments to study the solar atmosphere, helioseismology, and the solar wind. Information from the mission has allowed scientists to learn more about the Sun's internal structure and dynamics, the chromosphere, the corona, and solar particles. The SOHO and Cluster missions, part of ESA's Solar Terrestrial Science Programme (STSP), are ESA's contributions to the International Solar Terrestrial Physics (ISTP) program, which has involved the work of other spacecraft such as Wind and ACE, which, like SOHO, operate in the vicinity of the Sun–Earth L1 point. NASA contributed three instruments on SOHO as well as launch and flight operations support. About two months after launch, on 14 February 1996, SOHO was placed at a distance of 1.5 million kilometers from Earth in an elliptical Lissajous orbit around the L1 libration point where it takes approximately six months to orbit L1 (while the L1 itself orbits the Sun every 12 months). The spacecraft returned its first image on 19 December 1995 and was fully commissioned for operations by 16 April 1996. SOHO finished its planned two-year study of the Sun's atmosphere, surface, and interior in April 1998. Communications with the spacecraft were interrupted for four months beginning 24 June 1998, after which the spacecraft was apparently spinning, losing electrical power, and not pointing at the Sun. After intensive search efforts, by 25 September, controllers managed to regain control and return SOHO to "normal mode." Because of the failure of onboard gyros, ESA developed a special gyroless method of orientation (which used reaction wheels) that was successfully implemented beginning 1 February 1999. Barring three instruments, the spacecraft was functional and was declared operational once again by mid-October 1998. SOHO's original lifetime was three years (to 1998), but in 1997, ESA and NASA jointly decided to prolong the mission to 2003, thus enabling the spacecraft to compare the Sun's behavior during low dark sunspot activity (1996) to the peak (around 2000). One of SOHO's most important discoveries has been locating the origin of the fast solar wind at the corners of honeycomb-shaped magnetic fields surrounding the edges of large bubbling cells located near the Sun's poles. Another has been its discovery, as of September 2015, of over 3,000 comets (more than one-half of all known comets), by over 70 people representing 18 different nations. These discoveries were made possible because of the LASCO instrument that while observing the Sun, blocked out the Sun's glare, rendering comets visible. SOHO's mission at L1 has now been extended six times, most recently in June 2013, to at least December 2016. In December 2015, SOHO marked 20 years of continuous operation, having fundamentally changed our conception of the Sun "from a picture of a static, unchanging object in the sky to the dynamic beast it is," in the words of Bernhard Fleck, the ESA project scientist for SOHO. The longevity of the mission has allowed SOHO to cover an entire 11-year solar cycle and the beginning of a new one. One of the recent highpoints of the mission was SOHO's observation of a bright comet plunging toward the Sun on 3–4 August 2016 at a velocity of nearly 2.1 million kilometers/hour.

182

NEAR Shoemaker

Nation: USA (68)

Objective(s): asteroid (Eros) orbit and landing

Spacecraft: NEAR

Spacecraft Mass: 805 kg

Mission Design and Management: NASA / GSFC / APL

Launch Vehicle: Delta 7925-8 (no. D232)

Launch Date and Time: 17 February 1996 / 20:43:27 UT

Launch Site: Cape Canaveral Air Force Station / Launch Complex 17B

Scientific Instruments:

1. multi-spectral imager (MSI)
2. magnetometer (MAG)
3. near infrared spectrometer (NIS)
4. x-ray/gamma ray spectrometer (XGRS)
5. laser rangefinder (NLR)
6. radio science and gravimetry experiment

Results: Near Earth Asteroid Rendezvous (NEAR) was the first mission flown under NASA's Discovery program, a series of low-cost (less than c. $150 million in mid-nineties dollar amounts) planetary science projects that were selected competitively and led by a Principal Investigator rather than a NASA manager. NEAR's primary goal was to rendezvous with the minor planet 433 Eros (an S-class asteroid), approximately 355 million kilometers from Earth, and gather data on its physical properties, mineral components, morphology, internal mass distribution, and magnetic field. The spacecraft was the first to rely on solar cells for power during operations beyond Mars orbit. On the way to its primary mission, NEAR performed a 25-minute flyby of the asteroid 253 Mathilde on 27 June 1997. Closest approach to 1,200 kilometers was at 12:56

UT. During the encounter, the spacecraft photographed 60% of the minor planet from a range of 1,200 kilometers. The collected information indicated that the 4.5 billion-year-old asteroid is covered with craters and less dense than previously believed. After a mid-course correction on 3 July 1997, NEAR flew by Earth on 23 January 1998 at 07:23 UT for a gravity assist on its way to Eros. Closest approach was 540 kilometers. After the Earth flyby encounter, NEAR's previously planned mission profile had to be revised in the light of an aborted engine burn on 20 December 1998 that prevented a critical trajectory correction to meet up with Eros a month later. Instead, NEAR was put on a backup trajectory that afforded a different flyby than that originally planned. As part of this new plan, the spacecraft first flew past Eros on 23 December 1998 at 18:41:23 UT at a range of 3,827 kilometers (distance measured from the center of mass) during which it observed about 60% of the asteroid, and discovered that the minor planet was smaller than expected. NEAR also found that the asteroid has two medium-sized craters, a long surface ridge, and a density similar to Earth's crust. After several more trajectory adjustments, NEAR finally moved into orbit around Eros at 15:33 UT on 14 February 2000, roughly a year later than intended, becoming the first human-made object to orbit a minor planet. Orbital parameters were 321 × 366 kilometers. Through 2000, NEAR's orbit was shifted in stages to permit specific research programs. There were a few problems in the lead up to the landing on the asteroid. For example, on 13 May 2000, controllers had to turn off the Near Infrared Spectrometer due to an excessive power surge. By 30 April the spacecraft was in its operational orbit at an altitude of about 50 kilometers from Eros' center. Later, on 13 July,

NEAR entered an even lower orbit at 35 kilometers that brought the vehicle as close as 19 kilometers from the surface. After about 10 days, it moved back into a higher orbit. On 26 October, NEAR performed another close flyby of the surface, this time to just 5.3 kilometers. By the end of the year, the spacecraft had entered a circular 35-kilometer low orbit around the asteroid, and began to make a series of very close passes—on the order of 5 to 6 kilometers—to its surface. Following a slow controlled descent, during which it took 69 high-resolution photos of Eros, NEAR touched down on Eros at a gentle 6.4 kilometers/hour, just south of a saddle-shaped feature named Himeros, on 12 February 2001 at 19:44 UT. This was the first time a U.S. spacecraft was the first to land on a celestial body, having been beaten by the Soviets in landing on the Moon, Mars, and Venus. Remarkably, the orbiter survived contact and returned valuable data, especially from its gamma-ray spectrometer, for about two weeks. Last contact with NEAR was on 28 February 2001, the spacecraft having succumbed to the extreme cold (–173°C). NASA's attempt to contact the probe nearly two years later on 10 December 2002 was unsuccessful. NEAR data showed that Eros had no magnetic field. It mapped more than 70% of the surface using the near infrared spectrometer and provided important data about the asteroid's interior. The spacecraft returned about 10 times more data than originally planned, including 160,000 images. Earlier, on 14 March 2000, a month after entering asteroid orbit, NASA renamed the NEAR spacecraft NEAR Shoemaker in honor of Eugene M. Shoemaker (1928–1997), the renowned geologist.

183

Mars Global Surveyor

Nation: USA (69)
Objective(s): Mars orbit
Spacecraft: MGS

Spacecraft Mass: 1,030.5 kg
Mission Design and Management: NASA / JPL
Launch Vehicle: Delta 7925 (no. D239)
Launch Date and Time: 7 November 1996 / 17:00:49 UT
Launch Site: Cape Canaveral Air Force Station / Launch Complex 17A
Scientific Instruments:
1. Mars orbital camera (MOC)
2. Mars orbital laser altimeter (MOLA)
3. thermal emission spectrometer (TES)
4. magnetometer/electron reflectometer (MAG/ER)
5. radio science experiment (RS)
6. Mars relay antenna for future spacecraft (MR)

Results: Mars Global Surveyor was the first spacecraft in NASA's new Mars Surveyor Program, a new generation of American space probes to explore Mars every 26 months from 1996 to 2005, and formulated (in 1994) to economize costs and maximize returns by involving a single industrial partner with the Jet Propulsion Laboratory to design, build, and deliver a flight-worthy vehicle for Mars every two years. (A new Mars Exploration Program was inaugurated in 2000.) The Mars Global Surveyor spacecraft carried five instruments similar to those carried by the lost Mars Observer probe that fell silent in 1993. Among its instruments was a French-supplied radio relay experiment to serve as a downlink for future Mars landers, including for the then-planned Russian Mars 96 mission. After mid-course corrections on 21 November 1996 and 20 March 1997, Mars Global Surveyor entered a highly elliptical orbit around Mars on 12 September 1997 after engine ignition at 01:17 UT. Initial orbital parameters were 262 × 54,026 kilometers. Commencement of its planned two-year mission was delayed because one of its two solar panels (-Y) had not fully deployed soon after launch. The solar panels were designed to act as atmospheric brakes to alter its orbit. As a result, mission planners reconfigured the aerobraking process required to place the vehicle in its intended orbit: the modified aerobraking maneuver began on 17 September 1997 and lasted until 11 October. A second aerobraking phase lasted from November 1997 to March 1998 and a third one began in November 1998 whose goal was to reduce the high point of its orbit down to 450 kilometers. The revised maneuvers were finally completed on 4 February 1999 with a major burn from its main engine. A subsequent firing on 19 February finally put Mars Global Surveyor into a near-circular polar orbit at 235 kilometers—and on 9 March 1999, its mapping mission formally began. The orbit was Sun-synchronous, ensuring that all its images were taken by the spacecraft of the same surface features at different times under identical lighting conditions. Despite the early problems, Mars Global Surveyor, already during its movement to its new orbit, began to send back impressive data and high-resolution images of the surface of Mars. The spacecraft tracked the evolution of a dust storm, gathered information on the Martian terrain, found compelling evidence indicating the presence of liquid water at or near the surface (first announced by NASA on 22 June 2000). During its mission, Mars Global Surveyor also produced the first three-dimensional profiles of Mars' north pole using laser altimeter readings. The spacecraft's primary mission was concluded on 1 February 2001, by which time it had returned 83,000 images of Mars, more than all previous missions to Mars combined. In addition, the laser altimeter essentially mapped almost all of the planet, by firing approximately 500 billion pulses at the surface, providing topographical data that was more detailed than many places on Earth. On 1 February 2001, Mars Global Surveyor's mission was extended for a year, and then again on 1 February 2002, this time continuing 11 months. In the early 2000s, the spacecraft supported other missions to Mars, including that of Mars Odyssey (in 2001) and the Mars Exploration Rovers (in 2004) by providing either atmospheric data or relaying telemetry back to Earth. Between 2004 and 2006, it conducted experiments simultaneously with the European Mars Express. On 1 October 2006, mission planners, based on the recommendations of a Senior Review Board, once again extended its mission by another two years but only a month later, on 2 November, the spacecraft lost contact with Earth when attempting to orient a solar panel. Although weak signals were received three days later, on 21 November 2006, NASA announced that the mission of Mars Global Surveyor was over. The final problem was probably related to a flaw in the system's software.

184

Mars 8 / Mars 96

Nation: Russia (106)
Objective(s): Mars orbit and landing
Spacecraft: M1 (no. 520)
Spacecraft Mass: 6,795 kg
Mission Design and Management: NPO imeni Lavochkina
Launch Vehicle: Proton-K + Blok D-2 (8K82K no. 392-02 + 11S824F no. 3L)
Launch Date and Time: 16 November 1996 / 20:48:53 UT
Launch Site: GIK-5 / Site 200/39
Scientific Instruments:
Orbiter:
   1. Argus imaging complex
     – TV camera (HRSC)
     – spectroscopic camera (WAOSS)
     – Omega infrared and optical spectrometer
   2. infrared Fourier spectrometer (PFS)
   3. Termoskan mapping radiometer
   4. Svet mapping spectrometer
   5. Spikam multi-channel spectrometer
   6. ultraviolet spectrophotometer (UFS-M)
   7. long-wave radar (RLK)
   8. Foton gamma-ray spectrometer
   9. Neytron-S neutron spectrometer
  10. mass spectrometer (MAK)
  11. Aspera-S ion and particles power and mass analyzer

12. Fonema omni non-scanning energy-mass ion analyzer

13. Dimio omni ionosphere energy mass spectrometer

14. Mari-prob ionospheric plasma spectrometers

15. Maremf electron analyzer/magnetometer

16. Elisma wave complex experiment

17. Sled-2 low energy charged particle spectrometer

18. precision gamma-ray spectrometer (PGS)

19. Lilas-2 cosmic and solar gamma-burst spectrometer

20. Evris stellar oscilllation photometer

21. solar oscillation spectrometer (SOYa)

22. Radius-M dosimeter

23. tissue-equivalent dosimeter (TERS)

Small Autonomous Stations (MAS):

1. meteorology complex (MIS)

2. meteorology complex (DPI)

3. alpha-particle proton and x-ray spectrometer

4. Optimizm seismometer/magnetometer/inclinometer

5. panoramic camera (PanCam)

6. descent phase camera (DesCam)

7. soil oxidization capacity instrument (MOx)

Penetrators:

1. PTV-1 camera

2. Mekom meteorological unit

3. Pegas gamma-ray spectrometer

4. Angstrem x-ray spectrometer

5. Alfa alpha/proton spectrometer

6. Neytron-P neutron spectrometer

7. Grunt accelerometers

8. Termozond temperature probes

9. Kamerton seismometer

10. IMAP-7 magnetometer

Results: Mars 8, the only Soviet/Russian lunar or planetary probe in the 1990s, was an ambitious mission to investigate the evolution of the Martian atmosphere, its surface, and its interior. Originally planned as two spacecraft, Mars 94 and Mars 96, the missions were delayed and became Mars 96 and Mars 98. Subsequently Mars 98 was cancelled leaving Mars 96 as the first Russian deep space mission beyond Earth orbit since the collapse of the Soviet Union. The entire spacecraft comprised an orbiter, two small autonomous stations, and two independent penetrators. The three-axis stabilized orbiter carried two platforms for pointing several optical instruments for studying the Martian surface and atmosphere. After an initial period in low orbit lasting three to four weeks acting as a relay to the landers, the orbiter was designed to spend approximately two Earth years in a 250 × 18,000-kilometer orbit at 101° inclination mapping Mars. The orbiter would have released the two 88-kilogram small autonomous stations (*Malaya avtonomnaya stantsiya*, MAS), four to five days before entering orbit. The small stations would have landed on the Martian surface, cushioned by an inflatable shell that was to split open after landing. The stations were to have transmitted data daily (initially) and then every three days for about 20 minutes each session. The stations would have also studied soil characteristics and taken photos on the surface. The two 123-kilogram penetrators, each 2.1 meters long, would have impacted the Martian surface at a velocity of 76 meters/second to reach about 6 to 8 meters in depth. The plan was for the orbiter to release them between 7 and 28 days after entering orbit. During their one-year lifetimes, the penetrators would have served as nodes of a seismic network. In the event, the Proton-K rocket successfully delivered the payload to Earth orbit (after the first firing of the Blok D-2 upper stage). Initial orbit parameters were 150.8 × 165.7 kilometers at 51.53° inclination. At that point, the Blok D-2 was to fire once again to place Mars 8 into an elliptical orbit, after which the Fregat propulsion module (with its S5.92 engine) would have sent the spacecraft on a Martian encounter trajectory. The Blok D-2 was to have fired at 10:57:46 UT on 16 November for 528 seconds but either didn't fire or shut down very soon after ignition, thus putting its precious payload into an incorrect and similar orbit of 143.7 × 169.6 kilometers. Mars 8 and its Fregat module then automatically separated from the Blok D-2. The latter seems to have fired (as planned earlier), placing Mars 8 in an 80 × 1,500-kilometer orbit that deposited the planetary probe in Earth's atmosphere, with reentry between 12:30 and 01:30 UT on November 17, probably over southern Chile. Various parts of the vehicle, including 200 grams of plutonium-238, must have survived the reentry although there have been no reports of detection. Mars 8 was scheduled to arrive in Mars orbit on 23 September 1997.

185

Mars Pathfinder

Nation: USA (70)
Objective(s): Mars landing and roving operations
Spacecraft: Mars Pathfinder
Spacecraft Mass: 870 kg
Mission Design and Management: NASA / JPL
Launch Vehicle: Delta 7925 (no. D240)
Launch Date and Time: 4 December 1996 / 06:58:07 UT
Launch Site: Cape Canaveral Air Force Station / Launch Complex 17B
Scientific Instruments:
Pathfinder Lander:
1. IMP imager for Mars Pathfinder (including magnetometer and anemometer)
2. atmospheric and meteorology package (ASI/MET)

Sojourner Rover:
1. imaging system (three cameras)
2. laser striper hazard detection system
3. alpha proton x-ray spectrometer (APXS)
4. wheel abrasion experiment
5. materials adherence experiment
6. accelerometers

Results: Mars Pathfinder was an ambitious mission to send a lander and a separate remote-controlled rover to the surface of Mars, the second of NASA's Discovery missions. Launched one month after Mars Global Surveyor, Pathfinder was sent on a slightly shorter seven-month trajectory designed

On 21 July 1997, the Mars Pathfinder's Sojourner rover takes its Alpha Particle X-ray Spectrometer measurement on a rock near the landing site. *Credit: NASA/JPL*

for arrival earlier. The main spacecraft included a 10.5-kilogram six-wheeled rover known as Sojourner capable of traveling 500 meters from the main ship at top speeds of 1 centimeter/second. The mission's primary goal was not only to demonstrate innovative, low-cost technologies, but also to return geological, soil, magnetic property and atmospheric data. After a 7-month traverse and four trajectory corrections (on 10 January, 3 February, 6 May, and 25 June 1997), Pathfinder arrived at Mars on 4 July 1997. The spacecraft entered the atmosphere using an atmospheric entry aeroshell that slowed the spacecraft sufficiently for a supersonic parachute to deploy and slow the package to 68 meters/second. After separation of the aeroshell heatshield, the lander detached, and at about 355 meters above the surface, airbags inflated in less than a second. Three solid propellant retro-rockets reduced velocity further (firing about 100 meters above the surface) but were then discarded at 21.5 meters altitude—they flew up and away along with the parachute. The lander-within-the-airbag impacted on the surface at a velocity of 14 meters/second generating about 18 g's of acceleration. The package bounced at least 15 times before coming to rest, following which the airbags deflated revealing the lander. Landing time for Pathfinder was

16:56:55 UT on 4 July 1997 at 19° 7′ 48″ N / 33° 13′ 12″ W in Ares Vallis, about 19 kilometers southwest of the original target. The next day, Pathfinder deployed the Sojourner rover on the Martian surface via landing ramps. Sojourner was the first wheeled vehicle to be used on any planet. During its 83-day mission, the rover covered hundreds of square meters, returned 550 photographs and performed chemical analyses at 16 different locations near the lander. The latter meanwhile transmitted more than 16,500 images and 8.5 million measurements of atmospheric pressure, temperature, and windspeed. Data from the rover suggested that rocks at the landing site resembled terrestrial volcanic types with high silicon content, specifically a rock type known as andesite. Although the planned lifetime of Pathfinder and Sojourner were expected to be one month and one week respectively, these times were exceeded by 3 and 12 times respectively. Final contact with Pathfinder was at 10:23 UT on 27 September 1997. Although mission planners tried to reestablish contact for the next five months, the highly successful mission was officially declared over on 10 March 1998. After landing, Mars Pathfinder was renamed the Sagan Memorial Station after the late astronomer and planetologist Carl Sagan. In 2003, Sojourner was inducted into the Robot Hall of Fame. On 21 December 2006, NASA's Mars Reconnaissance Orbiter's HIRISE camera photographed the flood plain of the Ares and Tiu outflow channels; images clearly showed the Pathfinder and associated hardware.

1997

186

ACE

Nation: USA (71)
Objective(s): Sun–Earth L1 Lagrange Point
Spacecraft: ACE
Spacecraft Mass: 752 kg
Mission Design and Management: NASA / GSFC
Launch Vehicle: Delta 7920-8 (no. D247)
Launch Date and Time: 25 August 1997 / 14:39 UT
Launch Site: Cape Canaveral Air Force Station / Launch Complex 17A
Scientific Instruments:

1. solar wind ion mass spectrometer (SWIMS) and solar wind ion composition spectrometer (SWICS)
2. ultra-low energy isotope spectrometer (ULEIS)
3. solar energetic particle ionic charge analyzer (SEPICA)
4. solar isotope spectrometer (SIS)
5. cosmic ray isotope spectrometer (CRIS)
6. solar wind electron, proton, and alpha monitor (SWEPAM)
7. electron, proton, and alpha-particle monitor (EPAM)
8. magnetometer (MAG)
9. real time solar wind experiment (RTSW)

Results: The Advanced Composition Explorer (ACE) spacecraft was designed to study space-borne energetic particles from the Sun–Earth L1 libration point, about 1.4 million kilometers from Earth. Specifically, the spacecraft was launched to investigate the matter ejected from the Sun to establish the commonality and interaction between the Sun, Earth, and the Milky Way galaxy. In addition, ACE also provides real-time space weather data and advanced warning of geomagnetic storms. ACE's nine instruments have a collecting power that is 10 to 10,000 times greater than anything previously flown. After launch, the spacecraft's Delta 2 launch vehicle's second stage reignited (after 4 hours) to insert the satellite into a 177 × 1.37 million-kilometer orbit. After reaching apogee a month after launch, ACE inserted itself into its Lissajous orbit around the L1 point. The spacecraft was declared operational on 21 January 1998 with a projected two- to five-year lifetime. As of early 2015, it continues to provide near-real-time 24/7 coverage of solar wind parameters and measure solar energetic particle intensities. With the exception of the SEPICA instrument (data from which was no longer received after 4 February 2005), all instruments on ACE remain operational as of mid-2017, and the propellant on board could theoretically allow a mission until about 2024.

187

Cassini-Huygens

Nation: US and ESA (3)
Objective(s): Saturn orbit, Titan landing
Spacecraft: Cassini and Huygens
Spacecraft Mass: 5,655 kg
Mission Design and Management: NASA / JPL / ESA
Launch Vehicle: Titan 401B-Centaur (TC-21 / Titan 401 no. 4B-33)
Launch Date and Time: 15 October 1997 / 08:43 UT
Launch Site: Cape Canaveral Air Force Station / Launch Complex 40

Scientific Instruments:

Cassini:

1. cosmic dust analyzer (CDA)
2. visible and infrared mapping spectrometer (VIMS)
3. imaging science system (ISS)
4. radar
5. ion and neutral mass spectrometer (INMS)
6. radio and plasma wave spectrometer (RPWS)
7. plasma spectrometer (CAPS)
8. ultraviolet imaging spectrograph (UVIS)
9. magnetospheric imaging instrument (MIMI)
10. dual technique magnetometer (MAG)
11. composite infrared spectrometer (CIRS)
12. radio science subsystem (RSS)

Huygens:

1. atmospheric structure instrument (HASI)
2. gas chromatograph neutral mass spectrometer (GC/MS)
3. aerosol collector and pyrolyzer (ACP)
4. descent imager/spectral radiometer (DISR)
5. surface science package (SSP)
6. Doppler wind experiment (DWE)

Results: The Cassini-Huygens project was the result of plans at NASA dating from the early 1980s and formally approved in 1989 as a joint NASA-ESA mission. Having survived several attempts by Congress to cancel the mission, the mission that emerged was a cooperative project with ESA (as well as the Italian Space Agency, ASI) involving a NASA-supplied spacecraft, Cassini, that orbits Saturn, and an ESA-supplied lander, Huygens, which descended into the atmosphere of Titan, Saturn's largest moon, in 2005. ASI provided the high-gain and low-gain antenna assembly and a major portion of the radio system. The primary scientific goals of the mission included a diverse set of investigations of Saturn, its moons, and its near environment. The 3,132-kilogram orbiter with an original design life of 11 years was powered by three radioisotope thermoelectric generators (RTGs), of the same design as the RTGs carried aboard Ulysses, Galileo, and New Horizons. The

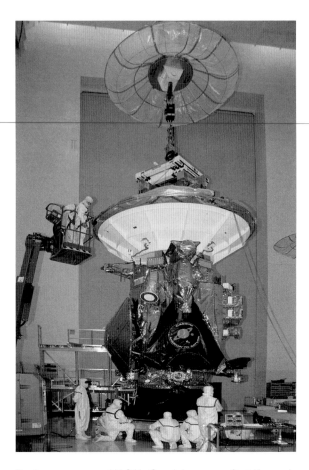

Engineers process NASA's Cassini spacecraft at Kennedy Space Center in Florida ahead of its launch to Saturn on 15 October 1997. *Credit: NASA*

335-kilogram Huygens, named after the Dutch physicist Christiaan Huygens (1629–1695), was designed to investigate Titan's atmosphere's chemical properties and measure wind, temperature, and pressure profiles from 170 kilometers down to the Moon's surface. The probe was not designed to survive past landing although scientists did not rule out the possibility entirely. Cassini's trip to Saturn included four gravity assists. Seven months after launch, the spacecraft passed Venus on April 26, 1988 at a range of 284 kilometers, gaining 26,280 kilometers/hour. Cassini performed a second flyby of Venus on June 24, 1999 at a range of 623 kilometers and one of Earth at 03:28 UT on 18 August 1999 at a range of 1,171 kilometers, before heading to Jupiter. During this portion of the traverse, Cassini passed by the asteroid 2685 Masursky on

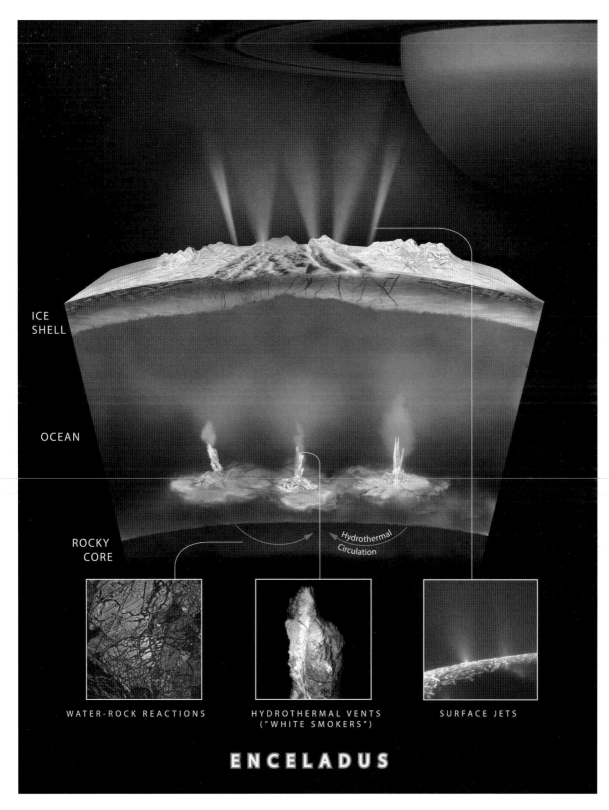

ICE
SHELL

OCEAN

ROCKY
CORE

Hydrothermal
Circulation

WATER-ROCK REACTIONS   HYDROTHERMAL VENTS
("WHITE SMOKERS")   SURFACE JETS

ENCELADUS

This graphic illustrates how scientists on NASA's Cassini mission think water interacts with rock at the bottom of the ocean of Saturn's icy moon Enceladus, producing hydrogen gas (H_2). *Credit: NASA/JPL-Caltech/Southwest Research Institute*

This artist's impression is based on images taken by ESA's Huygens probe which landed on the surface of Saturn's moon Titan. The parachute that slowed Huygens' reentry can be seen in the background, still attached to the lander. *Credit: ESA*

23 January 2000, flying as close as 1.5 million kilometers at 09:58 UT. During the encounter, Cassini used its remote sensing instruments to investigate the asteroid's size and dimensions and albedo. Nearly a year later, on 30 December 2000, the spacecraft passed by Jupiter at a distance of about 9.7 million kilometers. Among the data it returned was a detailed global color image of Jupiter, probably the most complete of the whole planet ever produced. In 2001–2002, controllers noticed a "haze" in images returned by the narrow-angle camera but these were eliminated following phases of heating the spacecraft. Its investigations en route to Saturn included an experiment in October 2003 in which scientists observed a frequency shift in radio waves to and from the probe as those signals traveled

close to the Sun. These results confirmed theoretical predictions based on Einstein's general theory of relativity. In May 2004, Cassini-Huygens entered the Saturn system, i.e., the gravitational pull of Saturn became higher than the pull from the Sun. After over five years of inactivity, Cassini's main engine was fired on 27 May as a test prior to orbital insertion. After a flyby of the Moon Phoebe on June 11 (at a distance of just 2,068 kilometers), Cassini performed one more correction five days later. Finally, on 1 July 2004, the spacecraft engine fired for 96 minutes, thus inserting Cassini-Huygens into a 0.02 × 9 million-kilometer orbit around Saturn. It was the first human-made object to enter orbit around Saturn. In its initial months, Cassini provided detailed data on Titan during three flybys (on 2 July, 27 October, and 13 December) and discovered two small new moons (Methone and Pallene). On Christmas Day 2004 at 02:00 UT, the Huygens lander, which had remained dormant for more than six years, separated from Cassini and began its 22-day coast to Titan. It entered Titan's atmosphere at 09:05:56 UT on 14 January 2005 and within 4 minutes had deployed its 8.5-meter diameter main parachute. A minute later, Huygens began transmitting a wealth of information back to Cassini for over 2 hours before impacting on the surface of Titan at 11:38:11 UT at a velocity of 4.54 meters/second. Landing coordinates were 192.32° W / 10.25° S, about 7 kilometers from its target point. A problem in the communications program limited the number of images that Huygens transmitted to Cassini, from about 700 to 376. Yet, to the excitement of planetary scientists back on Earth, it continued its transmissions for another 3 hours and 10 minutes during which it transmitted a view of its surroundings (224 images of the same view). Huygens appears to have landed in surface resembling "sand" made of ice grains; surface pictures showed a flat plain littered with pebbles as well as evidence of liquid acting on the terrain in the recent past. Subsequent data confirmed the existence of liquid hydrocarbon lakes in the polar regions of Titan. In April 2016, ESA announced that one of Titan's three large seas close to the north pole, known as Ligela Mare, is filled with pure liquid methane, with a seabed covered by a "sludge" of organic-rich material. The Cassini orbiter meanwhile continued its main orbital mission investigating the Saturn system, its voyage punctuated by repeated "targeted" flybys—flybys actively implemented by trajectory corrections—of various moons, particularly Titan, Enceladus, Tethys, Hyperion, Dione, Rhea, and Iapetus. Cassini ended its primary mission on 27 May 2008 with its 43rd flyby of Titan. During this period, the spacecraft discovered two new moons (Daphnis and Anthe), uncovered much valuable data about Titan, including the first radar images of the moon's

surface taken during its 27 October 2004 flyby and clear evidence of existing large lakes of liquid hydrocarbon in the northern latitudes of Titan, and performed a number of radio occultation experiments to study the size-distribution of particles in Saturn's rings and atmosphere. Perhaps the most exciting flyby was one on 12 March 2008 when Cassini flew within 50 kilometers of the surface of Enceladus, passing through the plumes from its southern geysers. The spacecraft detected water and carbon dioxide and also mapped surface features. In April 2008, NASA approved a two-year extension of its mission (i.e., 60 more Saturn orbits), which officially began on 1 July 2008 and was called the Cassini Equinox Mission, named as such because the mission coincided with Saturn's equinox. Further moons of Saturn were identified (Aegaeon and S/2009 S 1, the latter a "propeller moonlet" perhaps only 400 meters across) while additional encounters with Enceladus allowed Cassini to acquire very high-resolution images of its surface and directly sample its cryo-volcanic plumes that appear to contain complex organic chemicals. During the two-year Equinox Mission, which ended in September 2010, Cassini performed 26 targeted flybys of Titan, 7 of Enceladus, and 1 each of Dione, Rhea, and Helene. On 3 February 2010, NASA announced that Cassini's mission would continue beyond the original two-year extension into the new Cassini Solstice Mission that would last until September 2017, a few months past Saturn's summer solstice. The new mission was named after the Saturnian summer solstice occurring in May 2017, which marked the beginning of summer in the northern hemisphere and winter in the southern hemisphere. (The spacecraft had arrived at Saturn just after the planet's northern winter solstice. The extension thus allowed scientists to study a complete seasonal period of the planet.) The Cassini Solstice Mission was guided principally by its ability to continue close studies of Titan (particularly seasonal climate change such as storms, flooding, and changes in lakes) and Enceladus (particularly, its "astrobiological potential") but also some of the other icy moons (such as Dione and Rhea), the planet itself (and its magnetosphere), and its rings. The extension allowed 155 orbits around Saturn 155 times, plus 54 flybys of Titan, 11 of Enceladus, 2 of Rhea, and 3 of Dione. At the beginning of the Solstice Mission, on 2 November 2010, Cassini ran into a problem when a malfunction in the spacecraft's computer shut down all nonessential systems. Slowly, over a period of about three weeks, controllers were able to restore all of Cassini's instruments to working order. Only one targeted flyby of Titan was affected during the interim. On 6 March 2014, the spacecraft conducted its 100th flyby of Titan (at a range of 1,500 kilometers), conducting gravity measurements in order to explore the existence of a global subsurface ocean. By July 2014, Cassini had identified at least 101 distinct geysers erupting on the south polar region of Enceledaus. Researchers concluded that it is possible for liquid water to reach all the way to the surface from the moon's underground sea (whose existence had been announced in April 2014). The presence of this salty underground ocean, about 30–40 kilometers thick, under a 10-kilometer ice shell, raises the possibility that microbial life might exist there. Significant events in 2015–2016 included a close flyby (at 47,000 kilometers) of Rhea in February 2015, allowing very high-resolution images of the natural satellite, a flyby (at 34,000 kilometers) of the irregularly-shaped Hyperion in May 2015 showing a deeply impact-scarred surface, and two last flybys of Dione in June and August 2015, at just 516 and 474 kilometers range, respectively. Perhaps the most spectacular mission event of the year was Cassini's "deep-dive" to just 49 kilometers above the south polar region of the geologically active Enceladus in October 2015. During the encounter, the spacecraft's gas analyzer and dust detector instruments sampled the moon's plume of gas and dust-sized icy particles. A final flyby of Enceladus was carried out

in December at a range of 4,999 kilometers, concluding a chapter in the Cassini mission's encounters with Saturn's moons. In December 2015, Cassini initiated a number of delicate orbital maneuvers designed to tilt the spacecraft's orbit out of Saturn's ringplane. Each maneuver was followed by a gravity-assist from Titan ("Titan does all the heavy lifting," noted Earl Maize, Cassini project manager at JPL), thus sending the vehicle to an increasingly higher inclination (relative to Saturn's equator). These maneuvers set up Titan for its final dramatic year in 2016–2017, involving two distinct phases of the mission. On 30 November 2016, Cassini set off on a path that carried it high above and under Saturn's poles, diving every seven days through the hitherto unexplored region at the outer edge of the main rings. This phase of the mission, called "Cassini's Ring-Grazing Orbits" involved 20 "dives" through this region, that ended on 22 April 2017. During some of these passes, the spacecraft directly sampled ring particles and molecules of faint gases close to the rings. In March and April 2017, ring crossings had the spacecraft fly through the dusty outer regions of the F ring. After the last ring-grazing orbit concluded on 22 April 2017, a flyby of Titan reshaped Cassini's trajectory to send it on a new phase of the mission, the "Grand Finale," involving 22 plunges, the first on 26 April through the 2,400-kilometer gap between Saturn and its innermost ring. The mission concluded on 15 September 2017 on its 293rd orbit of Saturn when Cassini plunged into the Saturnian atmosphere ending one of the most ambitious and spectacular missions in the history of planetary exploration. By most estimates, the spacecraft burned up in the atmosphere and was destroyed about 45 seconds after the last transmission. During the final moments of the descent, data from eight of Cassini's science instruments beamed important data back to Earth, giving insight into the planet's formation and evolution. Cassini was named after Italian astronomer Giovanni Cassini (1625–1712).

188

Asiasat 3 / HGS 1

Nation: Asia Satellite Telecommunications Co. (1)

Objective(s): geostationary orbit, circumlunar mission

Spacecraft: Asiasat 1

Spacecraft Mass: 3,465 kg

Mission Design and Management: Asiasat (operator) + Hughes (design and manufacture)

Launch Vehicle: Proton-K + Blok DM3 (8K82K no. 394-01 / Blok DM3 no. 5L)

Launch Date and Time: 24 December 1997 / 23:19 UT

Launch Site: GIK-5 / Site 81/23

Scientific Instruments: [none]

Results: The lunar flyby accomplished by Asiasat 3 was not part of a science mission but rather the end result of a rescue mission of a satellite that had been stranded in an incorrect orbit. Asiasat 3 was a communications satellite, based on the Hughes HS-601HP bus, launched by the Russians for Asia Satellite Telecommunications Co. Because of the improper second firing of the Blok DM3 upper stage, the satellite ended up in a useless 203 × 36,000-kilometer orbit around the Earth and written off as a loss by Asiasat. Insurance underwriters subsequently signed an agreement with Hughes Global Systems who built the satellite to salvage the vehicle and bring it to its originally intended geostationary orbit by using as little propellant as possible. Using 11 carefully-planned burns beginning 12 April 1998, controllers raised the orbit's apogee to 321,000 kilometers. Then, with the 12th firing on 7 May 1998, the spacecraft was sent on a nine-day round trip around the Moon, approaching as close as 6,200 kilometers to its surface on 13 May. Using this gravity assist, Asiasat 3 hurled back into a usable orbit. By 16 May 1998, perigee had been raised to 42,000 kilometers and inclination reduced from 51° to 18°. A second circumlunar mission began on 1 June that culminated in a 34,300-kilometer flyby of the Moon on 6 June, a distance closer than the Soviet Luna 3. After four more engine firings, the satellite was finally in a 24-hour geosynchronous orbit by 17 June 1998 above 153°. The satellite, now owned by Hughes, was renamed HGS 1. In 1999, HGS-1 was bought by PanAmSat and renamed PAS 22 and moved to 60° W, and subsequently, in July 2002, it was deactivated and moved to a graveyard orbit.

1998

189

Lunar Prospector

Nation: USA (72)
Objective(s): lunar orbit
Spacecraft: Lunar Prospector
Spacecraft Mass: 300 kg
Mission Design and Management: NASA / ARC
Launch Vehicle: Athena-2 (LM-004)
Launch Date and Time: 7 January 1998 / 02:28:44 UT
Launch Site: Cape Canaveral Air Force Station / SLC-46
Scientific Instruments:

1. electron reflectometer and magnetometer (MAG/ER)
2. gamma-ray spectrometer (GRS)
3. neutron spectrometer (NS)
4. alpha particle spectrometer (APS)
5. Doppler gravity experiment (using S-band antenna) (DGE)

Results: Lunar Prospector was designed to collect data to compile the first complete compositional and gravity maps of the Moon during its planned one-year mission in lunar polar orbit. It was the third mission of NASA's Discovery Program of low-cost, highly-focused, and relatively frequent missions that were competitively selected. After two mid-course corrections, Lunar Prospector entered orbit around the Moon 105 hours after launch. Initial parameters were 92 × 153 kilometers. After two further corrections on 13 and 15 January, the spacecraft entered its operational 100 × 100-kilometer orbit at 90° inclination with a period of 118 minutes. Perhaps of most interest to scientists was to continue investigations into the signs of water ice on the Moon as found by the Clementine

probe. In March 1998, NASA announced that data from Lunar Prospector suggested the presence of water ice at both the lunar poles; the neutron spectrometer instrument detected hydrogen, assumed to be in the form of water. The information indicated that a large amount of water ice, possibly as much as 300 million (metric) tons was mixed into the regolith at each pole, the first direct evidence of water ice at the lunar poles. The spacecraft also detected strong localized magnetic fields, mapped the global distribution of major rock types, and discovered signs of a 600-kilometer diameter, iron-rich core. On 10 December 1998, Lunar Prospector's orbit was lowered to 40 kilometers to perform high-resolution studies. A subsequent maneuver on 28 January 1999 changed the orbit to 15 × 45 kilometers and ended the space probe's primary mission but began an extended mission for an additional seven months. Lunar Prospector was deliberately impacted onto the shadowed Shoemaker crater on the lunar surface at 09:52:02 UT on 31 July 1999. Observations of the resulting dust cloud by Earth-based telescopes showed no spectral signature characteristic of water vapor. The vehicle carried part of the cremated remains of geologist Eugene Shoemaker to the lunar surface.

190

Nozomi

Nation: Japan (4)
Objective(s): Mars orbit
Spacecraft: Planet-B
Spacecraft Mass: 536 kg
Mission Design and Management: ISAS
Launch Vehicle: M-V (no. 3)

Launch Date and Time: 3 July 1998 / 18:12 UT
Launch Site: Kagoshima / Launch Complex M-5
Scientific Instruments:

1. Mars imaging camera (MIC)
2. magnetic field measurement instrument (magnetometer) (MGF)
3. electron spectrum analyzer (ESA)
4. ion spectrum analyzer (ISA)
5. ion mass imager (IMI)
6. electron and ion spectrometer (EIS)
7. thermal plasma analyzer (TPA)
8. probe for electron temperature (PET)
9. plasma wave and sounder (PWS)
10. neutral mass spectrometer (NMS)
11. Mars dust counter (MDC)
12. extra ultraviolet scanner (XUV)
13. ultraviolet imaging spectrometer (UVS)
14. low frequency plasma wave analyzer (LFA)

Results: Nozomi, Japan's fourth "deep space" probe, was also its first planetary spacecraft and the first directed to Mars that was not from the United States or the Soviet Union/Russia. The spacecraft was slated to enter a highly elliptical orbit around Mars on 11 October 1999. Its mission was to conduct long-term investigations of the planet's upper atmosphere and its interactions with the solar wind and track the escape trajectories of oxygen molecules from Mars' thin atmosphere. It was also to have taken pictures of the planet and its moons from its operational orbit of 300 × 47,500 kilometers; during perigee, Nozomi would have performed remote sensing of the atmosphere and surface while close to apogee, the spacecraft would have studied ions and neutral gas escaping from the planet. Although designed and built by Japan, the spacecraft carried a set of 14 instruments from Japan, Canada, Germany, Sweden, and the United States. After entering an elliptical parking orbit around Earth at 340 × 400,000 kilometers, Nozomi was sent on an interplanetary trajectory that involved two gravity-assist flybys of the Moon

on 24 September and 18 December 1998 (at 2,809 kilometers), and one of Earth on 20 December 1998 (at 1,003 kilometers). The gravitational assist from Earth as well as a 7-minute engine burn put Nozomi on an escape trajectory towards Mars. Unfortunately, a problem valve resulted in loss of propellant, leaving the spacecraft with insufficient acceleration to reach its nominal trajectory. Subsequently, because two more mid-course corrections on 21 December used more propellant than intended, Nozomi's originally planned mission had to be completely reconfigured. The new plan involved four further years in heliocentric orbit, during which time it would conduct two more Earth flybys (in December 2002 and June 2003) leading to a Mars encounter in December 2003, four years after its original schedule. While heading towards Earth, on 21 April 2002, powerful solar flares damaged Nozomi's communications and power systems, causing the hydrazine to freeze in the vehicle's attitude control system. Contact was lost with the spacecraft on 15 May but two months later, controllers found the spacecraft's beacon. Mission scientists were able to thaw the frozen fuel as it approached Earth and the flybys were accomplished as intended: on 21 December 2002 at a range of 29,510 kilometers and once again on 19 June 2003 at a range of 11,023 kilometers. Soon after, the spacecraft's luck finally ran out: on 9 December 2003, in anticipation of the Mars orbit insertion (planned for five days later), the main thruster failed, essentially ending its mission. Instead, ground controllers commanded lower thrust attitude control thrusters to fire to ensure that Nozomi would not impact onto the Martian surface, which would have been a problem since the spacecraft had not been sterilized. The spacecraft passed by Mars at a range of 1,000 kilometers and remains in heliocentric orbit. Despite not accomplishing its primary mission, Nozomi provided important data from its suite of scientific instruments.

191

Deep Space 1

Nation: USA (73)

Objective(s): technology testing, comet flyby

Spacecraft: DS1

Spacecraft Mass: 486 kg

Mission Design and Management: NASA / JPL

Launch Vehicle: Delta 7326-9.5 (no. D261)

Launch Date and Time: 24 October 1998 / 12:08:00 UT

Launch Site: Cape Canaveral Air Force Station / Launch Complex 17A

Technology Instruments:

1. ion propulsion system
2. solar concentrator array with refractive linear element technology (SCARLET)
3. autonomous navigation system (AutoNav)
4. remote intelligent operations software (Remote Agent RAX)
5. Beacon monitor operations experiment
6. small deep-space transponder (SDST)
7. miniature integrated camera spectrometer (MICAS)
8. plasma experiment for planetary exploration instrument (REPE)
9. Ka-band solid-state power amplifier

Results: Deep Space 1 (DS1) was designed to test new innovative technologies appropriate for future deep space and interplanetary missions. It was the first in a new series of technology demonstration missions under NASA's New Millennium program. The spacecraft's main goals were to test 12 "high-risk" technologies as ion propulsion, autonomous optical navigation, a solar power concentration array, and a combination miniature camera/imaging spectrometer. As a bonus, the spacecraft also flew by the asteroid 9969 Braille. After a successful launch into parking orbit around Earth, a third stage burn at 13:01 UT on 24 October 1998 put DS1 on a heliocentric trajectory. On 10 November controllers commanded the ion thruster to fire for the first time but it operated for only 4.5 minutes

before stopping. Two weeks later, on 24 November 1998, controllers once again fired Deep Space 1's ion propulsion system (fueled by xenon gas) when the spacecraft was 4.8 million kilometers from Earth. This time, the engine ran continuously for 14 days and demonstrated a specific impulse of 3,100 seconds, as much as 10 times higher than possible with conventional chemical propellants. The mission tested its payload extensively to ensure that future users of such technologies would not take on unnecessary risks. DS1 passed by the near-Earth asteroid 9660 Braille at 04:46 UT on 29 July 1999 at a range of only 26 kilometers at a velocity of 15.5 kilometers/second. Although it was the closest asteroid flyby to date, it was only partially successful due to a problem that compromised data delivered to the onboard navigational system. These difficulties prevented a closer encounter, originally planned at 240 meters range. The few images returned from very long range were out of focus although much other data was useful. DS1 found Braille to be 2.2 kilometers at its longest and 1 kilometer at its shortest. Once the successful primary mission was over by 18 September 1999, NASA formulated an extended mission. Originally, the plan had been to have DS1 fly by the dormant Comet 107P/Wilson-Harrington in January 2001 and the Comet 19P/Borrelly in September 2001, but the spacecraft's star tracker failed on 11 November 1999. The continuation of the mission without the use of the star tracker—which initially was thought to be fatal to the mission since the spacecraft could not point its ion engine or sensors in the proper directions—required considerable ingenuity and effort on the part of controllers. Over two months, the operations team struggled to have the spacecraft point its antenna to Earth allowing it to download data on the tracker's failure (as well as other data collected by DS1). Over the subsequent five months, the team devised an innovative plan to revive the vehicle, by "building" a new attitude control system operating without the failed star tracker. Although it was no longer possible to visit the bonus targets (given its limited capability), DS1 was still healthy enough to be targeted to Borrelly with the hope of arriving in September 2001. By the end of 1999, DS1's ion engine had expended 22 kilograms of xenon to impart a total delta V of 1,300 meters/second. On its way to Borrelly, it set the record for the longest operating time for a propulsion system in space. By 17 August 2000, the engine had been operating for 162 days as part of an eight-month run. On 22 September 2001, DS1 entered the coma of Comet Borrelly, making its closest approach of 2,171 kilometers to the nucleus at 22:29:33 UT. Traveling at 16.58 kilometers/second relative to the nucleus at the time, it returned some of the best images of a comet ever as well as other significant data. The spacecraft's ion engine was finally turned off on 18 December 2001, having operated for 16,265 hours and provided a total delta-V of 4.3 kilometers/second, the largest delta-V achieved by a spacecraft with its own propulsion system. By this point, the spacecraft had operated far beyond its planned lifetime and was running low on attitude control hydrazine. A radio receiver was left on in case future contact with the spacecraft was desired, although an attempt in March 2002 to contact the spacecraft was unsuccessful.

192

Mars Climate Orbiter

Nation: USA (74)

Objective(s): Mars orbit

Spacecraft: MCO

Spacecraft Mass: 638 kg

Mission Design and Management: NASA / JPL

Launch Vehicle: Delta 7427-9.5 (no. D264)

Launch Date and Time: 11 December 1998 / 18:45:51 UT

Launch Site: Cape Canaveral Force Station / Launch Complex 17A

Scientific Instruments:

1. pressure modulated infrared radiometer (PMIRR)

2. Mars color imaging system (two cameras) (MARCI)

Results: Mars Climate Orbiter (MCO) was the second probe in NASA's Mars Surveyor program, which also included the Mars Global Surveyor (launched in November 1996) and Mars Polar Lander (launched in January 1999). Mars Climate Orbiter was designed to arrive at roughly the same time as Mars Polar Lander and conduct simultaneous investigations of Mars' atmosphere, climate, and surface. Arrival in orbit was dated for 23 September 1999; MCO would then attain its operational near-circular Sun-synchronous orbit at 421 kilometers by 1 December 1999. The satellite was also designed to serve as a communications relay for Mars Polar Lander. After the lander's mission lasting three months, MCO would have performed a two-year independent mission to monitor atmospheric dust and water vapor and take daily pictures of the planet's surface to construct an evolutionary map of climatic changes. Scientists hoped that such information would aid in reconstructing Mars' climatic history and provide evidence on buried water reserves. After the end of its main mapping mission on 15 January 2001, Mars Climate Orbiter would have acted as a communications relay for future NASA missions to Mars. After launch, the spacecraft was put into a Hohmann transfer orbit to intersect with Mars. It performed four mid-course corrections on 21 December 1998, 4 March, 25 July, and 15 September 1999. At 09:00:46 UT on 23 September 1999, the orbiter began its Mars orbit insertion burn as planned. The spacecraft was scheduled to reestablish contact after passing behind Mars, but, unfortunately, no further signals were received from the spacecraft. An investigation indicated that the failure resulted from a navigational error due to commands from Earth being sent in English units (in this case, pound-seconds) without being converted into the metric standard (Newton-seconds). The error caused the orbiter to miss its intended 140–150-kilometer altitude orbit and instead fall into the Martian atmosphere at approximately 57 kilometers altitude and disintegrate due to atmospheric stresses.

1999

193

Mars Polar Lander and Deep Space 2

Nation: USA (75)

Objective(s): Mars landing

Spacecraft: MPL

Spacecraft Mass: 576 kg total (including 290 kg lander)

Mission Design and Management: NASA / JPL

Launch Vehicle: Delta 7425-9.5 (no. D265)

Launch Date and Time: 3 January 1999 / 20:21:10 UT

Launch Site: Cape Canaveral Air Force Station / Launch Complex 17B

Scientific Instruments:

1. stereo surface imager (SSI)
2. robotic arm (RA)
3. meteorology package (MET)
4. thermal and evolved has analyzer (TEGA)
5. robotic arm camera (RAC)
6. Mars descent imager (MARDI)
7. light detection and ranging instrument (LIDAR)
8. Mars microphone

Results: The Mars Polar Lander (MPL) was one of NASA's Mars Surveyor missions that called for a series of small, low-cost spacecraft for sustained exploration of Mars. MPL's primary goal was to deploy a lander and two penetrators (known as Deep Space 2) on the surface of Mars to extend our knowledge on the planet's past and present water resources. The objective was to explore the never-before studied carbon dioxide ice cap, about 1,000 kilometers from the south pole. The mission also called for recording local meteorological conditions, analyzing samples of polar deposits, and taking multi-spectral images of local areas. MPL was to have performed its mission simultaneously with that of the Mars Climate Orbiter that would have acted as a communications relay during its surface operations. MPL itself comprised a bus section (for power, propulsion, and communications during the outbound voyage) and a 290-kilogram lander that stood 1.06 meters tall on the ground. The lander was equipped with a 2-meter long remote arm to dig into the terrain and investigate the properties of Martian soil (using the Thermal and Evolved Gas Analyzer). Having arrived at Mars on 3 December 1999, the spacecraft would enter the atmosphere, and about 10 minutes prior to landing, would jettison its cruise stage and solar panels and then release the two 3.572 kilogram (each) Deep Space 2 microprobes. Unlike Mars Pathfinder, MPL was scheduled to make a completely controlled landing using retro-rockets all the way to the surface. Landing was scheduled for 21:03 UT on 3 December 1999 with two-way communications planned to begin 20 minutes later. The two Deep Space 2 microprobes (renamed Amundsen and Scott on 15 November 1999), meanwhile, would impact the ground at a speed of 200 meters/second about 50–85 seconds prior to the lander and about 100 kilometers away. Each penetrator was designed to obtain a small sample of subsurface soil using an electric drill for analysis. The microprobes' mission was expected to last about 36 hours while the lander mission would continue until 1 March 2000. Mars Polar Lander successfully left Earth on a Mars transfer trajectory on 3 January 1999. During its traverse to Mars, the spacecraft was stowed inside an aeroshell capsule. The complete vehicle approached Mars in early December in apparently good health. Last contact with the vehicle was at 20:02 UT on

3 December 1999 as the spacecraft slewed to entry attitude. Then, traveling at 6.9 kilometers/second, the capsule entered the Martian atmosphere about 8 minutes later. Controllers expected to reestablish contact 24 minutes after landing (scheduled for 20:14 UT) but no signal was received. With no communications for over two weeks, on 16 December 1999, NASA used the Mars Global Surveyor orbiting Mars to look for signs of the lander on the Martian surface, but the search proved fruitless. On 17 January 2000, NASA finally terminated all attempts to establish contact with the lost lander. An independent investigation into the failure, whose results were released publicly on 28 March 2000, indicated that the most probable cause of the failure was the generation of spurious signals when the lander's legs deployed during the descent. These signals falsely indicated that the spacecraft had touched down on Mars when in fact it was still descending. The main engines prematurely shut down, and the lander fell to the Martian landscape. The demise of MPL undoubtedly set NASA's Mars exploration program back and also spelled the effective end of NASA's "Faster, Better, Cheaper" initiative for low-cost highly innovative missions. The Phoenix lander, which arrived on Mars in 2008, subsequently accomplished most of the original Mars Polar Lander's objectives. MPL carried a CD-ROM with the names of one million children from around the world as part of the "Send Your Name to Mars" program formulated to foster interest in space exploration among young people.

194

Stardust

Nation: USA (76)
Objective(s): comet sample return, comet flybys
Spacecraft: Stardust
Spacecraft Mass: 385 kg
Mission Design and Management: NASA / JPL
Launch Vehicle: Delta 7426-9.5 (no. D266)

Launch Date and Time: 7 February 1999 / 21:04:15 UT
Launch Site: Cape Canaveral Air Force Station / Launch Complex 17A
Scientific Instruments:
1. dust flux monitor instrument (DFMI)
2. cometary and interstellar dust analyzer (CIDA)
3. navigation camera (NC)
4. stardust sample collection (SSC)
5. dynamic science experiment (DSE)

Results: Stardust was the fourth of NASA's Discovery program of low-cost exploration missions (after NEAR, Mars Pathfinder, and Lunar Prospector), and the first American mission dedicated solely to studying a comet. It was also the second robotic mission (after Genesis) designed to bring extraterrestrial material from beyond lunar orbit back to Earth. Its primary goal was to fly by the Comet Wild 2 (pronounced "vilt 2"), collect samples of dust from the coma of the comet as well as additional interstellar particles, and then return the samples to Earth. Stardust comprised a 254-kilogram spacecraft that included a 45.7-kilogram return capsule shaped like a blunt-nosed cone. It had five major components: a heat shield, back shell, sample canister, parachute system, and avionics. The samples were to be collected using a low-density microporous silica-based substance known as aerogel, attached to panels on the spacecraft to "soft-catch" and preserve the cometary materials. The spacecraft was launched into heliocentric orbit that would bring it around the Sun and past Earth for a gravity-assist maneuver to direct it to Wild 2 after a flyby of the minor planet Annefrank in November 2002. After mid-course corrections on 28 December 1999, 18 January, 20 January, and 22 January 2000, its first interstellar dust collection operation was carried out between 22 February and 1 May 2000. After approximately a year in heliocentric orbit, Stardust flew by Earth (at a range of 6,008 kilometers) on 15 January 2001 for a gravity assist to send it on a second sample collection exercise between July and December 2002. On 2 November 2002 at 04:50 UT, Stardust flew by asteroid 5535 Annefrank at a range of 3,078 kilometers. During the encounter, the spacecraft's dust collectors collected samples while its camera returned 72 images. Over a year later, on 31 December 2003, the spacecraft entered the coma of Comet Wild 2 (or 81P/Wild) with the closest encounter (at a range of 250 kilometers) taking place at 19:22 UT on 2 January 2004. The sample collector, which had been deployed on 24 December was retracted about 6 hours after closest approach, stowed, and then sealed in the "sample vault." The imaging system also took 72 images of the comet's nucleus. Exactly as planned, after a 4.63 billion-kilometer trip lasting over two years, at 05:57 UT on 15 January 2006, Stardust's Sample Return Capsule (SRC) separated from the main vehicle and, 4 hours later, entered Earth's atmosphere. Slowed down by the drogue and main parachutes, the capsule landed at 10:10 UT within a 30 × 84-kilometer landing zone at the U.S. Air Force Test and Training Range in Utah. Because of high winds, the capsule drifted north of the ground track, but fortunately a locator beacon allowed rescuers to find the capsule 44 minutes after landing. The capsule had returned more than 10,000 particles larger than 1 micrometer from Wild 2. The main spacecraft, meanwhile was diverted so as not to reenter Earth's atmosphere. At 06:13 UT on 15 January, it fired its engines, flew past Earth and then the Moon before entering hibernation mode on 29 January 2004 and remains in a 3-year-long heliocentric orbit. In July 2007, NASA approved an extended mission for Stardust known as New Exploration of Tempel 1 (NExT) that envisaged a flyby of Comet Tempel 1 (or 9P/Tempel), which had been the target for Deep Impact's impact probe in 2005. The spacecraft, now known as Stardust/NExT, flew by Tempel 1 at 04:42:00 UT on 15 February 2011 at a range of 181 kilometers, returning 72 images of the nucleus. This was the first time a comet had been revisited. It was also during this flyby that investigators were able to conclusively identify the impact crater from Deep Impact's Impactor probe. Stardust carried out a final engine burn on 24 March 2011 exhausting all of its propellant. It sent its last transmission at 12:33 UT the same day, ending an 11-year mission. The analysis of the samples returned showed the presence of a wide range of organic compounds. In August 2014, NASA announced that seven rare, microscopic interstellar dust particles dating from the very origins of the solar system were among the samples collected by Stardust.

2001 Mars Odyssey

Nation: USA (77)
Objective(s): Mars orbit
Spacecraft: 2001 Mars Odyssey
Spacecraft Mass: 1,608.7 kg
Mission Design and Management: NASA / JPL
Launch Vehicle: Delta 7925-9.5 (no. D284)
Launch Date and Time: 7 April 2001 / 15:02:22 UT
Launch Site: Cape Canaveral Air Force Station / SLC-17A
Scientific Instruments:

1. thermal emission imaging system (THEMIS)
2. gamma ray spectrometer (GRS)
3. Mars radiation environment experiment (MARIE)

Results: As of mid-2017, 2001 Mars Odyssey holds the record for the longest surviving continually active spacecraft in orbit around a planet other than Earth, at 16 years and counting. This, the first launch in NASA's revamped Mars Exploration Program (which was originally approved in 1993 but restructured in October 2000 after the failures associated with "Faster, Better, Cheaper"), was designed to investigate the Martian environment, providing key information on its surface and the radiation hazards future explorers might face. The goal was to map the chemical and mineralogical makeup of Mars as a step to detecting evidence of past or present water and volcanic activity on Mars. It was also designed to act as a relay for future landers, and did so for the Mars Exploration Rovers (Spirit and Opportunity), the Mars Science Laboratory, and the Phoenix lander.

During the coast to Mars, in August 2001, the MARIE radiation instrument failed to respond but was successfully revived by March 2002. About 200 days after launch, at 02:38 UT on 24 October 2001, Mars Odyssey successfully entered orbit around Mars after a 20-minute, 19-second-long engine burn. The initial orbit was highly elliptical (272 × 26,818 kilometers), taking the spacecraft 18.6 hours to complete one circuit. The spacecraft then implemented an unusual aerobraking maneuver that used the planet's atmosphere to gradually bring the satellite closer to the Martian surface on every succeeding orbit. This process saved an estimated 200 kilograms of propellant. Once the aerobraking was over, by 30 January 2002, Mars Odyssey was in its nearly Sun-synchronous polar orbit of 400 × 400 kilometers at 93.1° inclination, allowing the initiation of its science and mapping mission on 19 February 2002. This phase lased 917 Earth days during which entire ground tracks were repeated every two sols. One of the most exciting findings of Mars Odyssey came early on in the mission. In May 2002, NASA announced that the probe had identified large amounts of hydrogen in the soil, implying the presence of ice possibly a meter below the planet's surface. Much later, in March 2008, mission scientists revealed that Mars Odyssey had found evidence of salt deposits in 200 locations in southern Mars. These chloride minerals were left behind as places where water was once abundant. Having fully completed its primary mission by August 2004, mission planers began a series of extended missions starting 24 August 2004. NASA approved seven two-year extensions of the Mars Odyssey mission, in 2004, 2006, 2008, 2010, 2012, 2014, and 2016. Each was dedicated to a specific set of objectives. For example, the fourth extension ending in August

2012 was dedicated to observing the year-to-year variations in polar ice, clouds, and dust storms. One of its instruments, the MARIE radiation experiment, stopped working on 28 October 2003, after a large solar event bombarded the spacecraft early in the mission, most likely because of a damaged computer chip. In addition, one of the spacecraft's reaction wheels failed in June 2012, but a spare, installed on board as a redundancy, was activated and spun into service a month later. A few months later, in August 2012, NASA used Mars Odyssey's THEMIS instrument to help select a landing site for the Mars Science Laboratory (MSL) and later acted as a relay for the MSL rover Curiosity. By July 2010, NASA was able to announce that Mars Odyssey's camera had helped construct the most accurate global map of Mars ever, using 21,000 images from the THEMIS instrument. These pictures have been smoothed, matched, blended, and cartographically controlled to make a giant mosaic available to users online. Later that year, on 15 December 2010, Mars Odyssey claimed the record for the longest operating spacecraft at Mars, with 3,340 days of operation. In December 2016, the spacecraft put itself into "safe mode" due to a problem with orientation relative to Earth and the Sun but by early January 2017 was restored to full operating status. During its many years in Martian orbit, Mars Odyssey globally mapped the amount and distribution of the numerous chemical elements and minerals in the Martian surface and also tracked the radiation environment in low Mars orbit, both necessary before humans can effectively explore the Martian surface. By mid-2016, the THEMIS instrument had returned more than 208,000 images in visible-light wavelengths and more than 188,000 in thermal-infrared wavelengths.

196

Microwave Anisotropy Probe (MAP)

Nation: USA (78)
Objective(s): Sun–Earth L2 Lagrange Point
Spacecraft: Explorer 80
Spacecraft Mass: 840 kilograms
Mission Design and Management: NASA / GSFC
Launch Vehicle: Delta 7425-10 (no. D286)
Launch Date and Time: 30 June 2001 / 19:46:46 UT
Launch Site: Cape Canaveral Air Force Station / SLC-17B
Scientific Instruments:

1. Pseudo-Correlation Radiometer (fed by two back-to-back reflectors)

Results: The Microwave Anisotropy Probe was designed to map the relative Cosmic Microwave Background (CMB) temperature with high angular resolution and sensitivity over the full sky. In order to achieve this, MAP used differential microwave radiometers that measured temperature differences between two points on the sky from its operational position at the Sun–Earth L2 Lagrange Point, about 1.5 million kilometers from Earth. The two back-to-back telescopes were designed to focus microwave radiation from two spots on the sky approximately 140° apart and feed it to 10 separate differential receivers. The MAP mission was a successor to the Cosmic Background Explorer (COBE) mission, also known as Explorer 46, launched in 1989. The spacecraft was launched into an initial orbit of 167 × 204 kilometers at 28.8° inclination. A third stage burn directed MAP into a highly elliptical orbit at 182 × 292,492 kilometers at 28.7° degrees. After three large elliptical loops around Earth, the spacecraft flew by the Moon on 30 July and arrived at the Sun–Earth L2 on 1 October 2001. Its position at L2 minimizes the amount of contaminating solar, terrestrial, and lunar emissions while also ensuring a stable thermal state. At L2, MAP was in a 6-month Lissajous orbit. In 2003, MAP was renamed the Wilkinson Microwave Anisotropy Probe (WMAP) in honor of cosmologist and mission scientist David Todd Wilkinson (1935–2002) who passed away the year before. Since its operational mission began, NASA has issued public "data releases" in 2003, 2006, 2008, 2010, and 2012 that have added an immense amount of rich information to our understanding of the origins of the universe. First and foremost, WMAP's data played a major role in precisely confirming the origin, content, age, and geometry of the universe. Among its many findings has been: the first fine-resolution (0.2 degree) full-sky map of the microwave sky; a more precise determination of the age of the universe (13.77 billion years); more accurate data on the curvature of space; data that has allowed scientists to reduce the volume of cosmological parameters by a factor of over 68,000; and the discovery that dark matter and dark energy (in the form of the cosmological constant) make up about 24.0% and 71.4% of the universe, respectively. All of these findings and others cumulatively provided more confirmation for the prevailing standard model of Big Bang cosmology, the Lambda-CDM model. After 9 years of operation, in October 2010, the WMAP spacecraft was moved to a derelict heliocentric graveyard orbit outside L2 where the probe will circle the Sun once every 15 years.

197

Genesis

Nation: USA (79)
Objective(s): solar wind sample return, Sun–Earth L1 Lagrange Point
Spacecraft: Genesis
Spacecraft Mass: 636 kg
Mission Design and Management: NASA / JPL
Launch Vehicle: Delta 7326-9.5 (no. D287)

Launch Date and Time: 8 August 2001 / 16:13:40 UT
Launch Site: Cape Canaveral Air Force Station / SLC-17A
Scientific Instruments:
1. solar wind ion monitor
2. electron monitor

Results: During its nearly two years in halo orbit around the Sun–Earth L1 point, the Genesis, the fifth Discovery-class spacecraft, was designed to collect samples of the solar wind and then subsequently return them to Earth. The collection device, fixed inside the return capsule, was made of a stack of four circular metallic trays, one that would be continuously exposed, and the other three deployed depending on particular solar wind characteristics. After insertion into a low parking orbit around Earth, the third stage fired exactly an hour after launch, at 17:13 UT on 8 August 2001 to send the spacecraft towards its destination at L1. On 16 November, a 4-minute. 28-second burn of its engine put Genesis into a halo orbit around L1 with a radius of about 800,000 kilometers and a period of six months. The first array was exposed for sample collecting a little over two weeks later on 30 November, while the others, depending on the particular array, were left exposed from 193 days (coronal mass ejection collector) to 887 days (the bulk arrays). The prior record for a solar wind collector (on Apollo 16 in 1972) had been a short 45 hours. All the trays were stowed away on 1 April 2004, and exactly three weeks later, on 22 April, the spacecraft fired its four thrusters and began its long trek back via an unusual trajectory that took it past the Moon (at a range of 250,000 kilometers), Earth (at 392,300 kilometers), and then to L2, which it reached in July 2004. Swinging around L2 it headed in the direction of Earth. About 5.5 hours prior to reentry on 8 September, the spacecraft bus ejected its return capsule and then fired its thrusters to enter a parking orbit around Earth partly as a precaution in case separation had not occurred. The capsule successfully separated and hit Earth's atmosphere, experiencing a force of 27 g's. The drogue parachute (which would have deployed the main chute) unfortunately did not deploy, and the capsule hit the ground at an estimated speed of 311 kilometers/hour at 15:58 UT on 8 September 2004. The "landing" was, as designed, in the Dugway Proving Ground in the Utah Test and Training Range, but obviously the capsule was severely damaged and contaminated. The shattered sample canister was taken to a clean room and, over the subsequent month, disassembled carefully. Teams eventually tagged 15,000 fragments of the return capsule. Despite the condition of the capsule, over a period of several years, project scientists were able to glean a significant amount of data from the recovered debris and published results detailing, for example, the identification of argon and neon isotopes in samples of three types of solar wind captured by the spacecraft. An investigation into the accident found that the drogue failed to deploy due to a design defect that allowed an incorrect orientation during assembly of gravity-switch devices that initiate deployment of the spacecraft's drogue parachute and parafoil. The main Genesis bus meanwhile headed back to L1 and then entered heliocentric orbit. Contact was maintained until 16 December 2004.

198

CONTOUR

Nation: USA (80)

Objective(s): comet flyby

Spacecraft: Contour

Spacecraft Mass: 970 kg

Mission Design and Management: NASA / APL

Launch Vehicle: Delta 7425-9.5 (no. D292)

Launch Date and Time: 3 July 2002 / 06:47:41 UT

Launch Site: Cape Canaveral Air Force Station / SLC-17A

Scientific Instruments:

1. remote imager/spectrograph (CRISP)
2. forward imager (CFI)
3. neutral gas ion mass spectrometer (NGIMS)
4. dust analyzer (CIDA)

Results: This sixth Discovery-class mission (after Mars Pathfinder, NEAR, Lunar Prospector, Stardust, and Genesis) was designed to fly by at least two cometary nucleii with the goal of compiling topographical and compositional maps, sending back imagery, and collecting data on the structure and composition of their comas. Named Comet Nucleus Tour (CONTOUR), the spacecraft would carry out its principal mission from heliocentric orbit with encounters with Comet 2P/Encke (on 12 November 2003), 73P/Schwassmann-Wachmann-3 (on 19 June 2006) and possibly, 6P/d'Arrest (16 August 2008). The spacecraft was successfully launched into a high apogee orbit (with a period of five-and-a-half days). Controllers implemented at least 23 orbital maneuvers over the next 43 days (and 25 orbits) to position CONTOUR properly for its planned burn to heliocentric orbit on 15 August. On that day, at 08:49 UT, its solid propellant apogee motor fired as the spacecraft was approaching perigee over the Indian Ocean and out of radio contact. Unfortunately, nothing was ever heard from CONTOUR again. Later investigation showed that the spacecraft had broken up during its burn. The spacecraft probably suffered structural failure due to "plume heating" as its main engine was firing, caused either by problems in the design of the probe or the solid rocket motor itself.

199

Hayabusa

Nation: Japan (5)
Objective(s): asteroid sample return
Spacecraft: MUSES-C + MINERVA
Spacecraft Mass: 510 kg
Mission Design and Management: ISAS / JAXA
Launch Vehicle: M-V (no. 5)
Launch Date and Time: 9 May 2003 / 04:29:25
Launch Site: Kagoshima / M-V
Scientific Instruments:

1. light detection and ranging instrument (LIDAR)
2. near infrared spectrometer (NIRS)
3. x-ray fluorescence spectrometer (XRS)
4. wide-range camera (ONC-W)
5. telescopic camera (AMICA)
6. four ion thrusters

Results: With this mission, Japan hoped to be the first nation to visit a minor planet and return samples from it. The plan was for the spacecraft to launch in 2003, encounter its target in 2005, and then return to Earth, with a landing in the Woomera Test Range in South Australia in 2007. NASA had originally planned to supply a nano-rover for the mission but backed out from the mission in November 2000. Instead, ISAS built its own rover, called MINERVA (Micro/Nano Experimental Robot Vehicle for Asteroid), a 16-sided "hopper" that weighed only 600 grams and was equipped with six thermometers, a pair of stereoscopic cameras, and a short-focus camera. MUSES-C, named Hayabusa (or "Peregrine falcon") after launch, lifted off with MINERVA and was inserted directly into a solar orbit of 0.860 × 1.138 AU. Its goal, to

rendezvous with asteroid 1998SF36 (now renamed Itokawa after the founding figure of the Japanese space program). During its outbound trip, controllers operated its ion engines, although one (of a total of four) failed soon after launch. As per its mission profile, the spacecraft returned towards Earth for a gravity assist flyby on 19 May 2004 at a range of 3,725 kilometers and then moved into a 1.01 × 1.73 AU heliocentric orbit. Unfortunately, right after this, Hayabusa was caught in the aftermath of a massive solar eruption that degraded its solar cells (and thus power to its ion engines). Mission managers scrambled to come up with a new schedule, delayed now for both the asteroid encounter and return to Earth. As it approached in the direction of Itokawa, the spacecraft's ion engines (one of which had operated for 10,400 hours) were shut off, and its main engine fired at 01:17 UT on 12 September 2005 to terminate its "approach phase." At the time, it was only 20 kilometers from its target. In its station-keeping mode, Hayabusa took hundreds of high resolution pictures, but by 3 October, two of its three reaction wheels controlling the attitude had failed. Yet, the spacecraft was able to conduct two "rehearsal" landings (on 4 and 9 November). During a third rehearsal, at a distance of about 55 meters, controllers commanded the mother ship to release MINERVA. Because tracking was being transferred from a Japanese antenna to a NASA one, key information was lacking about Hayabusa's vertical speed. By the time the command reached Hayabusa, it was actually moving away at 15 meters/second. As a result, MINERVA missed its target and flew into heliocentric orbit. Remarkably, the little rover operated as planned for 18 hours. On 19 November, the main spacecraft successfully executed a descent run and landed softly at 21:09:32 UT at 6° S / 39° E in the middle of a feature named "MUSES Sea." The probe rebounded but settled down in a stable position by 21:41 UT. (The spacecraft was preceded by one of three target markers that had the names of 880,00 people engraved on it). The spacecraft was, however, not properly aligned and took off at 22:15 UT in what was the first ever liftoff from a celestial body apart from Earth or the Moon. During a second landing at 22:07 UT on 25 November, the spacecraft was supposed to fire two small projectiles to generate a spray of soil for collection, but later data indicated this probably did not occur, although some material was clearly collected. Hayabusa immediately lifted off again, although it had suffered some damage during landing that caused a propellant leak in the attitude control system. This caused a cascade of problems including loss of solar orientation necessary for the spacecraft to have any power. With contact becoming rather intermittent, on 8 December tracking stations noted an abrupt change in attitude and a decrease in signal strength. At this point, Hayabusa was tumbling. Remarkably, controllers re-established contact (presumably after the tumbling had ceased when the leaking propellant had run out) on 23 January 2006. The spacecraft's condition was dire but JAXA was able to revive a number of systems. A year later, on 17–18 January 2007, the lid of the sample catcher was closed, and the catcher stowed into the return capsule. Finally, a plan to use its ion thrusters to return it to Earth was implemented: on 25 April 2007, two ion engines began to fire, continuing for about seven months. A second firing period began on 4 February 2009, designed to bring the spacecraft back to Earth. By this time, only one of the four thrusters was operable, which itself failed in November, leaving Hayabusa four months short of the firing time needed to get back to Earth. Japanese engineers came up with an ingenious solution, to use the ion source from one engine (engine B) with the neutralizer from another (engine A). The final phase of operation of the ion thruster concluded on 27 March 2010, followed by four short course corrections. On 13 June 2010, at a range of 40,000 kilometers from Earth, the spacecraft released its return capsule, just 3 hours before reaching Earth's atmosphere. Both entered the atmosphere at 13:51 UT and began to decelerate. While the mother ship burned up as expected, the return capsule was unscathed. It deployed its parachute and landed just 500 meters from its target point in the Woomera Test Range in Australia at 14:12 UT and was successfully recovered soon after. In November 2010, Japanese scientists announced that the return capsule had indeed returned 1,500 grains of rock (most smaller in size than 10 micrometers) from Itokawa, this being the second sample (after that from Genesis) returned from another celestial body other than the Moon.

200

Mars Express and Beagle 2

Nation: European Space Agency (2)
Objective(s): Mars orbit, Mars landing
Spacecraft: Mars Express
Spacecraft Mass: 1,186 kg
Mission Design and Management: ESA
Launch Vehicle: Soyuz-FG + Fregat (no. E15000-005 + 14S44 no. 1005)
Launch Date and Time: 2 June 2003 / 17:45:26 UT
Launch Site: GIK-5 / Site 31/6
Scientific Instruments:
Mars Express:
1. visible and infrared mineralogical mapping spectrometer (OMEGA)
2. ultraviolet and infrared atmospheric spectrometer (SPICAM)
3. sub-surface sounding radar altimeter (MARSIS)
4. planetary fourier spectrometer (PFS)
5. analyzer of space plasmas and energetic atoms (ASPERA)
6. high resolution stereo camera (HRSC)
7. Mars Express lander communications (MELACOM)
8. Mars radio science experiment (MaRS)
9. camera (VMC)

Beagle 2:

1. gas analysis package
2. environmental sensors
3. 2 stereoscopic cameras
4. microscope
5. Mössbauer spectrometer
6. x-ray spectrometer
7. planetary underground tool (PLUTO)

Results: ESA's first planetary mission (although launched by the Russians) had two parts, an orbiter and a lander. The orbiter was designed to image the entire surface of Mars in high resolution, produce a map of the mineral composition of its surface, map the composition of its atmosphere, determine the structure of the sub-surface (to a depth of a few kilometers), and study the effects of the atmosphere on its surface as well as the interaction of the atmosphere with the solar wind. It carried a number of instruments that were originally on the ill-fated Russian Mars 8 probe; several other nations also contributed, including the U.S., Poland, Japan, and China. The 33.2-kilogram British Beagle 2 lander (its name, an allusion to the HMS Beagle that carried the young Charles Darwin (1809–1882) on his historic voyage) was designed to conduct exobiology research and geochemistry research on the Martian surface. Its scientific suite included a gas analysis package that used a set of 12 gas heating ovens to heat soil to study released gases. After launch, the Fregat stage ignited twice, first to Earth orbit and second (at 19:03 UT) to send the spacecraft into heliocentric orbit. About 6 hours prior to entry into Mars orbit, at 08:31 UT on 19 December 2003, Mars Express successfully released Beagle 2, and without any active means of propulsion on board, it began its passive journey to the surface of Mars, landing as expected by 03:14 UT on 25 December. The American 2001 Mars Odyssey was programmed to relay the first signals from Beagle but heard nothing. Mars Express meanwhile fired its main engine at 02:47 UT on Christmas Day and entered an initial Mars orbit of $260 \times 187,500$ kilometers (with a period of 10 days), ESA thus achieving what had only been accomplished before by the United States and Soviet Union. After about a hundred days, in early May, the orbiter was in a $10,107 \times 298$-kilometer orbit with an orbital period of 6.7 hours. Attempts to contact Beagle 2 proved unsuccessful and the lander aspect of the mission was declared officially lost on 6 February 2004. In January 2015, ESA announced that high-resolution images taken by NASA's Mars Reconnaissance Orbiter (MRO) showed Beagle 2's wreckage on the Martian surface. The pictures revealed the lander partially deployed on the surface, confirming that entry, descent, and the landing sequence had gone well. Further improvement of these images by University College, London in the spring of 2016 led researchers to conclude by October 2016 that Beagle 2 did indeed touch down softly on Mars and had possibly deployed three of its four solar panels. The malfunction that killed the mission more likely happened soon after landing rather than before. Undoubtedly, the lander's development program was mismanaged with fatal shortcomings including insufficient testing of key systems. Mars Express meanwhile began to return valuable data on the Red Planet. It deployed the first of its two 20-meter long radar booms for the MARSIS experiment about a year later, on 4 May 2005, the second one being deployed on 14 June. The two booms together created a 40-meter-long dipole antenna to operate the MARSIS experiment. The spacecraft returned spectacular images of the planet's terrain; during its first Martian year, it had mapped one-quarter of the surface at a resolution of 20 meters per pixel in color and more than half of the surface at 50 meters. Within its first five years in orbit, the spacecraft had discovered relatively recent evidence of volcanic and glacial processes and the presence of water ice below the surface. Mars Express mapped the various types of ice in the polar regions and determined the history of water abundance on Mars. One of its most striking discoveries has been to confirm the existence of Methane in the atmosphere (announced on 30 March 2004), a presence independently confirmed by ground observations, but later clarified by data from Curiosity. The mission of Mars Express was first extended to October 2007, and then again to May 2009, far exceeding its original 687-Earth day lifetime. A further five extensions have followed since then, with the most recent one extending the mission to 2016. During these extended missions, the spacecraft aided the operation of future Mars probes such as NASA's Phoenix lander (in 2008) and also collected data on Phobos. On 3 March 2010, for example, Mars Express passed by Phobos at a range of just 67 kilometers, the closest any spacecraft had ever come to it by that time. This distance was beaten on 29 December 2013, when the spacecraft approached within just 45 kilometers of the Martian moon. Earlier, in August 2011, the probe ran into problems with its onboard computer memory and entered "safe mode" twice in response to emergencies. Fortunately, by 24 November 2011, controllers were able to bring the spacecraft back to full operating capacity. By the time of the 10th anniversary of its launch, ESA was able to release a near-complete topographical map of Mars showing the mountains, craters, ancient river beds, and lava flows that mark the planet. A potentially dangerous event for Mars Express was the flyby of Comet C/2013 A1 (also known as Siding Spring) near Mars on 19 October 2014. Because the range of the comet was only 385,000 kilometers, there were concerns about the possibility of the comet's coma enveloping Mars and along with it, the operational spacecraft in the vicinity of Mars (including at that time, the American Mars Reconnaissance Orbiter, 2001 Mars Odyssey, MAVEN, the European Mars Express, and the Indian Mars Orbiter Mission). In February 2015, scientists announced, based on data from Mars Express and NASA's MRO, that Phlegra Montes, a complex network of hills, ridges, and basins spanning 1,400 kilometers might hide large quantities of water-ice, perhaps only 20 meters below the surface. During three years, the Mars Express team continued with an interesting media experiment using the simple low-resolution camera originally designed to image the lost Beagle 2 lander. After the search for Beagle 2 concluded in 2003, that camera was switched off, but in 2007, ESA switched it back on, this time entirely for education and outreach. Images from the so-called Mars Webcam has been used by the public in dozens of countries, as enthusiasts downloaded, shared, and processed images, originally posted on a Flickr page. By May 2016, some 19,000 images, some of them based on requests from schools and youth clubs, had been viewed over two million times. In October 2016, Mars Express helped in the collection and transfer of data for the ill-fated Schiaparelli EDM lander that set down upon Mars. In November 2014, funding for the Mars Express mission was extended for the sixth time, to December 2016, and it continues to operate at full strength in mid-2017.

201

Spirit

Nation: USA (81)

Objective(s): Mars surface exploration

Spacecraft: Mars Exploration Rover 2 (MER 2) [became MER-A]

Spacecraft Mass: 1,062 kg

Mission Design and Management: NASA / JPL

Launch Vehicle: Delta 7925-9.5 (no. D298)

Launch Date and Time: 10 June 2003 / 17:58:47 UT

Launch Site: Cape Canaveral Air Force Station / SLC-17A

Scientific Instruments:

1. Panoramic Mast Assembly
   a. panoramic cameras (Pancam)
   b. navigation cameras (Navcam)
   c. miniature thermal emission spectrometer (Mini-TES)
2. Mössbauer spectrometer (MB)
3. alpha particle x-ray spectrometer (APXS)
4. magnets (to collect dust particles)
5. microscopic imager (MI)
6. rock abrasion tool (RAT)

Image showing three generations of Mars rovers developed at JPL in Pasadena, California. They are on display here at JPL's Mars yard testing area. Front center is the flight spare (called Marie Curie) for Sojourner which landed on Mars in 1977 as part of Mars Pathfinder. On the left is a Mars Exploration Rover Project test rover that is a working "sibling" to Spirit and Opportunity, which both landed in 2004. Finally, on the right is a ground model the size of Curiosity, which landed in 2012. The two JPL engineers shown are Matt Robinson (left) and Wesley Kuykendall. *Credit: NASA/JPL-Caltech*

Results: Spirit and Opportunity were two rovers that together represented the Mars Exploration Rover Mission (MER), itself part of NASA's Mars Exploration Program. The twin missions' main scientific objective was to search for a range of rocks and soil types and then look for clues for past water activity on Mars. Each rover, about the size of a golf cart and seven times heavier (185 kilograms) than the Sojourner rover on Mars Pathfinder, was targeted to opposite sides of the planet in locales that were suspected of having been affected by liquid water in the past. The plan was for the rovers to move from place to place and perform on-site geological investigations and take photographs with mast-mounted cameras (about 1.5 meters off the ground) providing 360° stereoscopic views of the terrain. A suite of instruments (MB, APXS, the magnets, MI,

and RAT) were deployed on a robotic arm (known as the Instrument Deployment Device, IDD). The arm would place the instruments directly against soil or rock and activate the instruments. The complete spacecraft was launched into an intermediate parking orbit around Earth of 163 × 4,762 kilometers at 28.5° inclination before the PAM-D upper stage fired to send it on to heliocentric orbit on a trajectory to intercept Mars. A mid-course correction followed 10 days later. After three more corrections, the spacecraft's "Cruise Stage" carrying the Spirit rover approached Mars for the landing on 4 January 2004. About 15 minutes prior to entry into the atmosphere, the lander (inside its protective aeroshell) separated from the Cruise Stage. At an altitude of 6 to 7.5 kilometers, a parachute deployed, followed 30 seconds later by release of the aeroshell's bottom heat shield. Within another 10 seconds, the rover unreeled down a "bridle" (or tether) while still descending at a rate of 70 meters/second. Four massive airbags, of the same type as the ones used on Mars Pathfinder, inflated soon after, followed by firing of the retrorockets at the base of the parachute until the lander was about 8.5 meters off the ground. The retrorockets were needed since the Martian atmosphere is less than 1% the density of Earth, and parachutes alone cannot reduce velocity. The entire package hit the Martian landscape at 04:26 UT at a velocity of 14 meters/second, bouncing a total of 28 times before rolling to a stop about 250–300 meters from point of first impact. Landing coordinates were 14.5692° S / 175.4729° E, about 13.4 kilometers from the planned target, inside the Gusev crater. The area was henceforth known as the Columbia Memorial Station. Through it all, the lander transmitted data via Mars Global Surveyor. About an hour-and-a-half after impact and after deflation of the airbags, MER-A deployed its petal solar panels, now relaying information to Earth via 2001 Mars Odyssey. Immediately after, Spirit began to transmit spectacular images back to Earth. The rover ran into a major problem on 21 January 2004 when NASA's Deep Space Network lost contact. Due to a problem in Spirit's flash memory subsystem, the rover entered a "fault mode." Fortunately, controllers were able to reformat the flash memory and send up a software patch (to preclude memory overload). Normal operations resumed again on 5 February and the day after, Spirit used its Rock Abrasion Tool (RAT) to ground down the surface of a rock (called "Adirondack"), a feat performed for the very first time on Mars. Investigating the exposed interior allowed scientists important insights into the composition of Martial soil. The original planned mission was to have lasted 90 Martian days (to approximately 4 April 2004). Yet, mission planners were able to repeatedly formulate extended missions well beyond the rover's original lifetime. Some of the subsequent highlights included a visit to Bonneville Crater, about 370 meters from its original landing point, then on to the base of Columbia Hills where it spent an extended period of time. By 2005, the rover began slowly making its way uphill to the apex of Husband Hill, over terrain that was both rocky and sandy. It stopped at many locations to investigate, often using the RAT. In March 2005, a peculiar and strange event, the passing of dust devils that swept dust from the top of the solar panels, increased power coming to Spirit from the usual 60% to 93%, thus significantly extending the lifetime of the mission. On 29 September 2005, the rover finally reached the summit of Husband Hill, a small flat plain, from which Spirit was able to take 360° panoramas in real color of the Gusev Crater. Early the following year, the rover was directed to the north face of McCool Hill where it was assumed that Spirit would receive sufficient sunlight to maintain operations through the impending Martian winter. The trip to McCool Hill was eventually canceled partly because a front wheel had stopped working. This malfunction proved to be beneficial since the inactive wheel scraped off the upper layer of Martian soil as the rover moved, exposing bright silica-rich dust that indicated contact between soil and water. In early 2007, controllers passed on new software to both Spirit and Opportunity. These new programs allowed the rovers to autonomously decide on a number of different actions, such as whether to transmit a particular image back to Earth or whether to extend the remote arm. Through much of the (Earth) summer of 2007, however, both Spirit and Opportunity faced massive dust storms that eroded their ability to operate effectively, mainly due to lack of power generated from the solar panels. These concerns did not abate into 2008 as another winter storm at the end of that year further reduced the output of Spirit's solar panel to about 89 watt-hours per Martian day (where a nominal amount would be about 700 watt-hours per day). At such low levels, the rover needed to resort to using its own batteries which, if ran dry, would basically end the mission. Through 2009, a series of fortuitous events—such as wind that blew dust off the panels—slowly increased power generated by the solar panels. By April 2009, the rover was back to about 372 watt hours per day, sufficient for "normal" science activities to resume. Unfortunately, soon after, on 1 May 2009, while driving south beside the western edge of a low plateau called Home Plate, Spirit was rendered immobile in soft soil, its wheels unable to generate traction against the ground. Subsequently, on 28 November, another of Spirit's six wheels, the right rear one, stopped working. By late January 2010, after many attempts to move Spirit had not bore fruit, mission planners reformulated the Spirit mission as a "stationary science platform." One of its goals would now be to study the tiny wobbles in Mars' rotation to determine the nature of the planet's core—whether it is liquid or solid. In order to do that, however, the rover had to be tilted slightly to the north to expose its panels to the Sun, since the winter sun would be in the northern sky. In the end, the desired tilt was not achieved, and after 22 March 2010, JPL was not able to regain contact with Spirit again. Despite more than 1,300 commands sent to Spirit, NASA officially concluded its recovery efforts on 25 May 2011. The most probable cause of the loss of contact was the excessive cold that made its survival heaters ineffective. By the time it stopped, Spirit had traveled 7.73 kilometers across the Martian plains. It had operated for 6 years, 2 months, and 19 days, more than 25 times its original intended lifetime.

202

Opportunity

Nation: USA (82)
Objective(s): Mars surface exploration
Spacecraft: Mars Exploration Rover 1 (MER 1) [became MER-B]
Spacecraft Mass: 1,062 kg
Mission Design and Management: NASA / JPL
Launch Vehicle: Delta 7925H (no. D299)

Launch Date and Time: 8 July 2003 / 03:18:15 UT
Launch Site: Cape Canaveral Air Force Station / SLC-17B
Scientific Instruments:
1. Panoramic Mast Assembly
    a. panoramic cameras (Pancam)
    b. navigation cameras (Navcam)
    c. miniature thermal emission spectrometer (Mini-TES)
2. Mössbauer spectrometer (MB)
3. alpha particle x-ray spectrometer (APXS)
4. magnets (to collect dust particles)
5. microscopic imager (MI)
6. rock abrasion tool (RAT)

Results: For description and background, see entry for Spirit. After launch, the MER-1 rover was dispatched on its six-month trek to Mars. After a final course correction on 16 January 2004, the spacecraft dived into the Martian atmosphere on 25 January 2004. The descent to the surface was uneventful with no anomalies. The lander, enclosed in the airbags, touched down at 04:54 UT and then bounced at least 26 times before coming to rest in Meridiani Planum at 1.9483° S / 354.47417° E, about 14.9 kilometers from the intended target. This area was now named the Challenger Memorial Station, in tribute to the Space Shuttle crew lost in 1986. Opportunity landed in a relatively flat plain but within an impact crater known as Eagle. After extensive studies within Eagle, on 22 March 2004, Opportunity climbed up the edge of the crater and rolled out and headed for a new phase of its mission in Endurance Crater, about 750 meters away. Having exited Eagle, the rover took some spectacular shots of the abandoned area where the lander, backshell, and parachute were still visible. Near its discarded heat shield, Opportunity discovered an unusual basketball-sized rock in January 2005 (known as "Heat Shield Rock") that turned out to be an iron-nickel meteorite. Later that year, the rover drove into an area where several of its wheels were buried in sand, rendering the vehicle immobile. JPL controllers were able to maneuver the vehicle a few centimeters at a time and free

Opportunity in June 2005 after six weeks of rest. Through the remainder of the year and into 2006, the rover headed slowly in a southward direction towards the 800-meter diameter Victoria crater, first arriving at Erebus, a highly eroded impact crater about 300 meters in diameter. In March 2006, it then began its 2- kilometer journey to Victoria, a crater wider and deeper than any yet examined by the two rovers. After a 21-month trip, Opportunity arrived at Victoria in September 2006 and sent back striking pictures of its rim. The following year, 2007, was an important test for Opportunity given the severe dust storms that plagued Mars. By 18 July, the rover's solar panels were reporting power at only 128 watt hours, the lowest for either rover at that point. All science activities were indefinitely suspended for Opportunity which faced much more severe conditions than Spirit. After about six weeks and abatement of the dust storms, Opportunity was back in action, and on 11 September 2007, it entered Victoria Crater, staying inside for almost a year sending back a wealth of information on its soil. Opportunity's next target was the enormous Endeavour Crater, 22 kilometers in diameter. On the way there, the rover found the so-called Marquette Island rock, "different in composition and character from any known rock on Mars or meteorite from Mars," according to Steve Squyres (1956–), the principal investigator for the rovers. The rock appeared to have originated deep in the Martian crust and someplace far away from the landing site, unlike almost all the rocks previously studied by Opportunity. On 24 March 2010, Opportunity passed the 20-kilometer milestone on Mars, more than double the distance recorded by Spirit, and far in excess of what was originally considered a nominal mission—600 meters. Two months later, on 20 May—with Spirit already inactive—Opportunity broke the record set by the Viking 1 Lander for the longest continuous operation on the surface of Mars, 6 years and 116 days. Another milestone was passed when Opportunity, still heading towards Endeavour Crater, passed the

30-kilometer mark on 1 June 2011. Finally, after a journey of nearly three years and about 21 kilometers, Opportunity arrived at Endeavour crater on 9 August 2011. In September 2011, NASA announced that an aluminum cuff that served as a cable shield on each of the RATs on the rovers was made from aluminum recovered from the World Trade Center towers, destroyed during the terrorist attacks on 11 September 2001. Honeybee Robotics, which helped build the tool, had its offices in New York that day not far from the attacks. As a memorial to the victims, JPL and Honeybee worked together to include the aluminum on the Mars rovers. Through late 2012 and into 2013, Opportunity worked around a geographic feature named Matijevic Hill (which overlooks the Endeavour crater), analyzing rocks and soil. On 16 May 2013, NASA announced that Endeavour had passed the previous record for the farthest distance traveled by any NASA vehicle on another celestial body, 35.744 kilometers, a record set by the Apollo 17 Lunar Roving Vehicle in December 1972. By August 2013, Opportunity was at Solander Point, an area of contact between a rock layer that was formed in acidic wet conditions long before and an older one from a more "neutral" environment. Both Cape York (location of Matijevic Hill) and Solander Point are raised segments near the western rim of the Endeavour crater. On 4 January 2014, Opportunity passed 10 years on the surface of Mars, now with relatively clean surfaces on the solar panels that had allowed increased power to the rover. A "selfie" from March 2014 showed a rover cleaned by wind events earlier in the month that raised hopes for continuing the mission. As it continued its exploration mission on the Martian surface, on 28 July 2014, NASA announced that Opportunity had passed the distance record set on another celestial body, set by Lunokhod 2, when the American rover's odometer showed 40.25 kilometers, exceeding the Soviet vehicle's record of 39 kilometers. However, Russian analysis from LRO images suggest that Lunokhod 2 may have traveled as much as 42 kilometers, rather than the revised

39 (itself a "revision" up from 37 kilometers). While the rover was generally in good health, because of the large number of computer resets in the preceding month, which interfered with its science goals, mission planners implemented a complete reformat of its flash memory on 4 September 2014. The same day, NASA announced a further (ninth) extension of the mission of Opportunity to another two years with a mission to nearby Marathon Valley. At the beginning of September, it had covered 40.69 kilometers. At launch, like its sister rover, Spirit, Opportunity was designed to have a lifetime of 90 sols (Martian days)—about three Earth months. In December 2014, NASA announced that the rover had been plagued with problems with saving telemetry information into its "non-volatile" (or flash) memory, a problem traced to one of its seven memory banks (Bank 7). By May 2015, NASA controllers configured the memory so the rover was operating only in RAM-only mode. On 25 March 2015, NASA announced that, having traveled 42.195 kilometers (or 26.219 miles), Opportunity became "the first human enterprise to exceed marathon distance of travel on another world." In June 2015, because Mars passed almost directly behind the Sun (from Earth's perspective) and therefore communications were curtailed. Later, through its seventh Martian winter (during Earth winter 2015–2016), at a time when it was kept at "energy-minimum" levels due to the relative lack of solar energy, Opportunity kept busy, using its Rock Abration Tool to remove surface dust from a target called "Private John Potts," the name a reference to a member of the Lewis and Clark Expedition. During this period, Opportunity continued to explore the western rim of the 22-kilometer wide Endeavour crater, particularly the southern side of Marathon Valley, which slices through Endeavour crater's rim from west to east. On 10 March 2016, while making its closest approach to a target near the crest of Knudsen Crater, it drove at a tilt of 32°, breaking the record for the steepest slope driven by any rover on Mars (a record previously set by Opportunity during a climb in

January 2004). In October 2016, Opportunity began a two-year extended mission that is to include investigations in the "Bitterroot Valley" portion of the western rim of the Endeavour Crater. The plan is for the rover to travel into a gully that slices Endeavor and is about two football fields in length. Opportunity Principal Investigator Steve Squyres noted that scientists were "confident [that] this is a fluid-carved gully, and that water was involved." On 7 February 2017, Opportunity passed the 44-kilometer mark on its odometer, as it made slow progress towards its next major scientific objective, a gully named Perseverance Valley, which it reached by the first week of May. Having collected several panoramas of high-value targets in the gully, on 4 June, the rover experienced a problem due to a stall on the left front wheel, which left the wheel "toed out" by 33 degrees. Fortunately, after several straightening attempts, the wheel appeared to be steering straight again, although controllers could identify any conclusive cause for the problem. For about three weeks during June and July, there was reduced communication with the rover due to a solar conjunction (when the Sun comes between Earth and Mars). In mid-July, Opportunity finally entered Perseverance Valley and began driving down into the gully during which time, rover energy levels dropped due to reduced Sun exposure. As of 31 October 2017, Opportunity's odometer read 45.04 kilometers.

203

SIRTF / Spitzer Space Telescope

Nation: USA (83)
Objective(s): solar orbit
Spacecraft: Space Infrared Telescope Facility (SIRTF)
Spacecraft Mass: 950 kg
Mission Design and Management: NASA / JPL / Caltech
Launch Vehicle: Delta 7920H (no. D300)

Image showing assembly of the Spitzer Space Telescope, the infrared telescope designed to study the early universe, young galaxies, and star formation. *Credit: Russ Underwood, Lockheed Martin Space Systems; courtesy NASA/JPL-Caltech.*

Launch Date and Time: 25 August 2003 / 05:35:39 UT

Launch Site: Cape Canaveral Air Force Station / SLC-17B

Scientific Instruments:

1. infrared array camera (IRAC)
2. infrared spectrograph (IRS)
3. multiband imaging photometer for Spitzer (MIPS)

Results: SIRTF was the fourth and last of NASA's "Great Observatories," after the Hubble Space Telescope (launched in 1990), the Compton Gamma Ray Observatory (1991), and the Chandra X-Ray Observatory (1999). It carried an 85-centimeter infrared telescope and three scientific instruments as part of the Cryogenic Telescope Assembly (CTA). Its planned two-and-a-half-year mission was designed to detect infrared radiation from its vantage point in heliocentric orbit. The CTA was cooled to only 5 degrees above absolute zero (or a temperature of –268°C) using 360 liters of liquid helium, ensuring that the observatory's "body heat" did not interfere with the observation of relatively cold cosmic objects. The Delta II Heavy (in a two-stage Delta 7925H configuration) inserted the second stage and payload into an initial orbit of 166 × 167 kilometers at 31.5° before the second stage ignited again at 06:13 UT on 25 August 2003 to send both the stage and the observatory into a hyperbolic orbit where SIRTF, by 3 September passed into an "Earth-trailing orbit" around the Sun. It ejected its dust cover on 29 August and then opened its aperture door the day after. In this orbit, at 0.996 × 1.019 AU, Earth (rather prominent in the infrared) does not hinder observation of potential targets of observation. On 18 December 2003, the SIRTF was renamed the Spitzer Space Telescope in honor of Lyman S. Spitzer, Jr. (1914–1997), one of the first to propose the idea of using telescopes in space. One of the early successes of the mission (in 2005) was to capture direct light from extrasolar planets for the first time. Many other findings followed in the subsequent four years, including seeing the light from the earliest objects in the universe, mapping the weather on an extrasolar planet for the first time, finding water vapor on another extrasolar planet, and identifying a new ring (the Phoebe ring) around Saturn. The observatory worked far longer than expected and its supply of liquid helium finally depleted at 22:11 UT on 15 May 2009, nearly six years after launch. At that point, mission scientists reformulated the mission as the Spitzer Warm Mission, which would use the two shortest-wavelength modules of the IRAC instrument, which did not require the cryogenic helium to operate, for future observations. More discoveries followed. In August 2010, for example, data from Spitzer revealed the identification of the first Carbon-rich planet (known as WASP-12b) orbiting a star. In October 2012, astronomers announced that data from the observatory had allowed a more precise measurement of the Hubble constant, the rate at which the universe is stretching apart. The following year, Spitzer celebrated 10 full years of operation in space and continued operation of its two instruments which, in August 2014, observed an eruption of dust around a star (NGC 2547-ID8), possibly caused by a collision of large asteroids. Such impacts are thought to lead to the formation of planets. Continuing discoveries based on results from Spitzer (as well as data integrated with information from other space-based observatories such as Swift) were announced in April 2015 (discovery of one of the most distant planets ever identified, about 13,000 light-years from Earth) and in March 2016 (discovery of the most remote galaxy ever detected, a high-redshift galaxy known as GN-z11). The latter was detected as part of the Frontiers Field project that combines the power of all three of NASA's Great Observatories, Spitzer, Hubble, and Chandra. In August 2016, mission planners at JPL announced a new phase of the Spitzer mission known simply as "Beyond," leveraged on a two-and-a-half-year mission extension granted by NASA earlier in the year. Because the distance between Spitzer and Earth has widened over time, during Beyond, its antenna must be pointed at higher angles towards the Sun to communicate with Earth. As a result, parts of the spacecraft will experience increasing amounts of heat. Simultaneously, its solar panels will be pointed away from the Sun in this configuration, thus putting onboard batteries under more stress. These challenges will be a part of the Beyond phase as Spitzer continues to explore planetary bodies both within and beyond the solar system. In October 2017, NASA announced that it was seeking information from potential funders who might be able to support operation of the telescope after NASA funding runs out in March 2019. With such possible funding, it might be possible to operate Spitzer beyond September of that year when operations are expected to cease with government funding.

204

SMART-1

Nation: European Space Agency (3)
Objective(s): lunar orbit
Spacecraft: SMART-1
Spacecraft Mass: 367 kg
Mission Design and Management: ESA
Launch Vehicle: Ariane 5G (no. V162) (L516)
Launch Date and Time: 27 September 2003 / 23:14:46 UT
Launch Site: Centre spatial Guyanais / ELA-3
Scientific Instruments:

1. advanced Moon micro-imager experiment (AMIE)
2. demonstration of a compact x-ray spectrometer (D-CIXS)
3. x-ray solar monitor
4. SMART-1 infrared spectrometer (XSM)
5. electric propulsion diagnostic package (EPDP)
6. spacecraft potential, electron and dust experiment (SPEDE)
7. K_a band TT&C experiment (KATE)

Results: The Small Missions for Advanced Research in Technology (SMART)-1 spacecraft was a technology demonstrator designed to test solar-electric propulsion and other deep space technologies on the way to the Moon. A second part of the mission would focus on studying polar mountain peaks that are in perpetual sunlight as well as the dark parts of the lunar parts that might contain ice. The ESA spacecraft, the first European spacecraft to enter orbit around the Moon, had a French-built Hall effect thruster (known as PPS®1350) derived from a Russian ion propulsion system originally designed by OKB Fakel, a Russian company that specializes in attitude control thrusters using ion and plasma sources. The thruster used xenon propellant to generate 88 mN of thrust (about the weight of a postcard) and a specific impulse of 1,650 seconds. The engine was powered by solar arrays which generated the 1,350 watts needed to power the ion engines. Initially launched into a geostationary transfer orbit of 7,035 × 42,223 kilometers by the Ariane 5 hypergolic EPS upper stage (with a 2,600 kgf thrust Aestus engine), SMART-1 used its electric propulsion system to slowly spin out into higher and higher elliptical orbits in what was a highly efficient mission profile. Two days every week, mission controllers at the European Space Operations Centre (ESOC) in Darmstadt, Germany, repeated burns of the ion engine, gradually expanding the spacecraft's spiral orbit. By the time it was 200,000 kilometers out, the Moon's gravity began to exert a significant influence on SMART-1. Nine days after its last perigee (on 2 November 2004), the spacecraft passed through the L1 Lagrange Point into a region dominated by the Moon's gravitational field. At 17:47 UT on 15 November, the vehicle then passed through its first perilune, having moved into polar orbit around the Moon. Initial orbital parameters were 6,704 × 53,208 kilometers, with an orbital period of 129 hours. During the following weeks, SMART-1's ion engine fired to gradually reduce orbital

parameters to allow closer views of the surface; it reached its operational orbit (with an orbital period of about 5 hours) by 27 February 2005. While in orbit, SMART-1's instruments studied the Moon's topography and surface texture as well as mapping the surface distribution of minerals such as pyroxenes, olivines, and feldspars, thus improving the data returned by Clementine. The mission was designed to end in August 2005 but was extended a year to August 2006 with plans for an impact. On 17 September 2005, the ion engine was fired for the last time, having exhausted all its propellant, leaving the vehicle in a natural orbit determined only by lunar gravity (and the gravitational influences of Earth and the Sun) and the occasional use of its attitude control thrusters. By that time, the ion engine had fired for 4,958.3 hours, a record length of operation in space for such an engine. The mission of SMART-1 finally ended at 05:42:22 UT on 3 September 2006 when the spacecraft was deliberately crashed onto the nearside of the Moon in Lacus Excellentiaie at 46.2° W / 34.3° S. Its impact (at 2 kilometers/second) created a dust cloud visible with Earth-based telescopes.

Rosetta and Philae

Nation: ESA (4)
Objective(s): comet orbit and landing
Spacecraft: Rosetta Orbiter / Rosetta Lander
Spacecraft Mass: 3,000 kg (includes 100 kg lander)
Mission Design and Management: ESA
Launch Vehicle: Ariane 5G+ (V158) (no. 518G)
Launch Date and Time: 2 March 2004 / 07:17:44 UT
Launch Site: CSG / ELA-3
Scientific Instruments:

Rosetta Orbiter:

1. ultraviolet imaging spectrometer (ALICE)
2. comet nucleus sounding experiment by radiowave transmission (CONSERT)
3. cometary secondary ion mass analyzer (COSIMA)
4. grain impact analyzer and dust accumulator (GIADA)
5. micro-imaging dust analysis system (MIDAS)
6. microwave instrument for the Rosetta orbiter (MIRO)
7. optical, spectroscopic and infrared remote imaging system (OSIRIS)
8. Rosetta orbiter spectrometer for ion and neutral analysis (ROSINA) Rosetta plasma consortium (RPC)
9. radio science investigation (RSI)
10. visible and infrared thermal imaging spectrometer (VIRTIS)

Philae:

1. alpha proton x-ray spectrometer (APXS)
2. cometary sampling and composition instrument (COSAC)

3. Ptolemy evolved gas analyzer
4. comet nucleus infrared and visible analyzer (CIVA)
5. Rosetta lander imaging system (ROLIS)
6. comet nucleus sounding experiment by radiowave transmission (CONSERT)
7. multi-purpose sensors for surface and sub-surface science (MUPUS)
8. Rosetta lander magnetometer and plasma monitor (ROMAP)
9. surface electric sounding and acoustic monitoring experiments (SESAME)
10. sample and distribution device (SD2)

Results: Rosetta was a European deep space probe launched on an originally projected 11.5-year mission to rendezvous, orbit, land, and study the 67P/Churyumov-Gerasimenko comet. Part of ESA's Horizon 2000 cornerstone missions, which includes SOHO (launched 1995), XMM-Newton (1999), Cluster II (2000), and INTEGRAL (2002), Rosetta consists of two parts—an orbiter (Rosetta) and a lander (Philae)—each equipped with a variety of scientific instruments. Originally, the mission was targeting comet 46P/Wirtanen but when the launch was delayed due to problems with the Ariane 5, the mission was redirected to Churyumov-Gerasimenko. Rosetta was launched into an escape trajectory with a 17-minute burn of Ariane's EPS second stage, putting the spacecraft on a trajectory that culminated in a 0.885×1.094 AU heliocentric orbit inclined at 0.4° to the ecliptic. Its voyage to its target comet was punctuated by a series of gravity-assist maneuvers, the first of which occurred at 22:09 UT on 4 March 2005 when Rosetta flew by Earth (over the Pacific, west of Mexico) at a distance of 1,954.7 kilometers. A most risky flyby of Mars followed on 25 February 2007, when Rosetta came a mere 250 kilometers close to the Red

Planet, experiencing a short and critical period out of contact with Earth and in Mars' shadow. Both these flybys produced spectacular photographs of Earth and Mars, respectively. The assist sent the spacecraft towards Earth for a second time, arriving and flying past our planet at a range of 5,295 kilometers on 13 November 2007. Before the final Earth flyby (on 12 November 2009), Rosetta performed a close flyby (just 800 kilometers) of asteroid 2867 Steins in the main asteroid belt at 18:58 UT on 5 September 2008, collecting a large amount of information. A second asteroid flyby, that of 21 Lutetia at 16:10 UT on 10 July 2010 at a range of 3,162 kilometers, produced spectacular images (using the OSIRIS instrument) of a battered minor planet riddled with craters. Resolution was as high as 60 meters in a body whose longest side is around 130 kilometers. Soon after, in June 2011, Rosetta was placed under "hibernation" as it made its way beyond the orbit of Jupiter—where there was no solar energy to power the vehicle— and back again close to the Sun. On 20 January 2014, its internal clock "awoke" the spacecraft and sent a signal back to Earth that all was well. Now only 9 million kilometers from its primary target, Rosetta began its final race to comet 67P/C-G. On 6 August 2014, at a distance of 405 million kilometers from Earth (about halfway between the orbits of Mars and Jupiter), Rosetta finally rendezvoused with the comet as it completed the last of 10 maneuvers (that began in May 2014) to adjust velocity and direction. During close operations near the comet, on 15 September, scientists identified a landing site for the spacecraft, "Site J" (later named "Agilkia"), located near the smaller of the comet's two "lobes." By this time (10 September 2014), the spacecraft was in a roughly 29-kilometer orbit around 67P/C-G, becoming the first spacecraft to orbit a cometary nucleus. Just prior to the planned landing, on 12 November, controllers identified a problem in Philae's active descent system thruster which provides thrust to avoid a rebound, but it was decided to move on with the landing and rely only on the harpoons instead of

the thruster to keep the spacecraft moored. At 08:35 UT on 12 November, the two spacecraft separated, initiating Philae's 7-hour descent to the comet at a relatively velocity of just 1 meter/second. A signal confirming the touchdown arrived at Earth at 16:03 UT (about 28 minutes, 20 seconds after the actual event). It later transpired the Philae had actually landed three times on the comet (at 15:34:04, 17:25:26, and 17:31:17 UT comet time) as the two harpoons did not fire as intended after each touchdown. Later analysis showed that all of the three methods to secure the lander had faced some problems: the ice screws, which were designed for soft materials, did not penetrate the hard surface of the Agilkia region; the thruster failed to fire due to a problem with a seal; and the harpoons also did not fire due to an electrical problem. As a result, Philae bounced on the surface several times before settling down about one kilometer away from its intended landing site in an area known as Abydos. All of its instruments were subsequently activated for data collection. For a short period, ESA controllers did not know the disposition of the lander as it went into hibernation, but on 14 November, contact was reestablished with Philae, following which all of its collected data was transferred to the mothership. Due to exhaustion of the primary battery, last contact with Philae was at 00:36 UT on 15 November, thus coming to about 64 hours of independent operation (and 57 hours on the surface). During its mission, Philae completed 80% of its planned "first science sequence," returning spectacular images of its surroundings, showing a cometary surface covered by dust and debris in size measuring anywhere from a millimeter to a meter. Philae also found complex molecules that could be the key building blocks of life, monitored the daily rise and fall of temperature, and assessed the surface properties and internal structure of the comet. ESA controllers hoped that the lander could be revived in August 2015 when sunlight fell on the lander and its solar panels, but assumed that Philae's mission was essentially over by

November 2014. As hoped, the Philae lander was awoken after seven months of hibernation. At 20:28 UT on 13 June 2015, controllers at ESA's European Space Operations Center in Darmstadt received signals (about 663 kbits of data over 85 seconds) from the lander, suggesting at least initially that Philae was "doing very well" and "ready for operations," according to DLR Philae Project Manager Stephan Ulamec. A second smaller burst was received at 21:26 UT on 14 June followed by six more bursts by 9 July 2015, after which time Rosetta was no longer in range to receive data from Philae. A year after landing, in November 2015, mission teams still remained hopeful that there would be renewed contact with the lander, especially as the Rosetta orbiter began to approach the lander again. But in February 2016, ESA announced that it was unlikely that Rosetta would ever pick up any more signals from Philae again, partly due to failures in a transmitter and a receiver on board. On 5 September 2016, ESA announced that they had conclusively identified the landing site of Philae in images taken by Rosetta's OSIRIS narrow-angle camera when the orbiter approached to just 2.7 kilometers of the surface. Rosetta, meanwhile, had continued its primary mission orbiting Comet 67P/Churyumov-Gerasimenko as the comet itself arced closer to the Sun. In November 2014, the orbiter adjusted its orbit several times to position it about 30 kilometers above the comet, interrupted by a brief "dip" down to 20 kilometers for about 10 days in early December. On 4 February 2015, Rosetta began moving into a new path for an encounter, timed for 12:41 UT on 14 February, at a range of just six kilometers. The flyby took the spacecraft over the most active regions of the comet, allowing scientists to seek zones where gas and dust accelerates from the surface. In June 2015, ESA extended Rosetta's mission to at least September 2016 (an extension of nine months from its original "nominal" mission). During this extension, Rosetta was party to Comet 67P/C-G's closest approach to the Sun, a distance of 186 million kilometers, on 13 August 2015. At

the perihelion, gases and dust particles around the comet reached peak intensity, clearly visible in the many spectacular images sent back by the orbiter. Finally, at 20:50 UT on 30 September 2016, Rosetta carried out a final maneuver sending it on a collision course with the comet from a height of 19 kilometers. During the descent, Rosetta studied the comet's gas, dust, and plasma environment very close to the surface and took numerous high-resolution images. The decision to end the mission was predicated on the fact that the comet was heading out beyond the orbit of Jupiter again, and thus, there would be little power to operate the spacecraft. Confirmation of final impact arrived at Darmstadt at 11:19:37 UT on 30 September 2016, thus ending one of ESA's most successful planetary missions. Besides collecting a vast amount of data on the properties of the comet, including its interior, surface and surrounding gas, dust, and plasma, Rosetta's key findings include the discovery of water vapor in comet 67P/G-C (vapor that is significantly different from that found on Earth), the detection of both molecular nitrogen and molecular oxygen for the first time at a comet, the existence of exposed water ice on the comet's surface, and the discovery of amino acid glycine (commonly found in proteins) and phosphorus (a component of DNA and cell membranes) in the comet.

206

MESSENGER

Nation: USA (84)

Objective(s): Mercury orbit

Spacecraft: MESSENGER

Spacecraft Mass: 1,107.9 kg

Mission Design and Management: NASA / APL

Launch Vehicle: Delta 7925H (no. D307)

Launch Date and Time: 3 August 2004 / 06:15:57 UT

Launch Site: Cape Canaveral Air Force Station / SLC-17B

This image acquired by NASA's Messenger spacecraft on 2 October 2013 by its Wide Angle Camera (WAC) shows the sunlit side of the planet Mercury. Messenger was the first spacecraft to go into orbit around Mercury. *Credit: NASA/Johns Hopkins University Applied Physics Laboratory/Carnegie Institution of Washington*

Scientific Instruments:

1. Mercury dual imaging system (MDIS)
2. gamma-ray spectrometer (GRS)
3. neutron spectrometer (NS)
4. x-ray spectrometer (XRS)
5. magnetometer (MAG)
6. Mercury laser altimeter (MLA)
7. Mercury atmospheric and surface composition spectrometer (MASCS)
8. energetic particle and plasma spectrometer (EPPS)
9. radio science experiment (RS)

Results: MESSENGER (Mercury Surface, Space Environment, Geochemistry, and Ranging) was the seventh Discovery-class mission, and the first spacecraft to orbit Mercury. Its primary goal was to study the geology, magnetic field, and chemical composition of the planet. It was the first mission to Mercury after Mariner 10, more than 30 years before. MESSENGER was launched into an initial parking orbit around Earth after which its PAM-D solid motor fired to put the spacecraft

on an escape trajectory into heliocentric orbit at 0.92 × 1.08 AU and 6.4° inclination to the ecliptic. The six-and-a-half-year road to Mercury was punctuated by several gravity-assist maneuvers through the inner solar system, including one flyby of Earth (on 2 August 2005), two flybys of Venus (on 24 October 2006 and 5 June 2007), and three flybys of Mercury (on 14 January 2008, 6 October 2008, and 29 September 2009). The gravity-assist maneuvers allowed the spacecraft to overcome the problem of massive acceleration that accompanies flight toward the Sun; instead, the flybys helped to decelerate MESSENGER's velocity relative to Mercury and also conserve propellant for its orbital mission (although it prolonged the length of the trip). The Earth flyby allowed mission controllers to properly calibrate all of the spacecraft's instruments while also returning spectacular images of the Earth–Moon system. During the second Venusian flyby (at a range of only 338 kilometers), MESSENGER relayed back a vast amount of data, including visible and near-infrared imaging data on the upper atmosphere. Some of the investigations, especially its study of the particle-and-fields characteristics of the planet, were coordinated with ESA's Venus Express mission. The three Mercury flybys further slowed down the spacecraft, although during the last encounter in September 2009, MESSENGER entered a "safe mode" and, as a result, collected no data on Mercury. Fortunately, the spacecraft revived 7 hours later. MESSENGER finally entered orbit around Mercury at 00:45 UT 18 March 2011, nearly seven years after launch and began formal data collection on 4 April. The vehicle's orbit was highly elliptical, approximately 9,300 × 200 kilometers with a 12-hour orbital period. One of MESSENGER's most remarkable images was its mosaic of the Solar System, obtained on 18 February 2011 with all the planets visible except Uranus and Neptune, a visual counterpart to the image of the solar system taken by Voyager 1 on 14 February 1990. The spacecraft completed its primary year-long mission on 17 March 2012, having taken nearly 100,000 images of the surface of Mercury. Among its initial discoveries was finding high concentrations of magnesium and calcium on Mercury's nightside, identifying a significant northward offset of Mercury's magnetic field from the planet's center, finding large amounts of water in Mercury's exosphere, and revealing evidence of past volcanic activity on the surface. In November 2011, NASA announced that MESSENGER's mission would be extended by a year, thus allowing the spacecraft to monitor the solar maximum in 2012. The extended mission lasted from 18 March 2012 to 17 March 2013. During this phase, by 20 April, with the help of three engine firings, the orbital period was reduced to 8 hours. It was also during this period, in early May 2012, that MESSENGER took its 100,000th photograph from orbit. By this time, the imaging instrument had globally mapped in both high-resolution monochrome and color, the entire surface of the planet. It was during this first extended mission that the spacecraft found evidence of water ice at Mercury's poles, frozen at locations that never see sunlight (made possible by the fact that the tilt of Mercury's rotational axis is almost zero.) A second extension was soon granted that extended the mission to March 2015, and on 6 February 2014, NASA reported that MESSENGER had taken its 200,000th orbital image, far exceeding the original expectation of at least 1,000 photographs. During the second extension, MESSENGER photographed two comets: Comet 2P/Encke and Comet C/2012 S1 (also known as Comet ISON). Beginning the summer of 2014, controllers began moving MESSENGER gradually, burn by burn, to a very low orbit for a new research program. By 12 September 2014, just after the 10th anniversary of its launch, the spacecraft's orbit was down to a mere 25 kilometers. Since then, mission controllers implemented at least two orbital maneuvers (on 12 September and 24 October) to raise its orbit and continue its latest extended mission. By Christmas Day 2014, it was clear that the spacecraft's propellants were running out and that MESSENGER would impact the planet in late March 2015. On 21 January 2015, mission controllers carried out one last maneuver to raise the spacecraft's orbit sufficient to continue more science activities to early in the spring. On 16 April 2015, NASA announced that the spacecraft would impact the surface of Mercury by 30 April after it ran out of propellant. As scheduled, on that day, at 19:26 UT, MESSENGER slammed into the planet's surface at about 14,080 kilometers/hour, creating a new crater on Mercury. Impact coordinates were probably close to 54.4° N / 149.9° W, near the Janácek crater in Suisei Planitia.

2005

207

Deep Impact

Nation: USA (85)

Objective(s): comet impact, comet flyby

Spacecraft: DIF + DI Impactor

Spacecraft Mass: 650 kg

Mission Design and Management: NASA / JPL

Launch Vehicle: Delta 7925-9.5 (no. D311)

Launch Date and Time: 12 January 2005 / 18:47:08 UT

Launch Site: Cape Canaveral Air Force Station / SLC-17B

Scientific Instruments:

Flyby Spacecraft:

1. high resolution instrument (HRI)
2. medium resolution instrument (MRI)

Impactor:

1. impact or targeting sensor (ITS)

Results: Unlike previous cometary flyby missions, such as Vega, Giotto, and Stardust, the Deep Impact spacecraft, the eighth mission in NASA's Discovery program, was intended to study the interior composition of a comet by deploying an impact probe that would collide with its target. The spacecraft comprised two distinct parts, a flyby bus and an impactor. The former, weighing 601 kilograms, was solar powered and carried two primary instruments. The HRI, the main science camera for Deep Impact, was one of the largest space-based instruments ever built for planetary science. It combined a visible-light multi-spectral CCD camera (with a filter wheel) and an imaging infrared spectrometer called the Spectral Imaging Module (SIM). The MRI was the functional backup for the HRI, and like the HRI, it also served as a navigation aid for Deep Impact. The 372-kilogram Impactor carried

the ITS, nearly identical to the MRI, but without the filter wheel, which was designed to measure the Impactor's trajectory and to image the comet from close range before impact. One of the more unusual payloads on board was a mini-CD with the names of 625,000 people collected as part of a campaign to "Send Your Name to a Comet!" After launch, Deep Impact was put into low Earth orbit, then an elliptical orbit (163 × 4,170 kilometers), and after a third stage burn, the spacecraft and its PAM-D upper stage departed on an Earth escape trajectory. There were some initial moments of anxiety when it was discovered that the spacecraft had automatically entered "safe mode" shortly after entering heliocentric orbit, but by 13 January, Deep Impact returned to full operational mode following a program to "tumble" the vehicle using its thrusters. The spacecraft traveled 429 million kilometers for nearly six months (including course corrections on 11 February and 4 May 2005) on an encounter with Comet 9P/Tempel. As the spacecraft approached its target, it spotted two outbursts of activity from the comet, on 14 June and 22 June 2005. At 06:00 UT (or 06:07 UT Earth-receive time) on 3 July 2005, Deep Impact released the Impactor probe, which, using small thrusters, moved into the path of the comet, where it hit the following day, 4 July at 05:44:58 UT at a relative velocity of 37,000 kilometers/hour. The impact generated an explosion the equivalent of 4.7 tons of TNT and a crater estimated to be about 150 meters in diameter. Minutes after the impact, the Flyby probe passed the nucleus at a range of about 500 kilometers at 05:59 UT on 3 July and took images of the resultant crater (although it was obscured by the dust cloud), ejecta plume, and the entire nucleus. Simultaneous observations of the impact were coordinated with ground-based observatories as well as space-based ones, such as the European Rosetta (which was about 80 million kilometers from the comet), Hubble, Spitzer, the Swift x-ray telescope, and XMM-Newton. The Impactor itself took images as little as 3 seconds before impact, which were transmitted via the flyby vehicle back to Earth. Controllers registered about 4,500 images from the three cameras over the next few days. Based on the results of Deep Impact's investigations, scientists concluded that Comet Tempel 1 had probably originated in the Oort Cloud. The data also showed that the comet was about 75% empty space. Although Deep Impact's primary mission was over, because the vehicle still had plenty of propellant left, on 3 July 2007, NASA approved a new supplemental mission for Deep Impact, known as EPOXI, derived from the combination of the two components of this extended flight: Extrasolar Planet Observations (EPOCh) and Deep Impact Extended Investigation (DIXI). This so-called "Mission of Opportunity" was originally focused on Comet 85P/Boethin; on 21 July 2005, Deep Impact was set on a trajectory to conduct a flyby of Earth in anticipation of the intercept of Boethin. Unfortunately, scientists lost track of Comet Boethin (possibly because the comet had broken up) and Deep Impact was instead directed towards Comet 103P/Hartley (or Hartley 2) beginning with a burn on 1 November 2007. EPOXI's new plan set Deep Impact on three consecutive Earth flybys, spread over two years (in December 2007, December 2008, and June 2010) before the final trek to meet Comet Hartley 2. These flybys essentially "stole some energy" from the spacecraft, thus dropping Deep Impact into a smaller orbit around the Sun. Before the second Earth flyby, Deep Impact performed its EPOCh mission using the HRI instrument to perform photometric investigations of extrasolar planets around eight distant stars, returning nearly 200,000 images. In the fall of 2010, Deep Impact began its investigations of Comet Hartley 2, conducting its flyby of the target at a range of 694 kilometers at 15:00 UT on 4 November 2010. As with the encounter with Comet Tempel 1, Deep Impact used its three instruments to study Hartley 2 for three weeks. Some of the images were so clear that scientists were able to identify jets of dust with particular features on the comet's nucleus. The data showed that the two lobes of Hartley 2 were different in composition. Once past this second cometary encounter, Deep Impact had little propellant for further cometary investigations, but there was a possibility that the spacecraft, if still in working condition, could be used for a flyby of Near Earth Asteroid 2002 GT in 2020. With that goal in mind, thrusters were fired in December 2011 and October 2012 for targeting purposes. In the meantime, the spacecraft was used for remote study of faraway comets such as C/200P1 (Garradd) in early 2012 and C/2012 S1 (ISON) in early 2013. Communications with Deep Impact were lost sometime between 11 August and 14 August 2013, and after "considerable effort" to contact the spacecraft, NASA announced on 20 September that it had officially abandoned efforts to contact Deep Impact.

208

Mars Reconnaissance Orbiter

Nation: USA (86)

Objective(s): Mars orbit

Spacecraft: MRO

Spacecraft Mass: 2,180 kg

Mission Design and Management: NASA / JPL

Launch Vehicle: Atlas V 401 (AV-007)

Launch Date and Time: 12 August 2005 / 11:43:00 UT

Launch Site: Cape Canaveral Air Force Station / SLC-41

Scientific Instruments:

1. high resolution imaging science experiment camera (HiRISE)
2. context camera (CTX)
3. Mars color imager (MARCI)
4. compact reconnaissance imaging spectrometer (CRISM)

An image from the HiRISE instrument on board NASA's Mars Reconnaissance Orbiter (MRO) shows a fissure, less than 500 meters across at its widest point, on Olympus Mons on Mars. *Credit: NASA/JPL-Caltech/Univ. of Arizona*

5. Mars climate sounder (MCS)
6. shallow subsurface radar (SHARAD)
7. Optical navigation camera
8. Electra communications package
9. Gravity field investigation package
10. Atmospheric structure investigation

Results: Mars Reconnaissance Orbiter (MRO) is a large orbiter, modeled in part on NASA's highly successful Mars Global Surveyor spacecraft, designed to photograph Mars from orbit for about two Earth years. Its primary goals were to map the Martian surface with a high-resolution camera (the HiRISE 0.5-meter diameter reflecting telescope, the largest ever carried on a deep space mission), at least partly to help select sites for future landing missions. Supplementary investigations included studies of the Martian climate, weather, atmosphere, and geology. Along with the basic six instruments, MRO also carried an optical navigation camera and Electra, a UHF telecommunications

package to provide navigation and communications support to other landers and rovers on the surface of Mars. After launch, MRO entered orbit around Earth. Soon after, the Centaur upper stage fired for a second time to dispatch its payload (and itself) to escape velocity on a trajectory to intercept with Mars. After a seven-month trip through interplanetary space and three mid-course corrections, MRO approached Mars and, on 10 March 2006, fired its six engines (which displayed slightly reduced thrust), and successfully entered into a highly elliptical orbit around the Red Planet with parameters of 426 × 44,500 kilometers with a period of 35.5 hours. A subsequent combination of aerobraking in the upper atmosphere and engine burns between 7 April and 11 September 2006 left MRO in its final operational orbit of approximately 250 × 316 kilometers. Two months later, it began its primary science mission, joining five other active spacecraft in orbit or on the surface of Mars: Mars Global

Surveyor, 2001 Mars Odyssey, the two Mars Exploration Rovers, and the European Mars Express. By December 2006, the operation of one of MRO's instruments, the Mars climate sounder, was suspended due to anomalies in its field of view. All other instruments, however, returned vast amounts of uninterrupted and valuable data during the first two years of MRO's operations, known as the Primary Science Phase, which extended from November 2006 to November 2008. One of the early findings from imagery collected by HiRISE was the presence of liquid carbon dioxide or water on the surface of Mars in its recent past. During the Extended Science Phase, from November 2008 to December 2010, MRO faced a number of technical obstacles, primarily related to seemingly spontaneous rebooting of its computer four times in 2009. At one point, the spacecraft was essentially shut down beginning 26 August. Finally, on 8 December, engineers commanded the orbiter out of "safe mode" and slowly began initiating science operations using its scientific instruments. As it was back on the job, MRO passed an important symbolic milestone on 3 March 2010 when it had reached 100 terabits of data transmitted back to Earth, which NASA said was "more than three times the amount of data from all other deep-space missions combined." MRO continued to return high quality data, despite another reboot event in September 2010. Many of its activities were coordinated with other Mars spacecraft. For example, in December 2010, researchers used data from the CRISM instrument to help the Opportunity rover study the distribution of minerals in Endeavour Crater on the ground. A new phase of MRO's mission began in December 2010, the Extended Mission, whose goal was to explore seasonal processes on Mars, search for surface changes, and also provide support for other Martian spacecraft including the Mars Science Laboratory (MSL). It was during this period, in March 2011, that MRO passed its five-year anniversary orbiting Mars. Later in August, NASA announced that MRO data indicated that water might actually be flowing on

Mars during the warmest months of the year; MRO images had shown dark finger-like features, known as Recurring Slope Linea (RSLs) that appear and disappear on some slopes during late spring through summer but disappear during winter. On 14 March 2012, MRO captured a 20-kilometer-high dust devil whirling its way across the Amazonis Planitia region of northern Mars. Later, in October 2012, NASA initiated MRO's second Extended Mission, which expired in October 2014. Late in 2013, MRO turned its gaze outwards, to Comet ISON, a comet racing in from the Oort Cloud, which passed by Mars on 29 September. During this second Extended Mission, MRO passed the point of transmitting 200 terrabits of science data back to Earth. Once again, there was a computer anomaly on board the spacecraft: on 9 March 2014, MRO put itself in safe mode after an unscheduled swap from one main computer to another. Four days later, the vehicle resumed normal science operations (along with its activities relaying data back to Earth from the Curiosity rover). Because of the impending flyby of Mars by Comet C/2013A1 (or Comet Siding Spring) on 19 October 2014, NASA began to shift the orbit of MRO (as well as its other operational orbiter, 2001 Mars Odyssey) to minimize risk of damage from material shed by the comet. Orbit adjustments were made by MRO on 2 July and then again on 27 August. In the event, MRO captured the best ever views of a comet from the Oort Cloud when Siding Spring flew by Mars on 19 October. The spacecraft also suffered no damage as a result of the flyby. For the seventh time in its time in orbit, MRO put itself in a precautionary standby mode on 11 April 2015 when there was an unplanned switch from one main computer to another. Within a week the spacecraft once again returned to full operational capability. Later, in January 2016, controllers completed a planned flash-memory rewrite in one of the spacecraft's redundant computers in order to load new data in the form of tables on the positions of Earth and the Sun. Earlier, in August 2015, MRO celebrated a decade since its launch, by which time it had orbited Mars 40,000 times and returned 250 tera-bits of data; NASA announced that every week, the spacecraft was still returning more information on Mars than the weekly total of all other active Mars missions. Soon after, in September 2015, scientists published evidence in the journal *Nature Geoscience* that data from MRO's imaging spectrometer pro-vided the strongest evidence yet that liquid water still flows intermittently on present-day Mars. Scientists later concluded that water ice makes up half or more of an underground layer in the Utopia Planitia region. In July 2016, research results were published indicating that gullies on modern day Mars—channels with an alcove at the top and deposited material at the bottom—were probably not formed by flowing liquid water, and instead perhaps by the freeze and thaw of carbon dioxide frost. The data from MRO also provided the basis for a large crowd-sourced experiment in 2016. Using the Planet Four: Terrains Web site, ten thou-sand volunteers used images (taken by the Context Camera) of the Martian south polar regions to identify targets for closer inspection (by the HiRISE camera), thus generating new insights on seasonal slabs of carbon dioxide and erosional fea-tures on Mars known as "spiders." On 28 September 2016, MRO was to have provided critical commu-nications support for the arrival of the InSight Mars lander mission (enabled in part by an orbital maneuver carried out more than a year before, on 29 July 2015). However, the InSight launch was postponed to 2018 due to development problems in one of its instruments as well as the relative infrequency of the short launch window chosen for the mission. At several points during its mission, MRO photographed artificial objects on the Martian surface. In January 2015, NASA announced that high resolution images taken by MRO had identified Beagle 2's wreckage on the Martian surface. Similarly, images taken in December 2014 and April 2015 by the HiRISE instrument also showed NASA's Curiosity rover inside Gale Crater. Later, in October 2016, images taken by both the Context Camera and the HiRISE

camera showed ESA's Schiaparelli test lander that stopped transmitting before final impact. In early 2017, nearly 11 years after arriving at Mars, MRO remains operational and the second longest-lived spacecraft to orbit Mars, after 2001 Mars Odyssey.

209

Venus Express

Nation: European Space Agency (5)
Objective(s): Venus orbit
Spacecraft: VEX
Spacecraft Mass: 1,270 kg
Mission Design and Management: ESA
Launch Vehicle: Soyuz-FG + Fregat (no. Zh15000-010 + 14S44 no. 1010)
Launch Date and Time: 9 November 2005 / 03:33:34 UT
Launch Site: GIK-5 / Site 31/6
Scientific Instruments:

1. analyzer of space plasma and energetic atoms (ASPERA)
2. Venus Express magnetometer (MAG)
3. planetary Fourier spectrometer (PFS)
4. ultraviolet and infrared atmospheric spec-trometer (SPICAV/SOIR)
5. Venus radio science experiment (VeRa)
6. visible and infrared thermal imaging spec-trometer (VIRTIS)
7. Venus monitoring camera (VMC)

Results: Venus Express was a spacecraft, similar in design to ESA's Mars Express, designed to conduct a global investigation of the Venusian atmosphere, its plasma environment, and surface character-istics, from a 24-hour near-polar elliptical orbit around Venus. The spacecraft was launched by a Soyuz-FG/Fregat combination owned by Starsem, a French company which markets the Russian Soyuz in its "European" version. The Soyuz-FG delivered the payload into a low Earth orbit, with the Fregat firing a second time 96 minutes after launch to send the entire stack out of Earth orbit towards Venus. The spacecraft carried out a single mid-course correction on the way to Venus, on 11 November 2005, before arriving at Venus on 11 April 2006 after a five-month journey. The main engine fired at 07:10:29 UT (spacecraft time) to insert Venus Express into orbit around the planet, thus becoming the first European spacecraft to orbit Venus. It achieved its operational orbit—250 × 66,000 kilometers—by 7 May 2007. The original mission of Venus Express was anticipated to last no more than 500 Earth days, but the mission was extended five times past its nominal mission (which ended on 19 September 2007); it was extended first to May 2009, then to December 2009, then to December 2012, then to 2014, and finally to 2015. Among its initial accomplishments was to generate a complete temperature map of the southern hemisphere of the planet by December 2006. Further major findings included evidence for past oceans on the surface of Venus, a higher prevalence of lightning on Venus than Earth, and the discovery of a huge "double atmospheric vortex" at the south pole of the planet. In 2011, scientists studying data from Venus Express reported the existence of a layer of ozone in the upper atmosphere of the planet. After eight years in orbit, as propellant supplies to maintain its elliptical orbit began running low, routine science experiments were concluded on 15 May 2014, and mission scientists decided undertake a series of aerobraking campaigns during which the spacecraft would "dip" deeper into the atmosphere than it had before. The lowest point of 129.1 kilometers was reached on 11 July. The duration of these "dips" was about 100 seconds long with maximum dynamic pressure at 0.75 Newton per square meter, probably a record for a spacecraft still operating in orbit around a planetary body. After about a month in late June and early July "surfing in and out" of the Venusian atmosphere, during which time critical data was collected on the effects of atmospheric drag and heating, the spacecraft performed a 15-day climb back up, beginning 12 July, which ended by reaching an orbit with a lowest point of 460 kilometers. Having reached this orbit, Venus Express decayed naturally the remainder of the year. There was an attempt in late November to arrest this decay but contact with the spacecraft was lost on 28 November 2014, with only intermittent telemetry and telecommand links after that point. On 16 December, ESA officially announced the end of the mission although a carrier signal was still being received. The last time that this X-band carrier signal was detected was on 19 January 2015, suggesting that the orbiter burned up in the atmosphere soon after.

2006

210

New Horizons

Nation: USA (87)

Objective(s): Pluto flyby

Spacecraft: New Horizons

Spacecraft Mass: 478 kg

Mission Design and Management: NASA / APL

Launch Vehicle: Atlas V 551 (AV-010)

Launch Date and Time: 19 January 2006 / 19:00:00 UT

Launch Site: Cape Canaveral Air Force Station / SLC-41

Scientific Instruments:

1. Ralph visible and infrared imager/ spectrometer
2. Alice ultraviolet imaging spectrometer
3. radio-science experiment (REX)
4. long-range reconnaissance imager (LORRI)
5. solar wind and plasma spectrometer (SWAP)
6. Pluto energetic particle spectrometer science investigation (PEPSSI)
7. student dust counter (SDC)

Results: New Horizons is a mission sent to study the dwarf planet Pluto, its moons, and other objects in the Kuiper Belt, a region of the solar system that extends from about 30 AU, near the orbit of Neptune, to about 50 AU from the Sun. The first mission of NASA's New Frontiers program—a medium-class, competitively selected, and Principal Investigator-led series of missions—that also includes Juno and OSIRIS-REx, New Horizons was the first spacecraft to encounter Pluto, a relic from the formation of the solar system. By the time of its Pluto system encounter, the spacecraft had to travel farther away and for a longer time period (more than nine years) than any previous deep space spacecraft ever launched. The design of the spacecraft was based on a lineage traced back to the CONTOUR and TIMED spacecraft, both also built by the Applied Physics Laboratory at Johns Hopkins University. Besides its suite of scientific instruments, New Horizons carries a cylindrical radioisotope thermoelectric generator (RTG), a spare from the Cassini mission, that provided about 250 W of power at launch (decaying to 200 W by the time of the Pluto encounter). After reaching initial Earth orbit at 167 × 213 kilometers, the Centaur upper stage fired (for a second time) for 9 minutes to boost the payload out to an elliptical orbit that stretched to the asteroid belt. A second firing of the Star 48B solid rocket accelerated the spacecraft to a velocity of 58,536 kilometers/hour, the highest launch velocity attained by a human-made object relative to Earth. It was now set on a trajectory to the outer reaches of the solar system. Controllers implemented mid-course corrections on 28 and 30 January and 9 March 2006, and a month later, on 7 April, New Horizons passed the orbit of Mars. A fortuitous chance to test some of the spacecraft's instrumentation—especially Ralph—occurred on 13 June 2006 when New Horizons passed by a tiny asteroid named 132524 APL at a range of 101,867 kilometers. The spacecraft flew by the solar system's largest planet, Jupiter, for a gravity assist maneuver on 28 February 2007 with a closest approach at 05:43:40 UT. The encounter increased the spacecraft's velocity by 14,000 kilometers/hour, shortening its trip to Pluto by three years. During the flyby, New Horizons carried out a detailed set of observations over a period of four months in early 2007. These were both designed to gather new data on Jupiter's atmosphere, ring system, and moons (building on research from Galileo) and to test out instruments.

NASA's New Horizon spacecraft captured this high-resolution enhanced color view of Pluto on 14 July 2015. The image combines blue, red, and infrared images taken by the Ralph/Multispectral Visual Imaging Camera (MVIC). Resolution is as high in places as 1.3 kilometers. *Credit: NASA/JHUAPL/SwRI*

Although observing the moons from distances much farther than Galileo, New Horizons was still able to return impressive pictures of Io (including eruptions on its surface), Europa, and Ganymede. Following the Jupiter encounter, New Horizons sped its way towards the Kuiper Belt, performing a mid-course correction on 25 September 2007. It was in hibernation mode from 28 June 2007 during which time the spacecraft's on-board computer kept tabs on mission systems, transmitting special codes indicating that operations were either nominal or anomalous. During hibernation, most major systems of New Horizons were deactivated, revived only about two months every year. The second, third, and forth hibernation cycles were activated on 16 December 2008, 27 August 2009, and 29 August 2014. It passed the halfway point to Pluto on 25 February 2010. The discovery of new

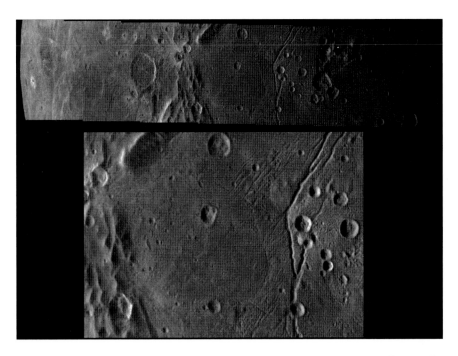

This mosaic of Pluto's largest moon Charon was taken by the Long Range Reconnaissance Imager (LORRI) on New Horizons just prior to closest approach on 14 July 2015. The scene at the bottom is approximately 200 kilometers. Resolution is as high as 310 meters. *Credit: NASA/JHUAPL/SwRI*

moons (Kerberos and Styx) around Pluto during the mission added to suspicions that there might be debris or dust around Pluto. Mission planners devised two possible contingency plans in case debris increased as the spacecraft approached Pluto, either using its antenna facing the incoming particles as a shield, or flying closer to Pluto where there might be less debris. On 6 December 2014, ground controllers revived New Horizons from hibernation for the last time to initiate its active encounter with Pluto. At the time, it took 4 hours and 25 minutes for a signal to reach Earth from the spacecraft. The spacecraft began its approach phase towards Pluto on 15 January 2015, its trajectory adjusted with a 93-second thruster burn on 10 March. Two days later, with about four months remaining before its close encounter, New Horizons finally became closer to Pluto than Earth is to the Sun. Pictures of Pluto began to reveal distinct features by 29 April, with detail growing literally week by week into its approach. A final 23-second engine burn on 29 June accelerated New Horizons towards

its target by 27 centimeters/second, also fine-tuning its trajectory. There was a minor concern on 4 July when New Horizons entered "safe mode" due to a timing flaw in the spacecraft command sequence. Fortunately, the spacecraft returned to fully nominal science operations by 7 July. Three days later, data from New Horizons was used to conclusively answer one of the most basic mysteries about the dwarf planet: its size. Mission scientists concluded that Pluto is 2,370 kilometers in diameter, slightly larger than prior estimates. Charon was confirmed to be 1,208 kilometers in diameter. Finally, at 11:49 UT on 14 July 2015, New Horizons flew by about 7,800 kilometers above the surface of Pluto. About 13 hours later, at 00:53 UT on 15 July, a 15-minute series of status messages was received at mission operations at Johns Hopkins University's APL (via NASA's Deep Space Network) confirming that the flyby had been fully successful. Besides collecting data on Pluto and Charon (flyby at 28,800 kilometers range), New Horizons also observed Pluto's other satellites, including Nix, Hydra, Kerberos, and Styx. The download of the entire set of data collected during the Pluto/Charon encounter—about 6.25 gigabytes of data—took over 15 months, and officially completed at 21:48 UT on 25 October 2016. Such a lengthy period was necessary because the spacecraft was roughly 4.5 light-hours from Earth and it could only transmit 1–2 kb/second. Data from New Horizons clearly indicated that Pluto and its satellites were far more complex than imagined, and scientists were particularly surprised by the degree of "current activity" on Pluto's surface. The atmospheric haze and lower-than-predicted atmospheric escape rate forced scientists to fundamentally revise earlier models of the system. Pluto, in fact, displays evidence of vast changes in atmospheric pressure and possibly past presence of running or standing liquid volatiles on its surface. There are hints that Pluto could have an internal water-ice ocean today. Photographs clearly showed a vast thousand-kilometer-wide heart-shaped nitrogen glacier (called Sputnik Planitia) on the surface, undoubtedly the largest known glacier in the solar system. On Charon, images showed an enormous equatorial extension tectonic belt, suggesting a long-past water ice ocean. In the fall of 2015, after its Pluto encounter, mission planners began to redirect New Horizons for a flyby on 1 January 2019 with 2014 MU$_{69}$, a Kuiper belt object that is approximately 6.4 billion kilometers from Earth. Four course corrections were implemented in the fall while a fifth was carried out on 1 February 2017. The goal of the encounter is to study the surface geology of the object, measure surface temperature, map the surface, search for signs of activity, measure its mass, and detect any satellites or rings. As of 3 April, the spacecraft was halfway from Pluto to its target. Soon after, on 10 April, New Horizons entered hibernation mode, when much of the vehicle remained in unpowered mode for "a long summer's nap" that lasted until 11 September. During this time, the flight computer broadcast a weekly beacon-status tone back to Earth, and another data stream once a month on spacecraft health and

safety data. On the first anniversary of its Pluto/Charon flyby, on 14 July 2017, the New Horizons team unveiled new detailed maps of both planetary bodies. As of 4 November 2017 New Horizons was 40.31 AU (6.03 billion kilometers) from Earth and traveling at approximately 14.22 kilometers/second (relative to the Sun) heading generally in the direction of the constellation Sagittarius. The mission is currently extended through 2021 to explore additional Kuiper belt objects.

211

STEREO A and STEREO B

Nation: USA (88)

Objective(s): solar orbit

Spacecraft: Stereo A / Stereo B

Spacecraft Mass: 623 kg / 658 kg

Mission Design and Management: NASA / GSFC / APL

Launch Vehicle: Delta 7925-10L (no. D319)

Launch Date and Time: 26 October 2006 / 00:52:00 UT

Launch Site: Cape Canaveral Air Force Station / SLC-17B

Scientific Instruments:

1. Sun Earth connection coronal and heliospheric investigation (SECCHI)
   a. extreme ultraviolet imager (EUVI)
   b. inner coronagraph (COR1)
   c. outer coronagraph (COR2)
   d. heliospheric imager (HI)
2. interplanetary radio burst tracker (SWAVES)
3. in-situ measurements of particles and CME transients (IMPACT)
4. plasma and suprathermal ion composition (PLASTIC)

Results: STEREO (Solar Terrestrial Relations Observatory), the third mission in NASA's Solar Terrestrial Probes (STP) program, consists of two space-based observatories to study the structure and evolution of solar storms as they emerge from the Sun and move out through space. The two spacecraft, one ahead of Earth in its orbit and the other trailing behind, are providing the first stereoscopic images of the Sun, and collecting data on the nature of its coronal mass ejections (CMEs), represented by large bursts of solar wind, solar plasma, and magnetic fields that are ejected out into space. Such CMEs can disrupt communications, power grids, satellite operations, and air travel here on Earth. Both spacecraft were inserted into an initial orbit around Earth at 165 × 171 kilometers. Following a second and third burn, the two spacecraft were sent into a translunar orbit, planned at 182 × 40,3810 kilometers at 28.5° inclination. Just after the final burn, at 01:19 UT, STEREO A separated from STEREO B. On the fifth orbit for both, on 15 December 2006, both spacecraft swung by the Moon, and using a gravitational assist maneuver, were sent to different orbits. STEREO A was in a solar orbit inside Earth's orbit (and "Ahead") while STEREO B remained in a high Earth orbit. STEREO B encountered the Moon again on 21 January 2007, and was accelerated into the opposite direction from STEREO A; it entered heliocentric orbit outside of Earth's orbit (and "behind"). The orbital periods of STEREO A and STEREO B are 347 days and 387 days, respectively. The two spacecraft separate from each other at a (combined) annual rate of 44°. A star tracker failed briefly on STEREO B but this had no impact on the mission. Later, in May 2009, the same spacecraft was successfully rebooted with a new guidance and control software. A similar reset was implemented with STEREO A in August of the same year. A transponder malfunction in July 2013 briefly interrupted science activities on STEREO B. More seriously, the spacecraft suffered a failure of its Inertial Measurement Unit in January 2014 but controllers managed to revive the spacecraft quickly. At various points, the spacecraft were separated from each other by 90° and 180°. The latter occurred on 6 February 2011 allowing the *entire* Sun to be seen at once for the first time by any set of spacecraft. On 23 July 2012, during an "extreme" solar storm more powerful than anything seen in the past 150 years, STEREO A was able to collect significant data on the phenomenon.

Unanticipated high temperatures in the high gain antenna feed horns of both spacecraft were detected in June 2014, effectively reducing the data return rate, thus curtailing the science program. Because of this problem, mission scientists formulated a reduced program of science operations for STEREO A in August 2014, one that was further thwarted by a massive proton storm (caused by a large solar flare on the far side of the Sun) on 3 September 2014. The high-energy particle fluxes were so high that star trackers on both STEREO spacecraft were reset. Later, on 1 October, communications were lost with STEREO B immediately after a planned reset of the spacecraft. All attempts to recover contact were in vain and it is thought that anomalies in the guidance and control system of the spacecraft might have rendered the spacecraft powerless as a result of drift away from direct exposure of the Sun to its solar panels. Controllers hoped at the time that eventually STEREO B would drift into "proper" orientation (much like SOHO in 1998) and would power up and resume its mission. Remarkably, 22 months after loss of contact, on 21 August 2016, NASA's DSN reestablished communications with STEREO B (having tried once a month through this period). Controllers concluded that STEREO B was probably spinning out of control around its principal axis of inertia. This uncontrolled orientation allowed some power generation but not enough time to upload a software fix. STEREO A meanwhile was entered into a "safe mode" deliberately in March 2015 for several months during a superior solar conjunction, i.e., a period when the spacecraft is on the opposite side of the Sun from Earth. Communication was reestablished with the spacecraft on 11 July 2015 when images were received again, although the science program remained at a low status until 17 November 2015 when STEREO A began to operate at full capacity again. The key element here was the transmission of real-time data, known as beacon data from coronagraph imagery. As of early 2017, STEREO A continues to operate without problems.

2007

Artemis P1 and Artemis P2

Nation: USA (89)

Objective(s): Earth–Moon L1 and L2 Lagrange points, lunar orbits

Spacecraft: THEMIS B / THEMIS C

Spacecraft Mass: 126 kg (each)

Mission Design and Management: NASA / University of California–Berkeley

Launch Vehicle: Delta 7925-10C (no. D323)

Launch Date and Time: 17 February 2007 / 23:01:00 UT

Launch Site: Cape Canaveral Air Force Station / SLC-17B

Scientific Instruments:

1. electric field instruments (EFI)
2. fluxgate magnetometer (FGM)
3. search coil magnetometer (SCM)
4. electrostatic analyzer (ESA)
5. solid state telescope (SST)

Results: The two Artemis lunar orbit missions were repurposed from the original Time History of Events and Macroscale Interactions during Substorms (THEMIS) mission that involved five NASA satellites, THEMIS A, B, C, D, and E, which studied a type of magnetic phenomena ("substorms") in Earth's magnetosphere that tend to intensify auroras near Earth's poles. Each of the five satellites carried identical instrumentation. After a burn of the third stage, the five THEMIS spacecraft—initially joined but soon separated—were deposited into a 469 × 87,337 km × 16.0° orbit around Earth. In its "string-of-pearls" configuration, the five THEMIS satellites carried out its initial mission without any significant anomalies. On 19 May 2008, Space Sciences Laboratories, developer of the spacecraft at University of California–Berkeley, announced that NASA had extended the THEMIS mission to 2012 and that two of the THEMIS satellites, B and C, would venture into lunar orbit as part of a new mission under the name ARTEMIS (Acceleration, Reconnection, Turbulence and Electrodynamics of the Moon's Interaction with the Sun). In this new mission, THEMIS B and C were renamed ARTEMIS P1 and ARTEMIS P2, respectively, and redirected to study the Earth–Moon Lagrange points, the solar wind, the Moon's plasma wake, and the interaction between Earth's magnetotail and the Moon's own weak magnetism. (The "P1" and "P2" designations were leftover terminology from the THEMIS mission which used "P1" and "P2" to denote the operational orbits of THEMIS B and C). On the 40th anniversary of the Apollo XI landing, on 20 July 2009, ARTEMIS P1 and P2 officially began low thrust maneuvers that, over the course of the following year-and-a-half, led them to the L2 and L1 Lagrange points, opposite the near and far sides of the Moon, respectively. (This phase included a lunar flyby on 28 March 2010 by ARTEMIS P2.) On 25 August 2010, an engine burn propelled ARTEMIS P1 into orbit around the Earth–Moon L2 Lagrange point, located on the far side of the Moon, about 61,300 kilometers above the lunar surface. This was the first time that a spacecraft had successfully entered orbit around an Earth–Moon libration point. The second spacecraft, ARTEMIS P2 arrived at L1 on 22 October 2010 by which time P1 had completed about four revolutions around L2. Although the station-keeping at the Lagrange points on the way to the Moon was motivated to avoid Earth's long shadows in its original orbits (thus keeping the spacecraft operational), here at the two Lagrange points, the two spacecraft collected magnetospheric data from opposite sides of the Moon, critical for simultaneous measurements of particles and electric and magnetic fields to build a three-dimensional map of the acceleration of energetic particles near the Moon's orbit. On 27 June 2011, ARTEMIS P1 successfully entered lunar orbit with an initial orbit of roughly 3,543 × 27,000 kilometers while its sister vehicle, ARTEMIS P2 arrived on 17 July 2011, after a two-year journey from Earth orbit. Over the next three months, mission controllers implemented a series of maneuvers to move the second spacecraft into an orbit with a period of 27.5 hours, similar to its companion, but moving in the opposite direction. The two spacecraft, orbiting in opposite directions around the Moon, began to provide the first 3D measurements of the Moon's magnetic field to determine its regional influence on solar wind particles. More specifically, the two spacecraft revealed new information on the lunar "wake" that extends about 12 lunar radii and in particularly how its void distorts the interplanetary magnetic field causing it to bulge moonward. As of January 2016, the two spacecraft remained in good health and operating in their stable but highly elliptical lunar orbits. Mission scientists marked the tenth anniversary of the launch in February 2017, with the spacecraft still in good health.

213

Phoenix

Nation: USA (90)
Objective(s): Mars landing
Spacecraft: Phoenix Lander
Spacecraft Mass: 664 kg (350 kg lander)
Mission Design and Management: NASA / JPL / University of Arizona
Launch Vehicle: Delta 7925-9.5 (no. D325)
Launch Date and Time: 4 August 2007 / 09:26:34 UT

Launch Site: Cape Canaveral Air Force Station / SLC-17A
Scientific Instruments:
1. robotic arm (RA)
2. microscopy, electrochemistry, and conductivity analyzer (MECA)
3. robotic arm camera (RAC)
4. surface stereo imager (SSI)
5. thermal and evolved gas analyzer (TEGA)
6. Mars descent imager (MARDI)
7. meteorological station (MET)

Results: The Phoenix mission was a landing mission to Mars, the first under NASA's new Mars Scout Program to send a series of small, low-cost, low complexity, and higher frequency robotic missions to Mars. (The second and last mission in the series was MAVEN launched in 2013; Mars missions were then folded into the Discovery Program where they would compete with missions to other planetary destinations). Its science goals included studying the history of water on Mars in all its phases, searching for evidence of habitable zones, and assessing the biological potential of the ice–soil boundary. More broadly, the lander was designed to determine whether life ever existed on Mars, characterize the climate and geology of the Red Planet, and help prepare for future human exploration of its surface. The spacecraft was essentially built on the basis of the abandoned and never-launched Mars Surveyor 2001 Lander and contained other instruments built in support of the unsuccessful Mars Polar Lander mission. It was the first NASA mission to Mars that was led directly from a public university, the University of Arizona, more specifically its Lunar and Planetary Laboratory. The primary mission was designed to last 90 sols (Mars days) or approximately 92 Earth days. After two burns of the Delta's second stage, the PAM-D upper stage (with a Star 48 motor) fired at 10:44 UT on 4 August 2007 to send the Phoenix lander towards Mars. It conducted mid-course corrections on 10 August and 30 October 2007, and

10 April and 17 May 2008, the latter two directing it toward the northern polar region of Mars. As it approached Mars, the orbits of three other spacecraft orbiting Mars—Mars Reconnaissance Orbiter (MRO), 2001 Mars Odyssey, and Mars Express—were adjusted so that they could observe Phoenix's entry into the atmosphere. In addition, MRO's HiRISE instrument was used to thoroughly scout out the landing area, with some images identifying rocks smaller than the lander itself. Phoenix entered the Martian atmosphere at nearly 21,000 kilometers/hour on 25 May 2008 and touched down safely on the surface at 23:38:38 UT in the Green Valley of Vastitas Borealis. It was the first successful landing of a stationary soft-lander on Mars since Viking 2, 32 years before. During its descent, MRO's HiRISE camera clearly photographed Phoenix suspended from its parachute, the first time one spacecraft photographed another during a planetary landing. The lander waited 15 minutes for the dust to settle before unfurling its solar panels. The first images showed a flat surface marred by pebbles and troughs, but no large rocks or hills as expected given its northern position. Within four days, Phoenix had transmitted a complete 360° panorama of the cold Martian surface, deployed the nearly 2.5-meter robotic arm, and started returning regular weather reports. On 31 May, the robotic arm scooped up dirt and began sampling Martian soil for ice. Already by 19 June 2008, mission scientists were able to conclude that clumps of bright material in the so-called "Dodo-Goldilocks" trench dug by the robotic arm were probably water ice: the material had vaporized in four days after the scoop. On 31 July 2008, NASA officially announced that, based on an analysis (by TEGA's mass spectrometer) of a sample collected by the lander, that there is water on Mars. William Boynton of the University of Arizona noted that such data adds to the claims from the 2001 Mars Odyssey orbiter whose data also indicated likewise. On 5 August, in response to media rumors about the possibility of life on Mars, the Phoenix team announced that they had found perchlorates on the surface of Mars that neither confirmed nor refuted the possibility of life on Mars. The results also led scientists to revisit the data from the Viking Landers. By the end of August, Phoenix had completed its originally planned 90-day mission, which was extended to 30 September. On 12 September, the lander scoop delivered a new soil sample to its Wet Chemistry Laboratory that mixed an aqueous solution from Earth to the soil as part of a process to identify soluble nutrients and other chemicals in the soil. Early results suggested that the soil was alkaline, composed of salts and other chemicals such as perchlorate, sodium, magnesium, chloride, and potassium. On 13 October, Phoenix weathered a dust storm and recovered another soil sample, the sixth, into the TEGA instrument. But as the Martian winter was upon the landing site, the lander went into safe mode on 28 October 2008 due to insufficient sunlight and poor weather conditions. During safe mode, non-critical activities were suspended while the spacecraft awaited further instructions from mission control. There was daily communication with the lander from 30 October to 2 November but no signals were received after that. On 10 November, NASA announced the lander had "finishe[d its] successful work on [the] Red Planet," and on 1 December, the Agency announced that NASA "had stopped using its Mars orbiters to hail the lander." The loss of the spacecraft was a combination of low power and the dust storm. During the cold harsh winter, CO_2 ice—some of it as thick as 19 centimeters—probably built up on the lander, sufficiently heavy to break the fragile solar arrays. Because of this kind of damage, subsequent communications attempts with the lander, in early 2010, were unsuccessful. On 24 May 2010, NASA announced that the project was formally ended. Images from MRO conclusively showed that Phoenix's solar panels were severely damaged by the freezing during the Martian winter.

214

Kaguya

Nation: Japan (6)

Objective(s): lunar orbit

Spacecraft: SELENE plus Okina (Rstar) and Ouna (Vstar)

Spacecraft Mass: 2,900 kg

Mission Design and Management: JAXA

Launch Vehicle: H-IIA 2022 (no. 13)

Launch Date and Time: 14 September 2007 / 01:31:01 UT

Launch Site: Tanegashima / Area Y1

Scientific Instruments:

1. x-ray spectrometer (XRS)
2. gamma-ray spectrometer (GRS)
3. multi-band imager (MI)
4. spectral profiler (SP)
5. terrain camera (TC)
6. lunar radar sounder (LRS)
7. laser altimeter (LALT)
8. lunar magnetometer (LMAG)
9. charged particle spectrometer (CPS)
10. plasma energy angle and composition experiment (PACE)
11. radio science experiment (RS)
12. upper atmosphere and plasma imager (UPI)
13. four-way Doppler measurements by relay satellite and main orbiter transponder (RSAT)
14. differential VLBI radio source experiment (VRAD)
15. high-definition television (HDTV)

Results: SELENE (Selenological and Engineering Explorer), named Kaguya ("Moon princess") after launch as a result of a public poll, was the second Japanese lunar probe, whose goal was to orbit the Moon and collect data on the origins and geological evolution of Earth's only natural satellite, study the lunar surface environment, and carry out radio science experiments. The Japanese noted it was "the largest lunar mission since the Apollo program."

Besides the main lunar satellite, SELENE, the mission also included two small spin-stabilized sub-satellites, each weighing 53 kilograms. These were the Relay Satellite (Rstar) and the VRAD satellite (Vstar). Upon launch, they were renamed Okina and Ouna, which mean "honorable elderly man" and "honorable elderly woman," respectively. Originally slated for launch in 2003 but delayed to 2007 due to problems with the H-II launch vehicle, the probe was launched into a highly elliptical parking orbit of 282 × 232,960 kilometers. On 3 October 2007, Kaguya entered into an initial polar orbit around the Moon at 101 × 11,741 kilometers, the first time that a Japanese spacecraft had done so. The two subsatellites, Okina and Ouna, were released on 9 October at 00:36 UT and 12 October at 04:28 UT into corresponding orbits: 115 × 2,399 kilometers and 127 × 795 kilometers. The orbiter itself attained its operational circular orbit at 100 kilometers by 19 October. Soon, on 31 October, Kaguya's two main HDTV cameras—each a 2.2 megapixel CCD HDTV camera—took the first high definition images of the Moon. A week later, on 7 November, the satellite took spectacular footage of an "Earthrise," the first since the Apollo missions in the 1970s. The fully operational phase of the mission began on 21 December following a successful checkout of all the onboard instruments. By 9 April 2008, JAXA was able to announce that Kaguya, using its Laser Altimeter, had been able to collect enough data to construct the topography of the entire lunar surface, with data points 10 orders larger than the previous model of the lunar surface, produced by the Unified Lunar Control Network in 2005, based largely on the American Clementine spacecraft. Its subsequent achievements include detecting gravity anomalies on both the near and far side of the Moon (based on Doppler data from both Kaguya and the Okina spacecraft) and the first optical observation of the permanently shadowed interior of the Shackleton Crater. Kaguya completed its original planned mission by late October 2008, with hopes to continue to March 2009 followed by impact in August 2009.

However, because of a faulty reaction wheel, the extended mission was ended early. On 1 February 2009, Kaguya's orbit was lowered to approximately 50 kilometers. The orbiter then impacted the Moon at 18:25 UT on 10 June 2009 at 65.5° S / 80.4° E near Gill Crater. Okina had already impacted the Moon at 10:46 UT on 12 February 2009.

215

Dawn

Nation: USA (91)

Objective(s): Vesta and Ceres orbit

Spacecraft: Dawn

Spacecraft Mass: 1,217.7 kg

Mission Design and Management: NASA / JPL

Launch Vehicle: Delta 7925H-9.5 (no. D327)

Launch Date and Time: 27 September 2007 / 11:34:00 UT

Launch Site: Cape Canaveral Air Force Station / SLC-17B

Scientific Instruments:

1. framing camera (FC)
2. visible and infrared mapping spectrometer (VIR)
3. gamma ray and neutron detector (GRaND)

Results: Dawn, the ninth mission in NASA's Discovery Program, was launched on a nearly-decade long mission to study two very different objects which both accreted early in the history of the solar system, the asteroid Vesta (arrival in 2011) and the dwarf planet Ceres (arrival in 2015). The investigation of these objects—the two largest in the asteroid belt—was driven by the three principal scientific motivations: to investigate the conditions at the origin of the solar system; to find the nature of the building blocks from which the inner planets were formed; and to contrast the very different evolutionary paths of Ceres and Vesta. Besides its complement of scientific equipment, Dawn also carried three xenon ion thrusters (derived from the technology used on the Deep Space 1 spacecraft),

This image from Dawn shows Kupalo Crater on Ceres. The crater, one of the youngest on the minor planet, measures 26 kilometers across. The image was taken on 21 December 2015 from Dawn's low-altitude mapping orbit from a distance of approximately 385 kilometers. *Credit: NASA/JPL-Caltech/UCLA/MPS/DLR/IDA*

each with a thrust of 91 mN and a specific impulse of 3,100 seconds. Some of the scientific equipment was provided by German and Italian institutions. After launch, Dawn was accelerated to escape velocity of 11.50 kilometers/second by the PAM-D solid propellant third stage which fired at 12:29 UT. The spacecraft passed lunar orbit at around 14:30 UT on 28 September and entered solar orbit at roughly 1.00 × 1.62 AU. Long-term cruise with the ion thrusters began on 17 December 2007, and completed on 31 October 2008, nearly 11 months later. Subsequently, a single trajectory correction on 20 November 2008 orchestrated a gravity assist flyby past Mars at a range of 542 kilometers on 17 February 2009. Over two years later, Dawn began to approach its first target, Vesta, returning progressively higher resolution images of the protoplanet. At around 05:00 UT on 16 July 2011, Dawn gently slipped into orbit around Vesta at an altitude of about 16,000 kilometers, thus becoming the first spacecraft to orbit any object in the main asteroid belt. Using its xenon-ion engine, it moved into a closer "survey" orbit at an altitude of about 2,700 kilometers by 2 August, staying at that orbit until the end of the month. Later, on 27 September, it moved into a closer orbit at an altitude of 680 kilometers (with an orbital period of 12.3 hours) staying there until 2 November, during which period it fully mapped Vesta six times, including in color and in stereo. By 8 December, the spacecraft was in a 4.3-hour orbit at an average altitude of just 210 kilometers. Original plans were

This mosaic image shows the mysterious mountain Ahuna Mons on Ceres. The images were taken by NASA's Dawn spacecraft from a low-altitude mapping orbit, about 385 kilometers above the surface, in December 2015. *Credit: NASA/ JPL-Caltech/UCLA/MPLS/DLR/IDA/PSI*

to carry out a 70-day mission at that low orbit but this was extended to 1 May, during which the spacecraft took 13,000 photos covering most of Vesta as well as more than 2.6 million visible and IR spectra. From 23 June to 25 July, Dawn was back up to 680 kilometers conducting more mapping, including of areas that had not been visible before. Data from the extended period of study in 2012 allowed scientists to estimate the size of its dense iron-nickel core (about 220 kilometers across), and conclusively identify Vesta as one of the few remaining remnants of large planetoids that formed the rocky planets of the solar system. In December 2012, investigators announced that Dawn had detected sinuous gullies on the surface of Vesta that might have been caused by liquid water—similar gullies on Earth are carved by liquid water. The data collected by Dawn suggested that Vesta is more closely related to terrestrial planets such as Earth (and its Moon) than to typical asteroids. Dawn's findings also showed that Vesta is the source of more meteorites on Earth than Mars or the Moon. Pictures also identified immense basins on the surface of Vesta such as the 400-kilometer diameter Veneneia and the 500-kilometer diameter Rheasilvia basins, created by impacts two and one billion years ago, respectively. At 07:26 UT on 5 September 2012, Dawn escaped the gravitational grip of Vesta and headed towards its second destination, the dwarf plant Ceres. During its stay at Vesta, the two bodies travelled around the Sun for 685 million kilometers. On the way to Ceres, Dawn stopped normal ion thrusting and spuriously entered "safe mode" on 11 September 2014. Fortunately, ground controllers were able to determine the source of the problem—a coincidental combination of high-energy particles that disabled the ion propulsion system and a previously unknown bug in the spacecraft software—and resumed normal ion firing by 15 September. Approach operations to Ceres began in January 2015 although during the approach phase, Dawn took fewer photographs of its target (than with Vesta) due to problems with two of its four reaction

wheels, the first of which failed on 17 June 2010. Needing at least three wheels to be operational, the mission team devised a plan to allow the spacecraft to operate using only two (in case another failed) in combination with hydrazine reaction control thrusters. A second wheel indeed failed, on 8 August 2012, just as Dawn was spiraling away from Vesta and beginning its trip to Ceres. The mission team implemented plans to conserve the much needed hydrazine (now more valuable than ever). All of these strategies allowed Dawn to approach Ceres slightly compromised but largely operational. At 00:39 UT on 7 March 2015, Dawn finally entered initial (polar) orbit around Ceres, thus becoming the first mission to study a dwarf plant, ahead of the New Horizons encounter with Pluto four months later. Dawn also became the first spacecraft to orbit two different celestial bodies (other than the Sun, of course). In planning for the mission, scientists had envisioned four different circular mapping orbits—called RC3, Survey, HAMO, and LAMO—around Ceres, from one as high as 13,600 kilometers (RC3) to as low as 385 kilometers (LAMO). Dawn remained in its first mapping orbit, RC3, from 23 April to 9 May 2015 carrying out photography, taking spectra at infrared and visible wavelengths, and searching for lofted dust as evidence of water vapor. On 9 May the spacecraft began using its ion engine to spiral down to reach its second mapping orbit by 3 June at about 4,400 kilometers, with its science mission beginning two days later. The orbital period at this point was 3.1 days. During this phase, lasting eight orbits, Dawn carried out extensive scientific observations over the sunlit side of Ceres. This phase, completed by 30 June, was punctuated by a minor alarm on 27 June due to anomalies in two of the scientific instruments, and a problem on 30 June when the spacecraft went into safe mode due to a problem with its orientation (later traced to a mechanical gimbal system that swivels one of its three ion engines). On 13 August 2015, it arrived in its third mapping orbit and stopped using its ion thrusters. Orbital altitude was 1,470 kilometers with a period of 19 hours. Its science mission in this orbit began four days later. Over the next six months, the spacecraft mapped Ceres six times. Mission Director at JPL Marc Rayman noted that this would be "some of the most intensive observations of its entire mission." At 23:30 UT on 23 October, Dawn turned one of its ion engines (no. 2) to move to its next orbit, a passage that took about seven weeks. On 7 December, the engine was turned off as the spacecraft reached its final mapping orbit at about 385 kilometers altitude. After some further orbital tweaking on 11–13 December, Dawn's orbit was synchronized with Ceres' rotation around its axis. Finally, on 18 December, the spacecraft began its next science phase in its new orbit, which it maintained for nearly nine months until 2 September 2016. During this period, the spacecraft obtained extensive data with its combined gamma-ray and neutron spectrometer as well as its infrared mapping spectrometer. Circling the dwarf planet every 5.4 hours, the standard high-resolution mapping programs were also in effect. By 3 May, Dawn's time in orbit around Ceres exceeded its time in orbit around Vesta in 2011–2012. A couple of months later, on 30 June 2016, JPL announced that Dawn had concluded its fully completed prime mission. By this point, it had taken 69,000 images and completed 2,450 orbits around both Vesta and Ceres. In addition, the ion engines had fired for 48,500 hours. NASA approved an extended mission at the time, opting not to have Dawn travel to a large asteroid known as Adeona but to continue to explore Ceres. On 2 September 2016, Dawn began a new five-week journey to a higher orbit after a highly successful stay at lower altitudes. The spacecraft reached a new orbit at 1,480 kilometers with an orbital period of 18.9 hours by 6 October, having begun an extended science mission 10 days later. On 4 November, it began climbing higher again and reached its sixth science orbit (known as "extended mission orbit 3" or XMO3) on 5 December at about 7,520 × 9,350 kilometers altitude. Here it measured the cosmic ray noise to

calibrate the data on Ceres' nuclear radiation that was collected when it was at 385 kilometers. Dawn was basically in good condition in early 2017 despite a temporary switch to "safe mode" on 17 January. On 23 April 2017, mission controllers discovered that two of the remaining reaction wheels on board the spacecraft had stopped working, thus jeopardizing attitude control. By using hydrazine, controllers were able to return Dawn to standard flight configuration. A few days later, on 29 April, Dawn successfully observed Ceres at opposition, i.e., from a position between Ceres and the Sun, allowing the spacecraft to view the bright Occator Crater from a new perspective. Soon after celebrating the tenth anniversary of its launch, NASA announced on 19 October 2017 that it had authorized a second extension to the mission. During the extension, Dawn will descend to lower than before—possibly 200 kilometers—to continue studies of the dwarf planet, focused on measuring the number and energy of gamma rays and neutrons. One of the most important discoveries made by Dawn was the existence of widespread ice just below the surface of Ceres, announced in December 2016. Dawn carries a memory chip with the names of more than 360,000 people who submitted their names as part of an outreach effort in 2005 and 2006.

216

Chang'e 1

Nation: China (1)

Objective(s): lunar orbit

Spacecraft: Chang'e yihao

Spacecraft Mass: 2,350 kg

Mission Design and Management: China National Space Administration

Launch Vehicle: Chang Zheng 3A (no. Y14)

Launch Date and Time: 24 October 2007 / 10:05:04 UT

Launch Site: Xichang / LC 3

Scientific Instruments:

1. stereoscopic CCD camera
2. Sagnac-based interferometer spectrometer imager
3. laser altimeter
4. microwave radiometer
5. gamma and x-ray spectrometer
6. space environment monitor system (a high-energy particle detector and 2 solar wind detectors)

Results: Chang'e 1 was the first deep space mission launched by China, part of the first phase of the so-called Chinese Lunar Exploration Program, divided into three phases of "circling around the Moon," "landing on the Moon," and "returning from the Moon" that would be accomplished between 2007 and 2020. The goal of this first mission, besides proving basic technologies and testing out several engineering systems, was to create a three-dimensional map of the lunar surface, analyze the distribution of certain chemicals on the lunar surface, survey the thickness of the lunar soil, estimate Helium-3 resources, and explore the space environment (solar wind, etc.) in near-lunar space. The spacecraft itself was based on the design of the reliable DFH-3 satellite bus. After launch, the spacecraft entered a 205 × 50,900-kilometer orbit for a day before firing its 50 kgf thrust main engine at 09:55 UT on 25 October to raise perigee to 593 kilometers. Subsequent burns (this time near perigee) were performed on 26 October (at 09:33 UT), 29 October (at 09:49 UT), and 31 October (at 09:15 UT) increasing apogee to 71,600, 119,800, and finally 400,000 kilometers, respectively. On its way to the Moon, Chang'e 1 (or CE-1, as it was often named in the Chinese English-language press) made one mid-course correction before entering lunar orbit with a 22-minute burn that began at 02:15 UT on 5 November 2007, thus becoming the first Chinese spacecraft to orbit the Moon. Initial orbital parameters were 210 × 860 kilometers. Two maneuvers on 6 and 7 November lowered perigee to 1,716 and 200 kilometers, respectively. Its final working orbit—a 200-kilometer polar orbit with a period of 127 minutes—was reached soon after on the same day. On 20 November, CE-1 returned the first raw image of the lunar surface, and by 28 November, all its scientific instruments were fully operational. A composite of 19 strips of raw images was issued by the Chinese media on 26 November at a ceremony attended by Chinese Premier Wen Jiabao. There was some controversy regarding this image which some believed was a fake or a copy of an image returned by Clementine but this proved not to be so: the high-quality image was indeed quite real. Through December 2007, CE-1 continued to photograph the Moon (including in stereo), and began imaging the polar regions in January 2008. The spacecraft successfully fulfilled its 1-year mission after which it continued extended operations. On 12 November 2008, Chinese space authorities issued a full-Moon image map produced using CE-1 images taken over 589 orbits covering 100% of the lunar surface with a resolution of 120 meters. In December 2008, over a period of two weeks, the spacecraft's perigee was progressively lowered to 15 kilometers to test operations for future orbiter and lander spacecraft. Finally, on 1 March 2009, CE-1 was commanded to impact on to the lunar surface, making contact at 08:13:10 UT at 52.27° E and 1.66° S, thus becoming the first Chinese object to make contact with the Moon. Its most significant achievement was to produce the most accurate and highest resolution 3D map of the lunar surface.

2008

217

Chandrayaan-1 and MIP

Nation: India (1)
Objective(s): lunar orbit, lunar impact
Spacecraft: Chandrayaan-1 / MIP
Spacecraft Mass: 1,380 kg
Mission Design and Management: ISRO
Launch Vehicle: PSLV-XL (no. C11)
Launch Date and Time: 22 October 2008 / 00:52:11 UT
Launch Site: Sriharikota / SLP
Scientific Instruments:

Main Satellite:

1. terrain mapping camera (TMC)
2. hyper spectral imager (HySI)
3. lunar laser ranging instrument (LLRI)
4. high energy x-ray spectrometer (HEX)
5. Moon impact probe (MIP)
6. Chandrayaan-1 x-ray spectrometer (CIXS)
7. near infrared spectrometer (SIR-2)
8. Sub keV atom reflecting analyzer (SARA)
9. miniature synthetic aperture radar (Mini SAR)
10. Moon mineralogy mapper (M3)
11. radiation dose monitor (RADOM)

MIP:

1. radar altimeter
2. video imaging system
3. Chandra's altitudinal composition explorer (mass spectrometer) (CHASE)

Results: Chandrayaan-1, the first Indian deep space mission, was launched to orbit the Moon and dispatch an impactor to the surface. Scientific goals included the study of the chemical, mineralogical, and "photo-geologic" mapping of the Moon. Besides five Indian instruments, the spacecraft carried scientific equipment from the United States, the U.K., Germany, Sweden, and Bulgaria. Chandrayaan-1 was launched into an initial geostationary transfer orbit of 225 × 22,817 kilometers at 17.9° inclination. Over a period of 13 days, the apogee of the orbit was increased by five burns of its 44.9 kgf Liquid Engine that successively raised orbit on 23 October (to 37,900 kilometers), 25 October (to 74,715 kilometers), 26 October (to 164,600 kilometers), 29 October (to 267,000 kilometers), and 4 November (to 380,000 kilometers). Finally, the probe successfully entered lunar orbit after a burn that began at 11:21 UT on 8 November and lasted about 13.5 minutes. Initial lunar orbital parameters were 7,502 × 504 kilometers. Between lunar orbit insertion on 8 November and 12 November, Chandrayaan-1's orbit was reduced gradually so that it ended up finally in its operational polar orbit at about 100 kilometers above the lunar surface. Two days later, at 14:36 UT, Chandrayaan released its 29-kilogram Moon Impact Probe (MIP) which fired a small deorbit motor and went into freefall, sending back readings from its three instruments until it crashed onto to the lunar surface at 15:01 UT near the Shackleton crater at the lunar south pole. Indian scientists reported that data from the CHASE instrument, which took readings every 4 seconds during its descent, suggested the existence of water in the lunar atmosphere, although the data remains inconclusive absent further verification. Chandrayaan-1 experienced abnormally high temperatures beginning late November 2008, and for a time, it could only run one scientific instrument at a time. In May 2009, the spacecraft was delivered to a higher 200-kilometer orbit, apparently in an attempt to keep the temperatures aboard the satellite to tolerable levels. Chandrayaan-1 also suffered a star sensor failure after nine months of operation in lunar orbit. A backup sensor also failed soon after, rendering inoperable the spacecraft's primary attitude control system. Instead controllers used a mechanical gyroscope system to maintain proper attitude. Last contact with the spacecraft was at 20:00 UT on 28 August 2009, thus falling short of its planned two-year lifetime, although ISRO noted that at least 95% of its mission objectives had been accomplished by then. The most likely cause of the end of the mission was failure of the power supply due to overheating. Perhaps Chandrayaan-1's most important finding was related to the question of water on the Moon. In September 2009, scientists published results of data collected by the American M3 instrument which had detected absorption features on the polar regions of the surface of the Moon usually linked to hydroxyl- and/or water-bearing molecules. This finding was followed later, in August 2013, by a further announcement of evidence of water molecules locked in mineral grains on the surface of the Moon, i.e., "magmatic water," or water that originates from deep in the Moon's interior. Magmatic water had been found in samples returned by Apollo astronauts but not from lunar orbit until the operation of the M3 instrument, although Cassini, during its flyby of the Moon in August 1999, had detected (using its VIMS instrument) water molecules and hydroxyl. Later, NASA's Deep Impact-EPOXI mission, which flew by the Moon in June 2009 also returned the same type of data.

2009

218

Kepler

Nation: USA (92)

Objective(s): solar orbit

Spacecraft: Kepler

Spacecraft Mass: 1,039 kg

Mission Design and Management: NASA / ARC / JPL

Launch Vehicle: Delta 7925-10L (no. D339)

Launch Date and Time: 7 March 2009 / 03:49:57 UT

Launch Site: Cape Canaveral Air Force Station / SLC-17B

Scientific Instruments:

1. photometer (Schmidt telescope)

Results: Kepler, the tenth in the series of low-cost, low-development-time, and highly-focused Discovery class science missions, is designed to discover Earth-like planets orbiting other stars in our region of the Milky Way. More specifically, Kepler has been equipped to look for planets whose size spans from one-half to twice the size of Earth ("terrestrial planets") in the habitable zone of their stars where liquid water might exist in the natural state on the surface of the planet. Its scientific goals include determining the abundance of these planets and the distribution of sizes and shapes of their orbits, estimating the number of planets in multiple-star systems, and determining the properties of stars that have planetary systems. Kepler detects planets by observing transits, tiny dips in the brightness of a star when a planet crosses in front of it. The spacecraft is basically a single instrument—in this case, a specially designed 0.95-meter diameter aperture telescope and image sensor array—with a spacecraft built around it. (The diameter of the telescope's mirror is 1.4 meters, one of the largest mirrors beyond Earth orbit). As originally planned, it was designed to monitor about 100,000 main-sequence stars over a period of three-and-a-half years. Kepler was initially launched into Earth orbit at 185 × 185 kilometers at 28.5° inclination. Subsequently, after another first stage burn, the second stage fired to set Kepler on an escape trajectory into solar orbit. It passed lunar orbit at 04:20 UT on 9 March, eventually entering heliocentric orbit at 0.97 × 1.041 AU at 0.5° inclination to the solar ecliptic. In order to improve resolution, on 23 April 2009, mission planners optimized the focus of the telescope by moving the primary mirror 40 micrometers toward the focal plane and tilting it by 0.0072°. Less than a month later, on 13 May, Kepler finished its commissioning and began its operational mission. Already during its first six weeks of operation, Kepler discovered five exoplanets (which were named Kepler 4b, 5b, 6b, 7b, and 8b), which NASA announced in January 2010. Later, in April 2010, mission scientists published results that showed that Kepler had discovered the first confirmed planetary system with more than one planet transiting the same star, Kepler-9. That discovery was the result of surveying more than 156,000 stars over a period of seven months. The planetary system orbiting Kepler-11, a yellow dwarf star about 2,000 light years from Earth, included six planets. NASA announced in February 2011 that these planets were larger than Earth, with the largest ones comparable in size to Uranus and Neptune. In 2011, Kepler suffered at least two "safe mode" events, when the spacecraft essentially shut down science operations as a result of a suspected anomaly. In both cases, in February and March, the Kepler project team were able to revive the vehicle relatively quickly, within two to three days. In September, mission scientists announced the discovery of a planet (Kepler-16b) orbiting two stars, where we might expect a double sunset, much like the fictional planet Tatooine depicted in the film *Star Wars*. (A subsequent double-star system was announced in January 2012 and multiple planets orbiting multiple stars—the Kepler-47 system—was announced in August 2012). Finally, in December 2011, NASA announced that Kepler had found its first planet (Kepler-22b) in the "habitable zone" of a star where liquid water could exist on the planet's surface. In April 2012, the mission, closing in on its three-and-a-half-year lifetime, was formally extended through fiscal year 2016 after a review of its operations, with the extended mission beginning on 15 November 2012. By that time, Kepler had identified more than 2,300 planet candidates and confirmed more than 100 planets. Based on data collected by Kepler, scientists were able to announce in January 2013 that about 17% of stars (about one-sixth) have an Earth-sized planet in an orbit closer than Mercury is to our Sun. Given that the Milky Way has about 100 billion stars, this would suggest at least 17 billion Earth-sized worlds in our galaxy. (In November 2013, this number was revised up to 40 billion). Following two brief lapses into "safe mode" in May, one of the spacecraft's four reaction wheels (no. 4) was found to have failed. Given that an earlier one failed in July 2012 and that at least three such wheels were needed to accurately aim the telescope, there was anxiety that the mission might be jeopardized. Subsequent to that point, and after another safe mode event in late May, Kepler operated in Point Rest State (PRS) mode—where the spacecraft used thrusters and solar pressure to control pointing—while controllers devised a way to reactivate the wheels necessary for accurate pointing of the spacecraft. After several months of activity, on 15 August 2013, NASA officially announced that it would be ending efforts to fully recover Kepler. NASA solicited proposals from the public on how to reformulate a new mission for Kepler given its obvious limitations. During this period, in October 2013, Kepler mission scientists announced that

they had conclusively identified the first Earth-sized rocky planet, Kepler-78b, which circles its host star every eight-and-a-half years, making it a very hot planet. A further announcement in April 2014 confirmed the discovery of the first Earth-sized planet (Kepler-186f) in the "habitable zone" of a star. At the end of the year, the Kepler team proposed a new mission, known as K2 ("Second Light"), using the two remaining reaction wheels to investigate smaller and dimmer red dwarf stars. Mission definition of the K2 proposal continued into 2014, with the mission finally approved by NASA in May 2014 and data collection beginning on 30 May. Observations continued through the year with several "campaigns" of data collection. As of January 2015, Kepler had found 1,004 confirmed exoplanets in about 400 star systems. By November 2016, Kepler, now in its K2 mission (which included "relaxed fault or sensitivity limits") was in its eleventh "campaign" of scientific observation, which began on 24 September. There was some concern on the eve of Campaign 9, slated to begin on 8 April, when controllers found the spacecraft in a "fuel-intensive coma," a kind of emergency mode much more serious than a "safe mode" and closer to complete systems failure. Fortunately, controllers were slowly able to fully revive the spacecraft by 22 April. On 9 June 2016, NASA announced that Kepler would continue science operations through to the end of Fiscal Year 2019 by which time, on-board propellant would probably be depleted. The spacecraft was named after the famed German astronomer Johannes Kepler (1571–1630).

219

Herschel

Nation: European Space Agency (6)
Objective(s): Sun–Earth L2 Lagrange Point
Spacecraft: Herschel
Spacecraft Mass: 3,400 kg

Mission Design and Management: ESA

Launch Vehicle: Ariane 5ECA (no. V188)

Launch Date and Time: 14 May 2009 / 13:12 UT

Launch Site: Kourou / ELA 3

Scientific Instruments:

1. infrared telescope
2. heterodyne instrument for the far infrared (HIFI)
3. photoconductor array camera and spectrometer (PACS)
4. spectral and photometric imaging receiver (SPIRE)

Results: Both Herschel and Planck were launched by the same Ariane launch vehicle and were both ESA missions (with significant NASA contributions) but they had different science missions. Herschel, the largest infrared telescope ever launched into space (3.5-meter mirror), was designed to study the origin and evolution of stars and galaxies, the chemical composition of atmospheres and surfaces of solar system bodies, and molecular chemistry across the universe, to help understand the evolution of the universe. Herschel's mirror, one-and-a-half times bigger than the one on Hubble, was made almost entirely of silicon carbide, 12 such segments being brazed together. Following launch, Ariane's ESC-A stage sent both Herschel and Planck into a highly elliptical transfer orbit of 270 × 1,197,080 kilometers at 6° inclination to enable the spacecraft to reach the Sun–Earth L2 Lagrange Point, the local gravitationally stable point that is fixed in the Sun–Earth System, about 1.5 million kilometers directly "behind" Earth as viewed from the Sun. Herschel did not have a dedicated engine for major course changes but used its own small thrusters for minor corrections. Herschel's operating lifetime was expected to be about three-and-a-half years, determined by the amount of coolant available for its instruments. In mid-July 2009, about two months after launch, Herschel entered a Lissajous orbit of 800,000 kilometers (average) radius around L2 and soon, on 21 July, began active operations. (Herschel's distance from Earth varied, depending on its orbital position, between 1.2 and 1.8 million kilometers). The observatory's operations were organized in 24-hour cycles where it communicated with ground control for 3 hours every day with the remainder of the time dedicated to scientific observations. Less than a year later, at a symposium to discuss the first results of Herschel, scientists reported a number of major findings: Herschel had found high-mass protostars around two ionized regions in the Milky Way, showing an early phase in the evolution of stars; the HIFI instrument (which had actually been inoperable due to a glitch between August 2009 and February 2010) had investigated the trail of water in the universe over a wide range of scales, from the solar system to extragalactic sources; and Herschel found a previously unresolved population of galaxies in the GOODS (Great Observatories Origins Deep Survey) fields identified by the Hubble, Spitzer, and Chandra spacecraft. In August 2011, Herschel mission scientists reported that they had identified molecular oxygen in the Orion molecular cloud complex, previously reported by the Swedish Odin satellite. Herschel data also suggested that much of Earth's water could have come from comets, results suggested by observations of Comet Hartley 2—although this notion has been dispelled since. Observations by Herschel, in fact, confirmed that Comet Shoemaker-Levy's impact into Jupiter in 1994 had actually delivered water to the gas giant. One of the oldest objects in the universe was located in 2013; scientists published results in April that showed the existence of a starburst galaxy which had produced over 2,000 solar masses of stars a year, originating only 880 million years after the Big Bang. On 29 April 2013, Herschel finally ran out of the liquid helium coolant required to maintain the operational temperature of the instrument detectors. On 13–14 May the spacecraft conducted a maneuver with its thrusters to boost it out of orbit around L2 and into its final resting place in heliocentric orbit. (A possible end on the lunar surface was also contemplated but not chosen because of cost). Following a last maneuver to deplete the propellant on board, at 12:25 UT on 17 June 2013, the final command to terminate communications was sent to Herschel, rendering the spacecraft dead. After a highly successful mission, the inert spacecraft will remain in heliocentric orbit. Scientists continued analyzing the vast amount of data returned from the observatory. For example, in January 2014, scientists announced that data from Herschel indicated the definitive detection of water vapor on the dwarf planet Ceres, a discovery which came prior to the arrival of NASA's Dawn mission at Ceres. This research was part of the so-called Measurements of 11 Asteroids and Comets Using Herschel (MACH-11) program. The observatory is named after British astronomers William Herschel (1738–1822) and his sister Caroline Herschel (1750–1848).

220

Planck

Nation: European Space Agency (7)
Objective(s): Sun–Earth L2 Lagrange Point
Spacecraft: Planck
Spacecraft Mass: 1,950 kg
Mission Design and Management:
Launch Vehicle: Ariane 5ECA (no. V188)
Launch Date and Time: 14 May 2009 / 13:12 UT
Launch Site: Kourou / ELA-3
Scientific Instruments:
1. low frequency instrument (LFI)
2. high frequency instrument (HFI)
3. telescope

Results: Both Herschel and Planck were launched by the same Ariane launch vehicle as ESA missions (although Herschel, especially, had significant NASA contributions) but they had different science objectives. Planck, named after German physicist Max Planck (1858–1947) was the first European space observatory whose primary objective was to study the Cosmic Microwave Background (CMB). The spacecraft used sensitive radio receivers operating at very low temperatures

to determine the black body equivalent temperature of the background radiation. Such measurements were then used to produce detailed maps of directional (anisotropic) temperature differences in the CMB radiation field, improving upon observations made by NASA's Wilkinson Microwave Anisotropy Probe (WMAP)—Planck had a higher resolution and sensitivity than WMAP. Besides mapping CMB anisotropies, Planck would also provide data to test inflationary models of the early universe, measure the amplitude of structures in the CMB, and perform measurements of the Sunyaev-Zeldovich effect (the distortion of CMB by high energy electrons through inverse Compton scattering). The Planck spacecraft was made up of two primary components, the payload and service modules. The former contained a telescope with primary and secondary mirrors that collected microwave radiation and directed it into the focal plane units. The octagonal service module (SVM) was common to both Herschel and Planck, both being cornerstone missions in ESA's science program (along with Rosetta and Gaia). Like Herschel, Planck was put into an elliptical orbit that eventually led to Sun–Earth L2 where it entered a Lissajous orbit with a 400,000-kilometer radius on 3 July 2009. ESA announced that Planck's High Frequency Instrument reached their low temperatures of −273.05°C, making them the "coolest" known objects in space. (This temperature is only 0.1°C above absolute zero, the coldest temperature theoretically possible in our universe.) Such low temperatures are necessary to study CMB, the "first light" released by the universe only 380,000 years after the Big Bang. This visible light gradually faded and moved to the microwave wavelengths due to the expansion of the universe. By studying, with Planck's two instruments, patterns imprinted in that light today, scientists sought to understand the Big Bang and the very early universe. The spacecraft began its first "all-sky" survey on 13 August working for two continuous weeks, generating excellent preliminary results within a month. On 15 January 2010, ESA extended the mission by 12 months (from its original end point of late 2011). In July 2010, ESA reported that Planck had returned its first all-sky image, "the moment that Planck was conceived for," as ESA Director of Science and Robotic Exploration David Southwood (1945–) noted. The image spanned the closest portions of the Milky Way to the "furthest reaches of space and time." As expected, the High Frequency Instrument's sensor ran out of coolant on 14 January 2012, concluding its ability to detect this faint energy, but not before fully completing its survey of the early universe, i.e., the remnant of light from soon after the Big Bang itself. Based on data from Planck collected over its initial work spanning 15.5 months, in March 2013, ESA released the most detailed map ever created of the CMB. The map suggested that the universe is slightly older than earlier thought; the data points to an age of 13.798±0.037 billion years. In August 2013, having completed its mission, Planck was "nudged" away from its L2 orbit towards a more stable orbit around the Sun. Through September and October, mission controllers prepared the spacecraft for shutdown by using up its remaining fuel. Finally, at 12:10:27 UT, on 23 October 2013, ESA sent the final command to shut down Planck, ending a highly successful mission.

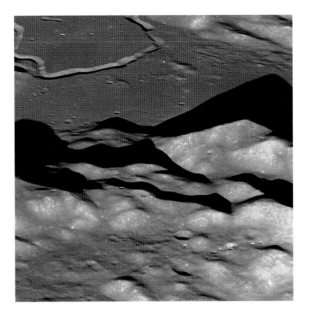

NASA's Lunar Reconnaissance Orbiter (LRO) captured this oblique view of the Moon, looking east-to-west over the Apennine Mountains towards Hadley Rille (upper left). Mount Hadley, at center right, casts a long shadow. The crew of Apollo 15 landed between the Apennines and Hadley Rille in 1971. *Credit: NASA/Arizona State University*

221

Lunar Reconnaissance Orbiter (LRO)

Nation: USA (93)
Objective(s): lunar orbit
Spacecraft: LRO
Spacecraft Mass: 1,850 kg
Mission Design and Management: NASA / GSFC
Launch Vehicle: Atlas V 401 (no. AV-020)
Launch Date and Time: 18 June 2009 / 21:32:00 UT
Launch Site: Cape Canaveral Air Force Station / SLC-41
Scientific Instruments:

1. cosmic ray telescope for the effects of radiation (CRaTER)
2. diviner lunar radiometer experiment (DLRE)
3. Lyman-Alpha mapping project (LAMP)
4. lunar exploration neutron detector (LEND)
5. lunar orbiter laser altimeter (LOLA)
6. lunar reconnaissance orbiter camera (LROC)
7. Mini-RF miniature radio frequency radar

Results: The Lunar Reconnaissance Orbiter (LRO) was part of NASA's now-cancelled Lunar Precursor Robotic Program (which also included LCROSS) and was the first U.S. mission to the Moon in over 10 years. LRO's primary goal was to make a 3D map of the Moon's surface from lunar polar orbit as part of a high-resolution mapping program to identify landing sites and potential resources, investigate the radiation environment, and prove new technologies in anticipation of future automated and human missions to the surface of the Moon. LRO was launched together with LCROSS; the Centaur upper stage boosted them both into high apogee orbits soon after launch. At 11:27 UT on 23 June 2009, LRO successfully entered orbit around the Moon, having fired its rocket motor

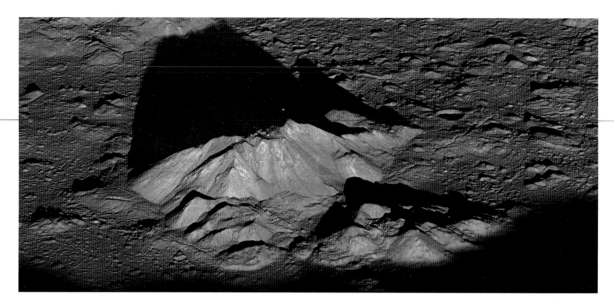

On June 10, 2011, Lunar Reconnaissance Orbiter (LRO) took this dramatic view of the Tycho crater. The summit of the central peak is 2 kilometers above the crater floor. *Credit: NASA/Arizona State University*

on the far side of the Moon. Initial orbital parameters were roughly 30 × 216 kilometers. A series of four engine firings over the next four days left LRO in its optimal orbit—roughly circular at 50 kilometers—allowing the satellite to begin its primary mission on 25 September 2009, expected to last one year and overseen by NASA's Exploration Systems Mission Directorate (ESMD). During this period, LRO gathered information on day-night temperature maps, contributed data for a global geodetic grid, and conducted high-resolution imaging. The spacecraft paid particular emphasis to the polar regions, where constant solar illumination might be possible and where there is the possibility of water in the permanently shadowed regions. In September 2010, LRO operations were handed over to NASA's Science Mission Directorate (SMD) to continue the science phase of the mission (rather than activities purely related to exploration and future missions) for another five years. Among LRO's achievements was to take extremely high-resolution photographs of landing sites of several older lunar landers and impact vehicles, such as landing sites from all the Apollo landing missions (plus Surveyor III near the Apollo 12 site) and the Apollo 13, 14, 15, and 17 Saturn IVB upper stages.

Other targets included the later Ranger impact probes (VI, VII, VIII, and IX), and the Soviet Luna 16, 17, 20, 23, and 24 soft-landers, and the Chinese Chang'e 3 lander/rover. In November 2011, NASA released the highest resolution near-topographical map of the Moon ever created, showing surface features over nearly the entire moon. An interactive mosaic of the lunar north pole was published in March 2014. LRO also carried out the first demonstration of laser communication with a lunar satellite when, in January 2013, NASA scientists beamed an image of the Mona Lisa from the Next Generation Satellite Laser Ranging (NGSLR) station at NASA's Goddard Space Flight Center in Greenbelt, Maryland. to the LOLA instrument on board LRO. One of the LRO instruments, the Mini-RF partially failed in January 2011, although fortunately, it had already completed its primary science objectives by that time. On 4 May 2015, controllers at Goddard Space Flight Center sent commands to LRO to fire its engines twice to change its orbit, taking it closer to the Moon than before—a polar orbit of about 20 × 165 kilometers. Perilune was close to the lunar South Pole. The new orbit allowed LRO's LOLA instrument to produce better return signals and also allow it

to better measure specific regions near the South Pole that have unique illumination conditions. One of the more interesting finds of the orbiter was its identification, in December 2015, of the hitherto unknown impact site of Apollo 16's S-IVB upper stage that was deliberately impacted on the lunar surface in 1972. In early 2017, LRO was still in excellent shape with propellant use limited to a few kilograms per year. Total remaining in early 2017 was about 30 kilograms. It is not unlikely that the spacecraft will remain operational in its low elliptical orbit—which was about 30 × 150 kilometers in late 2016—for several more years.

222

Lunar Crater Observation and Sensing Satellite (LCROSS)

Nation: USA (94)
Objective(s): lunar orbit
Spacecraft: S-S/C
Spacecraft Mass: 621 kg
Mission Design and Management: NASA / ARC
Launch Vehicle: Atlas V 401 (no. AV-020)
Launch Date and Time: 18 June 2009 / 21:32:00 UT
Launch Site: Cape Canaveral Air Force Station / SLC-41
Scientific Instruments:
1. visible camera
2. 2 near infrared cameras
3. 2 mid-infrared cameras
4. visible spectrometer
5. 2 near infrared spectrometers
6. total luminescence photometer (TLP)

Results: The mission of Lunar Crater Observation and Sensing Satellite (LCROSS), like LRO, part of NASA's now-cancelled Lunar Precursor Robotic Program, was to confirm the presence of water ice in a permanently shadowed crater at the Moon's south pole. Notably, both missions were originally funded as part of NASA's human spaceflight Constellation program to return humans to

the Moon. Ancillary goals for LCROSS included the testing of new modular subsystems for potential use in future mission architectures. Its mission profile involved impacting a Centaur upper stage on the surface of the Moon and then flying LCROSS through the debris plume about four minutes later to collect data on the soil, and then itself impact a little later. LCROSS was launched along with the Lunar Reconnaissance Orbiter (LRO) and traveled to the Moon as a "co-manifested" payload aboard the launch vehicle. The Centaur upper stage entered a 180 × 208-kilometer parking orbit before firing again at 22:15 UT on 18 June to reach a 194 × 353,700-kilometer orbit at 28.2° inclination. At that point, LRO separated from the Centaur-LCROSS combination, and the Centaur then vented some remaining propellant, which slightly altered its orbit to 133 × 348,640-kilometer orbit at 28.0° inclination, ensuring a lunar flyby. The combined Centaur-LCROSS then passed the Moon at a distance of 3,270 kilometers at 10:29 UT on 23 June and entered into an Earth polar orbit at approximately 357,000 × 582,000 kilometers at 45° inclination with an orbital period of 37 days. The combined stack reached apogees near the Moon on 10 July, 16 August, and 22 September until its trajectory intersected with that of the Moon on 9 October. A serious problem was discovered earlier, on 22 August, when mission controllers found that a sensor problem had caused the spacecraft burning through 140 kilograms of propellant, more than half of the amount remaining at the time. The loss meant that the original mission could still be accomplished but with very little margin. At 01:50 UT on 9 October, Centaur and LCROSS separated. The former then crashed onto the surface of the Moon at 11:31 UT in the Cabeus crater at the lunar South Pole. The impact excavated roughly 350 (metric) tons of lunar material and created a crater estimated to be about 20 meters in diameter. Four minutes later, LCROSS flew through the resulting debris plume that rose above the lunar surface, collecting data before it itself struck the Moon at 11:36 UT at a velocity of 9,000 kilometers/hour. The LRO's Diviner instrument obtained infrared observations of the LCROSS impact point as it flew by about 90 seconds after the impact at a range of about 80 kilometers. On 13 November 2009, NASA formally announced that data from LCROSS "indicates that the mission successfully uncovered water…near the Moon's south pole." Nearly a year later, on 21 October 2010, mission scientists announced new data, including evidence that the lunar soil within Cabeus was rich in useful materials (mercury, magnesium, calcium, silver, sodium) and that the Moon is chemically active and has a water cycle. They also confirmed that in some places the water on the south pole is in the form of pure ice crystals.

2010

223

Venus Climate Orbiter (VCO) / Akatsuki

Nation: Japan (7)
Objective(s): Venus orbit
Spacecraft: PLANET-C
Spacecraft Mass: 517.6 kg
Mission Design and Management: JAXA

Launch Vehicle: H-IIA 202 (no. 17)
Launch Date and Time: 20 May 2010 / 21:58:22 UT
Launch Site: Tanegashima / Area Y1
Scientific Instruments:

1. 1-micron camera (IR1)
2. 2-micron camera (IR2)
3. ultraviolet imager (UVI)
4. longwave infrared camera (LIR)
5. lightning and airglow camera (LAC)
6. ultra-stable oscillator (USO)
7. radio science experiment (RS)

Artist's impression of Japan's Venus Climate Orbiter. *Credit: JAXA/Akihiro Ikeshita*

Results: This was Japan's first interplanetary mission after Nozomi (launched in 1998), that spacecraft having failed to enter orbit around Mars. The goal of Venus Climate Orbiter, or Akatsuki ("dawn") as it was known after launch, was to investigate atmospheric circulation in Venus by globally mapping clouds and minor constituents with four cameras at ultraviolet and infrared wavelengths, detect lightning, and observe the vertical structure of the atmosphere. A nominal mission would last two or more years in an elliptical orbit around Venus at 300 × 80,000 kilometers. The scientific goals of VCO were closely related to those of ESA's Venus Express launched in 2005. VCO was launched with a fleet of satellites, including three Japanese cubesats (KSAT, Negai*, and Waseda-Sat 2). The H-IIA rocket's second stage restarted in Earth orbit to accelerate three other spacecraft to escape velocity: Akatsuki, UNITEC 1, and IKAROS. On 21 May 2010, Akatsuki conducted a mid-course correction to sharpen its trajectory to Venus orbit. This was the first time that a ceramic thruster (with 51 kgf thrust), made of silicon nitride, was used in space conditions. The spacecraft was supposed to enter Venus orbit by firing its orbital maneuvering engines at 23:40 UT on 6 December 2010. Although the engine fired on time, it apparently cut off early. When communications were reestablished with Akatsuki (after a planned blackout due to occultation), the spacecraft was found to be in "safe mode" and spin-stabilized, slowly rotating once every 10 minutes. A later investigation (whose results were issued on 30 June 2011) showed that the engine had fired for less than 3 minutes (instead of 12 minutes), thus providing insufficient delta-V to enter Venusian orbit. Apparently, salt deposits jammed a valve that delivered fuel to the combustion chamber, making the combustion oxidizer-rich. The thruster nozzle was probably damaged in the firing. Soon, JAXA scientists began devising a backup plan to enter Venusian orbit in November 2015, which would be possible if subsequent tests of thruster were successful. On 7 and 14 September 2011, test firings

were carried out of the beleaguered engine which showed that the thrust was about one-ninth of normal (about 4.1 kg). As a result, a new plan was formulated to discard all the unused oxidizer (65 kilograms), lightening the spacecraft, and using the smaller reaction control system (RCS) thrusters to attempt Venus orbital insertion on 22 November 2015. Three firings of the RCS system (587.5, 544, and 342 seconds in length) in November 2011 were successful in altering the trajectory of VCO for rendezvous with Venus in either 2015 (preferable in terms of spacecraft lifetime) or 2016 (preferable in terms of the original science mission). The main problem at this time was VCO's exposure to incredibly high temperatures as it sped through perihelion—six times by April 2014—conditions it was not designed to survive, described by JAXA as "three times hotter than that of Earth." Yet telemetry showed that the spacecraft was still functioning. In January 2015, the planned Venus orbit insertion was shifted to early December 2015 with a possible apogee of 300,000 to 400,000 kilometers. As per the new plan, Akatsuki successfully entered orbit around Venus on 7 December 2015 by using its attitude control thrusters for 20 minutes, following earlier spacecraft from the former Soviet Union, the United States, and the European Space Agency. Initial orbital parameters were approximately 440,000 × 400 kilometers at 3° inclination and an orbital period of 13 days 14 hours, fortuitously much lower than mission planners had hoped to achieve in original pre-launch planning. (Originally, the plan was to have an apoapsis of 79,000 kilometers and an orbital period of 30 hours, partly to match the flow of Venusian winds for part of the spacecraft's orbit.) The new orbit had one drawback—it meant that Akatsuki would be in Venus' shadow for part of each day, longer than the 90-minute limit set for the spacecraft. Remarkably, despite the stress from overheating, three of the five cameras on Akatsuki remained in perfect condition and sent back impressive imagery of the planet. On 26 March 2015, the vehicle lowered the high point of its orbit to about 330,000 kilometers, shortening its orbital period to about nine days. After several months of checkout, on 28 April 2016, the spacecraft finally began its standard science mission. In March 2017, JAXA announced that two of the cameras, the 1-μm and 2-μm cameras) on Akatsuki had "pause[d] scientific observations" as of 9 December 2016 due to an electrical problem. Perhaps the most significant discovery of Akatsuki was the detection, in December 2015, of a bow-shaped feature in the Venusian atmosphere stretching about 9,600 kilometers from almost pole to pole, a feature which some have called a "sideways smile." In a paper published in January 2017 in *Nature Geoscience*, scientists suggested the "smile" was the result of a gravity wave, a kind of disturbance in Venusian winds that propagates upwards as a result of the topography on the surface. In September 2017, Japanese scientists announced, based on Akatsuki data, the discovery of an equatorial jet in the Venusian atmosphere. Like a number of other deep space vehicles, Akatsuki carried the names of people (260,214 names) printed in fine letters on an aluminum plate.

224

Shin'en

Nation: Japan (8)
Objective(s): Venus flyby
Spacecraft: UNITEC 1
Spacecraft Mass: 20 kg
Mission Design and Management: UNISEC
Launch Vehicle: H-IIA 202 (no. 17)
Launch Date and Time: 20 May 2010 / 21:58:22 UT
Launch Site: Tanegashima / Area Y1
Scientific Instruments:

1. radiation counter
2. camera

Results: The UNISEC Technology Experiment Carrier 1 (UNITEC 1) was a Japanese student-built spacecraft designed for a flyby of Venus to study the effects of deep space travel on computers.

Intended to be the first student spacecraft to operate beyond geocentric orbit, it was built and operated by UNISEC (University Space Engineering Consortium), a collaborative program involving 20 Japanese universities developing nano-satellites. Known as Shin'en ("abyss") after launch, the satellite carried six computers, a camera, a radiation counter, a low-power communications system, and solar cells for power. The launch and transplanetary injection occurred without incident (see Venus Climate Orbiter). Shin'en was the last spacecraft to separate from the multi-satellite stack and enter heliocentric orbit. The plan was for it to fly past Venus in December 2010. Although signals from the spacecraft were received on the first day after launch, there was no further communication after 15:43 UT on 21 May 2010 when Shin'en was 320,000 kilometers from Earth. A beacon was tracked until 31 May when it also stopped. The inert spacecraft flew past Venus sometime in December 2010.

225

IKAROS

Nation: Japan (9)
Objective(s): Venus flyby
Spacecraft: IKAROS
Spacecraft Mass: 310 kg
Mission Design and Management: JAXA
Launch Vehicle: H-IIA 202 (no. 17)
Launch Date and Time: 20 May 2010 / 21:58:22 UT
Launch Site: Tanegashima / Area Y1
Scientific Instruments:

1. instrument to measure variation in dust density (ALADDIN)
2. gamma-ray burst polarimeter (GAP)

Results: One of the most unique and innovative missions in the history of deep space exploration, IKAROS (Interplanetary Kite-craft Accelerated by Radiation of the Sun) was the world's first spacecraft to use solar sailing as the main propulsion.

The mission was designed to demonstrate the deployment of a large (20 m diameter), square-shaped, and thin (7.5 micron) solar sail membrane, which had integrated into it thin-film solar cells to generate power for the main payload. It had instruments to measure the acceleration generated by radiation pressure. While radiation pressure constituted the main form of propulsion, the spacecraft used variable reflectance LCD panels (80 of them) for smaller attitude control movements. The sail also carried the ALLADIN (Arrayed Large-Area Dust Detectors in Interplanetary Space) instrument, which was a network of eight "channels" or separate (polyvinylidene difluoride) PVDF sensors attached to the sail. Collectively they had a detection area of 0.54 m², making it, according to JAXA, "the world's largest dust detector" ever sent into space. ALLADIN's goal was to count and time hyper-velocity impacts by micrometeoroids larger than a micron in size during its 6-month voyage to the vicinity of Venus. Launched along with the Venus Climate Orbiter, IKAROS was sent on a trajectory to Venus with its co-payloads, VCO and Shin'en. After separation from the launch vehicle, IKAROS, shaped like a drum, was spun up to 5 rpms, then spun down to 2 rpm to deploy four "tip mass" objects that uncoiled from the drum. At this point, on 3 June 2010, the "quasi-static" stage of the sail deployment began. The spacecraft spun up to 25 rpm and the membrane gradually unfurled through centrifugal force, slowing the central drum down to about 5–6 rpm once the four booms had reached to 15-meter-diameter lengths by 8 June. The second stage, the "dynamic" phase of the deployment began at that point when a stopper was dislodged at the center, releasing in 5 seconds the entire membrane (which after deployment, took about 100 seconds to stop vibrating). By 10 June, the full membrane was deployed with the spacecraft spinning at 1–2 rpms. At this point, IKAROS was 7.7 million kilometers from Earth and actively generating power from its thin-film solar cells. Two small separation cameras (DCAM2 and DCAM1) were deployed on 15 and 19 June, respectively, to take pictures of the solar sail. On 9 July, JAXA announced that IKAROS was, indeed, being accelerated by the solar sail (at a value 0.114 grams). On 13 July, the spacecraft successfully implemented attitude control using the LCDs on its solar panel, another first in the history of space exploration. IKAROS flew by Venus on 8 December 2010 at a range of 80,800 kilometers, essentially completing its originally planned mission, which JAXA extended into the indefinite future. Slowly, attitude control degraded on IKAROS, making it more difficult to communicate with. In March 2013, the IKAROS project team was disbanded but the project was reactivated three months later, on 20 June 2013 and telemetry was received from that point until 12 September 2013. There was intermittent detection of transmissions in 2014, punctuated by long periods of hibernation. In mid-March 2015, IKAROS appeared to wake up from hibernation. On 23 April 2015, ground controllers found the spacecraft about 120 million kilometers from Earth. However, after last data reception on 21 May 2015, IKAROS entered hibernation mode again, for the fifth time, when the spacecraft was about 110 million kilometers from Earth.

226

Chang'e 2

Nation: China (2)

Objective(s): lunar orbit, Sun–Earth L2, asteroid flyby

Spacecraft: Chang'e erhao

Spacecraft Mass: 2,480 kg

Mission Design and Management: CNSA

Launch Vehicle: Chang Zheng 3C

Launch Date and Time: 1 October 2010 / 10:59:57 UT

Launch Site: Xichang / LC2

Scientific Instruments:

1. CCD camera (TDI)
2. descent camera
3. directional antenna surveillance camera
4. solar wing surveillance camera

5. engine surveillance camera
6. gamma-ray spectrometer
7. x-ray spectrometer
8. laser altimeter

Results: The original mission of Chang'e 2 (or CE-2) was as backup to Chang'e 1 (CE-1), to basically repeat that mission with an improved suite of instruments. After Chang'e 1's highly successful mission, additional tasks were attached to CE-2, such that this mission essentially became a pathfinder mission to Chang'e 3 (CE-3), a landing mission. Unlike CE-1, CE-2 was launched on a more ambitious direct translunar trajectory (at 212 × 356,996 kilometers at 28.5° inclination), which required the more powerful Chang Zheng 3C launch vehicle. A midcourse correction on 2 October 2010 was so accurate that further adjustments were unnecessary on the way to the Moon. The spacecraft successfully entered lunar orbit after 4 days and 16 hours of flight (as opposed to 12 days for CE-1) at an orbit of 120 × 80,000 kilometers. Three adjustments followed on 7, 8, and 9 October that resulted in CE-2 being in its operational circular orbit at 100 kilometers. All of its instruments, activated during the coast to the Moon, continued operations in lunar orbit without problems. On 26 October at 13:27 UT, CE-2 fired its four 1 kgf thrusters for over 18 minutes to bring down perilune to 15 kilometers so that the spacecraft could photograph the planned landing site of CE-3 in Sinus Iridium. It returned to its nominal orbit two days later after obtaining a large number of high resolution images of the surface, some down to 1.2 to 1.5 meters resolution. In February 2012, Chinese authorities released a full map of the Moon at 7 meters resolution, claimed at the time as the highest resolution map of the Moon. Of all recent probes to the Moon, only the photographs from Lunar Reconnaissance Orbiter (LRO) had higher resolution. CE-2's main lunar orbital mission concluded on 1 April 2012 but because the spacecraft still had a relatively large amount of maneuvering propellant still left, mission planners decided in early 2013 to formulate

an extended mission, one that would culminate with a flight to the Sun–Earth L2 Lagrange Point. By 23 May 2013, CE-2 completed a second survey of Sinus Iridium from a low altitude and filled in some surface details where earlier data was of relatively low detail. On 8 June 2011, CE-2 raised its apolune to 3,583 kilometers. The next day, CE-2's main engine (50 kgf thrust) fired for 18 minutes, boosting the spacecraft out of lunar orbit. Finally, on 25 August 2011 at 15:24 UT, the 1 kgf thrusters fired to place CE-2 in a 180-day period Lissajous orbit around L2, about 1.5 million kilometers from Earth. China thus became only the third country or entity (after the United States and ESA) to send a spacecraft to a Lagrange Point. There, Chang'e 2 studied charged particles near Earth's magnetic tail and observed possible x-ray and gamma-ray bursts from the Sun. The spacecraft departed L2 on 15 April 2012 (although this was not announced by the Chinese until 14 June) and headed for a flyby encounter with the asteroid 4179 Toutatis, about 7 million kilometers from Earth. On 13 December 2012 at 08:30:09 UT, Chang'e 2 flew by Toutatis at a distance of just 1.9 kilometers (much better than the 30 kilometers hoped), making China the fourth nation or entity after the U.S., ESA, and Japan to perform an asteroid flyby. The encounter occurred at a relative velocity of 10.73 kilometers/second, giving very little time (about a minute) for useful imaging but some excellent pictures were returned. After the Toutatis encounter, CE 2 remains in heliocentric orbit, with Chinese controllers maintaining contact in 2014 when the probe was as far as 100 million kilometers from Earth. On 23 October 2016, chief scientist for China's Lunar Exploration Project, Ouyang Ziyuan, announced that Chang'e 2 had "fulfilled its mission," that it remained in heliocentric orbit (apparently as "the smallest man-made asteroid in the solar system," which was not true as some of NASA's early Pioneers were smaller), and that the spacecraft would be returning "somewhere closer to the earth around 2029." There was no word on whether the Chinese still maintained any contact with the spacecraft.

227

Juno

Nation: USA (95)

Objective(s): Jupiter orbit

Spacecraft: Juno

Spacecraft Mass: 3,625 kg

Mission Design and Management: NASA / JPL

Launch Vehicle: Atlas V 551 (AV-029)

Launch Date and Time: 5 August 2011 / 16:25:00 UT

Launch Site: Cape Canaveral Air Force Station / SLC-41

Scientific Instruments:

1. gravity science system (GS)
2. microwave radiometer (MWR)
3. vector magnetometer (MAG)
4. JADE and JEDI plasma and energetic particle detectors
5. Waves radio/plasma wave sensor
6. ultraviolet imager/spectrometer (UVS)
7. infrared imager/spectrometer (JIRAM)
8. JunoCam

Results: Juno, NASA's second New Frontiers mission (after New Horizons), was designed to study Jupiter from polar orbit around the gas giant. Its specific science goals include studying the planet's composition, gravity field, magnetic field, and polar magnetosphere, as well investigating the nature of the planet's core, the amount of water in its atmosphere, mass distribution, and the winds in its clouds. Juno is the first mission to Jupiter not to use radioisotope thermoelectric generators (RTGs) for power and relies on three giant solar arrays symmetrically arranged around the spacecraft that provide 450 watts of power in orbit around Jupiter. At launch, the optimal mission was planned to last about 14 Earth months in Jovian orbit. Juno entered parking orbit around Earth (at 194 × 226 kilometers at 28.8° inclination) and then a hyperbolic escape orbit less than 45 minutes after launch. As it headed outwards toward the asteroid belt, on 1 February 2012, Juno carried out its first mid-course correction, a firing lasting 25 minutes. Further course corrections were implemented on 30 August and 14 September 2012. The spacecraft passed the halfway point to Jupiter at 12:25 UT on 12 August 2013. Juno returned to the vicinity of Earth, using a gravity assist maneuver to pick up more speed, on 8 October 2013. Several of its instruments were activated as preliminary tests to see if they were in working condition. Twice after the flyby, Juno entered "safe mode" and shut down all inessential systems but ground controllers were quickly able to return the spacecraft to normal operating mode. On 13 January 2016, Juno broke the record for the furthest distance from the sun—793 million kilometers—where a spacecraft has been powered by solar power. The record had previously been held by ESA's Rosetta space probe and set in October 2012. After a mid-course correction on 3 February, on 27 May, Juno crossed from the primary gravitational influence of the Sun to that of Jupiter. A month later, on 30 June, Juno entered Jupiter's magnetosphere. Finally, after an almost five-year trip, Juno fired its Leros-1b engine for 35 minutes and 1 second (from 03:18 to 03:53 UT Earth receive time) and entered an elliptical and polar orbit—known as a "capture orbit"—around Jupiter of 8.1 million × 4,200 kilometers with an orbital period of 53.5 days. Five of its scientific instruments were powered up on 6 July with the rest activated by the end of the month. According to the original plan, Juno's dedicated science mission

This stunning view of Jupiter was acquired by NASA's Juno spacecraft on 19 May 2017 using its JunoCam from an altitude of 12,857 kilometers. Although the small bright clouds that dot the entire south tropical zone appear tiny, they are actually towers approximately 50 kilometers wide and 50 kilometers high that cast shadows below. *Credit: NASA/SWRI/MSSS/Gerald Eichstadt/Sean Doran*

was designed to begin only after two complete capture orbits, nearly two months after orbital insertion, and after a minor orbital correction on 13 July. In its first capture orbit, Juno made its first pass through perijove (lowest point) on 27 August when it approached the gas giant at a range of only 4,200 kilometers at 13:44 UT. This would be the closest the spacecraft would get to the planet during its primary mission, and it was the first time that its entire suite of scientific instruments was activated to study the planet as it flew by at a velocity of 208,000 kilometers/hour. Just prior to the perijove flyby on 19 October, at about 06:47 UT, Juno suddenly entered "safe mode" when a software performance monitor caused a reboot of the spacecraft's main computer. The switch to safe mode necessitated a postponement of exit from capture orbit to a new orbit with a period of 14 days; the burn was postponed to allow engineers to verify the performance of two helium check valves in the spacecraft's fuel pressurization system (associated with its main engine). On 24 October, the spacecraft exited safe mode and carried out a 31-minute burn using its smaller thrusters to slightly adjust its orbit. Juno successfully carried out its third flyby of Jupiter on 11 December at 17:04 UT, at a range of 4,150 kilometers, this time focusing on investigations on Jupiter's interior structure via its gravity field. (Only the JIRAM instrument was inactive during the flyby). During the flyby, Juno's JunoCam (a visible-light camera) captured spectacular images of one of Jupiter's so-called eight "string of pearls"—large counterclockwise rotating storms in the planet's southern hemisphere. For the fourth close flyby of Jupiter, at 12:57 UT on 2 February, NASA opened up an online vote for the public to choose in selecting what pictures should be taken. Juno's perijove for this encounter was 4,300 kilometers, while it was traveling at 57.8 kilometers/second. About two weeks later on 17 February 2017, NASA announced that Juno would remain in its 53-day capture orbit for the remainder of its primary mission and still be able to accomplish its science goals. Mission planners took this decision to avoid firing Juno's main engine due to concerns about the valves on the engine that did not operate as expected the previous October. There was concern that firing the engine would put the vehicle in a "less-than-desirable" orbit. In its capture orbit, scientists believe that Juno can still complete its primary science mission in 12 orbits, performed through 2018. If the spacecraft is operational at that time, there will remain the option of an extended mission. The orbiter performed its fifth close pass of Jupiter on 18 May 2017, swooping down to about 3,500 kilometers above the planet's clouds. All science instruments operated as planned. An even more spectacular flyby, its sixth, occurred on 10 July, when Juno flew over the Great Red Spot at an altitude of 9,000 kilometers, just 11 minutes and 33 seconds after reaching perijove (at 3,500 kilometers). The spacecraft remained in fine operation in the same orbit through its seventh (on 1 September) and eighth (on 24 October) pass to perijove. As part of an educational program with the LEGO group, Juno carries three aluminum figurines into outer space, each about 3.8 cm large, that of Galileo Galilei, the Roman god Jupiter, and his wife Juno. There is also a plaque, provided by the Italian Space Agency, dedicated to Galileo and including a facsimile of handwritten text by him. In June 2016, musicians Trent Reznor (of Nine Inch Nails) and Atticus Ross shared a nearly 9-minute piece of music to celebrate the Juno mission. The track originally scored NASA's short film about Juno entitled "Visions of Harmony."

228

Ebb and Flow

Nation: USA (96)
Objective(s): lunar orbit
Spacecraft: GRAIL-A / GRAIL-B
Spacecraft Mass: 202.4 kg (each)
Mission Design and Management: NASA / JPL
Launch Vehicle: Delta 7920H (no. D356)

Launch Date and Time: 10 September 2011 / 13:08:52 UT
Launch Site: Cape Canaveral Air Force Station / SLC-17B
Scientific Instruments:
1. lunar gravity ranging system (LGRS)
2. MoonKAM (Moon Knowledge Acquired by Middle school student) lunar-imaging system
3. radio science beacon (RSB)

Results: Gravity Recovery and Interior Laboratory (GRAIL), the eleventh of NASA's Discovery Program, was a dual-spacecraft mission that involved placing two identical spacecraft in orbit around the Moon to use high-quality gravitational field mapping to determine its internal structure. As the two spacecraft flew over areas of greater and greater gravity, the probes moved slightly toward and away from each other, while an instrument measured changes in their relative velocity, providing key information on the Moon's gravitational field. The nominal mission was planned to be three months. (The process used was very similar to the one employed by NASA's Gravity Recovery and Climate Experiment or GRACE since 2001, which used a similar instrument to GRAIL's LGRS). The spacecraft itself is a design evolution of the U.S. Air Force's Experimental Satellite System-11 (XSS-11) launched in 2005 while the avionics were derived from NASA's Mars Reconnaissance Orbiter (MRO). The names "Ebb" and "Flow" were given to GRAIL-A and GRAIL-B, respectively after a national contest (won by the fourth grade students at Emily Dickinson Elementary School in Bozeman, Montana). GRAIL made use of a low-energy translunar cruise that involved passing near the Sun–Earth L1 Lagrange Point and then heading for a rendezvous with the Moon. The two spacecraft arrived in lunar orbit about 25 hours apart, on 31 December 2011 (Ebb) and 2 January 2012 (Flow). The primary science phase of the two lunar satellites extended from 7 March to 29 May 2012. A second science phase, as part of the extended mission, was initiated on 8 August

2012. Both spacecraft were then decommissioned and powered down in anticipation of deliberate impact onto the lunar surface which occurred on 17 December 2012. Both Ebb and Flow impacted at 75.62° N / 26.63° W, crashing into the ground at 1.68 kilometers/second. The outcome of the mission was a gravitational map of the Moon unprecedented in its detail. In addition, the MoonKAM (Moon Knowledge Acquired by Middle school students), a digital video imaging system, was used as part of the education and public outreach activities of GRAIL. Each MoonKAM consisted of a digital video controller and four camera heads, one pointed slightly forward, two pointed below, and one pointed slightly backward. During the Ebb and Flow missions, MoonKam was operated by undergraduate students at the University of California–San Diego under the supervision of faculty, as well as Sally Ride Science, the foundation organized by America's first woman astronaut, Sally K. Ride (1951–2012) to encourage young people to enter careers in science, technology, engineering, and mathematics.

229

Fobos-Grunt

Nation: Russia (107)
Objective(s): Mars orbit, Phobos flyby, landing, soil sample return
Spacecraft: Fobos-Grunt
Spacecraft Mass: 13,505 kg (including 115 kg for Yinghuo-1)
Mission Design and Management: NPO imeni Lavochkina / IKI RAN
Launch Vehicle: Zenit-2SB41.1
Launch Date and Time: 8 November 2011 / 20:16:03 UT
Launch Site: Baikonur Cosmodrome / Site 45/1
Scientific Instruments:
1. gas chromatograph package (TDA analyzer + KhMS-1F chromatograph + MAL-1F mass spectrometer)
2. gamma-ray spectrometer (FOGS)
3. neutron spectrometer (KhEND)
4. Mössbauer spectrometer (MIMOS)
5. laser mass spectrometer (LAZMA)
6. mass spectrometer for secondary ions (MANAGA)
7. Fourier spectrometer (AOST)
8. gravimeter (GRAS-1)
9. seismometer (SEYSMO)
10. thermo-detector (TERMOFOB)
11. long-wave radar (DPR)
12. micrometeoroid detector (METEOR)
13. plasma complex (FPMS)
14. ultra-stable oscillator (USO)
15. optical sun star sensor (LIBRATSIYA)
16. microscope (MikroOmega)
17. dosimeter (LYULIN)
18. TV system (TSNN)
19. stereo and panoramic TV cameras

Results: Fobos-Grunt was a highly ambitious mission to the Martian system that had the goal of returning a sample (about 200 g) from Phobos. It was also the first Russian deep space/interplanetary mission since the failed mission of Mars 8 in 1996, 15 years earlier. Besides the main Russian spacecraft (which included significant scientific contributions from ESA and several European countries), Fobos-Grunt also carried a small passenger payload, the Chinese Yinghuo-1 Mars orbiter. The scientific goals of the mission included studying the physical and chemical characteristics of Phobos soil, the environment around Phobos, the seasonal and climatic variations of the Martian atmosphere and surface, and testing out several new technologies. The spacecraft was divided into three parts: a Flight Stage (PM), a Return Vehicle (VA)—which included a Reentry Vehicle (SA)—and the Main Propulsion Unit (MDU). After a voyage lasting about 10 months, Fobos-Grunt would have entered a highly elliptical Martian orbit with an 80,000-kilometer apogee, sometime in August/September 2012. At this point, the MDU would have separated from the main spacecraft, along with the truss structure

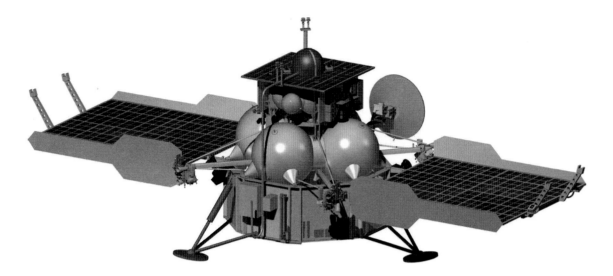

A simplified illustration of the Russian Fobos-Grunt lander package. The half-spherical object at the top was the descent capsule that would carry samples from the Martian moon Phobos back to Earth. *Credit: ESA/Lavochkin Association*

and the Chinese payload. These three objects would have remained in a 900 × 77,000 equatorial kilometer orbit while the main Fobos-Grunt spacecraft (with the Flight Stage and Return Vehicle) would have made two major burns to move into a 9,900-kilometer circular orbit. Following several more months in Mars orbit, the probe would have rendezvoused with Phobos, and then after a few more months, landed on its surface sometime in January 2013. A robotic arm would have performed 15–20 scoops totaling about 85–156 grams of soil, which would have been loaded into the Reentry Vehicle. The 287-kilogram Return Vehicle would then have taken off in April 2013 and headed back to Earth while the lander would have continued its surface experiment program for about a year. On approach to Earth in August 2014, the Return Vehicle would have released the 7-kilogram Reentry Vehicle, which would have reentered Earth's atmosphere, performing a hard landing (without a parachute) in the Sary Shagan test range in Kazakhstan. None of this happened. Within 2.5 hours after launch, at 22:55:48 UT, the MDU propulsion stage (derived from Fregat) was

supposed to fire to insert the payload into an elliptical orbit. A subsequent burn would then send the probe towards Mars. At the time that the first burn would have finished, ground tracking stations were, however, unable to find Fobos-Grunt. Later, at 20:25 UT on 22 November, a tracking station belonging to ESA received a signal from the probe (after a command had been sent to turn on one of the transmitters on the spacecraft). After a further brief communications session (with ESA) the next day and one with Russian stations on 24 November, no further contact was established with Fobos-Grunt. ESA gave up attempts to contact the probe on 2 December. American space assets tracked the probe in a 209 × 305-kilometer orbit in early December. The spacecraft and upper stage combination made an uncontrolled reentry on 15 January 2012, with wreckage apparently falling into the Pacific or parts of South America. An investigation later showed the probable cause of failure was a programming error that caused the simultaneous reboot of two channels of the onboard computer (TsVM22). The MDU never actually fired.

230

Yinghuo-1

Nation: China (3)

Objective(s): Mars orbit

Spacecraft: Yinghuo-1

Spacecraft Mass: 113 kg

Mission Design and Management: CNSA

Launch Vehicle: Zenit-2SB41.1

Launch Date and Time: 8 November 2011 / 20:16:03 UT

Launch Site: Baikonur Cosmodrome / Site 45/1

Scientific Instruments:

1. plasma package (including electron analyzer, ion analyzer, mass spectrometer)
2. fluxgate magnetometer
3. radio-occultation sounder
4. optical imaging system (2 cameras)

Results: This was a passenger payload on board the Russian Fobos-Grunt spacecraft, and a first attempt by the Chinese to test out an initial mission in Mars orbit. Nominal orbital parameters were planned to be highly elliptical—the low point would vary from 400 to 1,000 kilometers, while the high point would stretch out to 74,000 to 80,000 kilometers. This would be an equatorial orbit (5° inclination) with an orbital period of 72.8 hours. The spacecraft had a design lifetime of about one year that it would use to study the surface, atmosphere, ionosphere, and magnetic field of Mars. The payload was carried on a Russian spacecraft as a result of an agreement signed by the Chinese and the Russians in March 2007. The plan was for Yinghuo-1 to be released from Fobos-Grunt once the latter entered Martian orbit. As described in the entry for Fobos-Grunt, although launched successfully, the Russian Martian probe never left Earth orbit. As such, there was no way for the Chinese to test the spacecraft in Mars orbit. Both Fobos-Grunt and Yinghuo-1 reentered Earth's atmosphere on 15 January 2012.

231

Curiosity

Nation: USA (97)

Objective(s): Mars landing and rover

Spacecraft: MSL

Spacecraft Mass: 3,893 kg

Mission Design and Management: NASA / JPL

Launch Vehicle: Atlas V 541 (AV-028 + Centaur)

Launch Date and Time: 26 November 2011 / 15:02:00 UT

Launch Site: Cape Canaveral Air Force Station / SLC-41

Scientific Instruments:

1. mast camera (Mastcam)
2. Mars hand lens imager (MAHLI)
3. Mars descent imager (MARDI)
4. alpha x-ray spectrometer (APXS)
5. chemistry and camera (ChemCam)
6. chemistry and mineralogy x-ray diffraction/x-ray fluorescence instrument (CheMin)
7. sample analysis at Mars instrument suite (SAM)
8. radiation assessment detector (RAD)
9. dynamic albedo of neutrons (DAN)
10. rover environment monitoring station (REMS)
11. Mars science laboratory entry descent and landing instrument (MEDLI)

Results: The Mars Science Laboratory (MSL), part of NASA's Mars Exploration Program, consists of a large (899 kilogram) rover called Curiosity and a "sky crane" descent stage to bring the rover down to the Martian surface. Both were fixed inside an aeroshell for entry into the Martian atmosphere. The principal goal of the mission is to assess whether Mars ever had an environment hospitable for lifeforms such as microbes. To do this, Curiosity carries the most advanced complement of instruments ever sent to the surface of Mars. The rover is designed to scoop up soil and rocks and investigate their formation, structure, and chemical composition in order to look for the chemical building blocks of life (carbon, hydrogen, nitrogen, oxygen, phosphorus, and sulfur). Curiosity is able to travel up to 90 meters per hour on its six-wheeled rocker-bogie system, and is powered by a radioisotope power system to generate power from the heat of plutonium's radioactive decay, which ensured an operating lifetime of at least a full Martian year (687 Earth days). Curiosity operations depend heavily on the orbital communications capabilities of 2001 Mars Odyssey and Mars Reconnaissance Orbiter. The robot was named "Curiosity" after a nation-

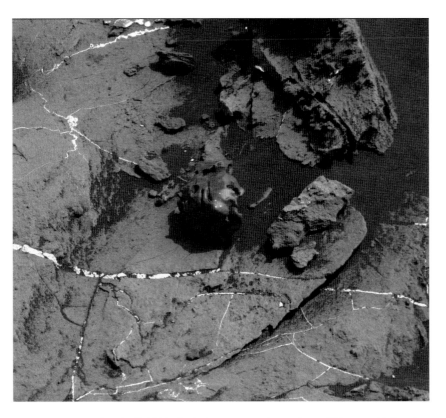

This image was taken by NASA's Curiosity rover on the surface of Mars on 30 October 2016. Taken by the Mast Camera (Mastcam), the photo shows a smooth-surfaced object about the size of a golf ball which was informally named "Egg Rock." The grid of shiny points visible in the object resulted from laser pulses produced by Curiosity's Chemistry and Camera (ChemCam) instrument. *Credit: NASA/JPL-Caltech/MSSS*

wide student contest involving more than 9,000 entries. The winning entry was submitted by Clara Ma, a sixth-grade student from Sunflower Elementary School in Lenexa, Kansas. Curiosity's expense, estimated at more than $1 billion, as well as the two-year delay in its launch, led to the institution of stronger requirements to maintain baseline cost and baseline schedules for future missions. After launch by an Atlas V 541 (powered by a Russian RD-180 engine), MSL was delivered into a 165 × 324-kilometer orbit around Earth at 35.5° inclination. At 15:33 UT, the Centaur upper stage fired to send the payload—the cruise stage and MSL—on a trajectory to Mars. MSL used a unique entry, descent, and landing (EDL) profile, designed to accommodate for the fact that the Martian atmosphere is too thin for regular parachutes and standard aerobraking but too thick for deceleration

with rockets. MSL's relatively heavy weight—much heavier than anything ever landed on Mars—presented engineers with some serious challenges. The EDL used by MSL was entirely autonomous without ground intervention and involved four separate stages: guided entry, parachute descent, powered descent, and sky crane landing, all of which took a total of only 7 minutes on 6 August 2012. The guided entry within the aeroshell was helped by small attitude control jets that narrowed the landing ellipse for MSL to a 20 × 7-kilometer area. Having slowed down to Mach 1.7, a supersonic parachute deployed (similar to those on Viking, Mars Pathfinder, and MER). The heat shield (from the aeroshell) was then discarded, and at about 1.8 kilometers altitude, with velocity down to 100 meters/second, the actual descent stage with Curiosity underneath it was released from the aeroshell, and eight variable thrust monopropellant hydrazine thrusters fired to slow the payload down further until Curiosity was slowly lowered from the descent stage with a 7.6-meter tether known as the sky crane system. The rover was then gently brought down to the surface at 05:17 UT on 6 August 2012. Just 2 seconds after the rover touched down, the sky crane/descent stage was freed and flew away and crashed about 650 meters away. Curiosity landed at Gale Crater, a 154-kilometer diameter impact crater estimated to be 3.5–3.8 billion years old. The precise landing coordinates are 4.5895° S / 137.4417° E. NASA named the landing site Bradbury Landing site after Ray Bradbury (1920–2012), author of *The Martian Chronicles*. On 15 August, Curiosity began initial instrument and mobility checks (including a test of the laser on a rock using the ChemCam instrument on 19 August). The rover began its first drive on 29 August, slowly taking about two months to cross

about 400 meters east to a location named Glenelg. During the fall of 2012, Curiosity identified several interesting rocks that it investigated using both the MAHLI and APXS instruments. An area called "Rocknest" was also identified as an area to test out the scoop on the rover's remote arm. On 27 September 2012, NASA announced that Curiosity had found hints of an ancient streambed, indicating that there might have been a "vigorous flow" of water on Mars. On 27 October 2012, Curiosity conducted its first x-ray diffraction analysis of Martian soil, and on 3 December, NASA announced the results of Curiosity's first extensive soil analysis, which had revealed the presence of water molecules, sulfur, and chlorine. The small amounts of carbon detected could not be properly sourced and might have been from instrument contamination. More conclusively, in March 2013, NASA announced the data from the rover (based on an investigation of the so-called "John Klein" rock)

This low-angle self-portrait of NASA's Curiosity rover shows the vehicle at the site from which it reached down to drill into a rock target called "Buckskin." Bright powder resulting from that drill, carried out on 30 July 2015, can be seen in the foreground. *Credit: NASA/JPL-Caltech/MSSS*

This plaque on board Curiosity bears the signatures of several U.S. officials including that of then-President Barack Obama and Vice-President Joe Biden. The image was taken by the rover's Mars Hand Lens Imager (MAHLI) on 19 September 2012, the rover's 44th Martian day on the Red Planet. *Credit: NASA/JPL-Caltech/MSSS*

suggested that Gale Crater was once suitable for microbial life. The result of further analysis showed the existence of water, carbon dioxide, sulfur dioxide, and hydrogen sulfide. In terms of rover operations, on 28 February 2013, the rover's active computer's flash memory developed a problem that made the computer reboot in a loop. As a result, controllers switched to the backup computer that became operational on 19 March. After drilling a rock in February, Curiosity drilled its second rock ("Cumberland") on 19 May 2013, generating a hole about 6.6 centimeters deep and delivering the material to its laboratory instruments. In early July 2013, Curiosity exited the Glenelg area and began a long trek to the mission's main destination, Mount Sharp. A long drive on 17 July of 38 meters meant that Curiosity had now traveled a total distance of 1 kilometer since it landed. As Curiosity continued toward Mount Sharp, NASA announced further findings. On 19 September 2013, scientists

revealed that data from Curiosity of samples of the atmosphere taken six times from October to June 2012, confirmed that the Martian environment lacks methane, suggesting that there is little chance that there might be methanogenic microbial activity on Mars at this time. At the end of the year, on 7 November, the rover abruptly reverted to "safe mode" (apparently due to a software error) but controllers revived the vehicle within three days to resume nominal surface operations. Later, on 17 November, there was a spurious voltage problem that suspended work for a few days. By 5 December, the rover's ChemCam laser instrument had been used for more than 100,000 shots fired at more than 420 different rock or soil targets. Through the next few months, Curiosity returned many spectacular images (including of Earth in the Martian night sky). One image returned in April 2014 showed both Ceres and Vesta. In May 2014, Curiosity drilled into a sandstone slab rock ("Windjana"), the third time on its traverse, this time at a waypoint along the route towards the mission's long-term destination on the lower slopes of Mount Sharp. Curiosity passed its two-year mark on Mars on 5 August 2014, having already far exceeded its original objectives. In August 2014, Curiosity was about to make its fourth drilling experiment but mission planners decided not to at the last minute as it was thought that the rock ("Bonanza King") was not stable enough. Finally, on 11 September 2014, Curiosity arrived at the slopes of Mount Sharp (or Aeolis Mons), now 6.9 kilometers away from its landing point. In less than two weeks, on 24 September 2014, the rover's hammering drill was used to drill about 6.7 centimeters into a basal-layer outcrop on Mount Sharp and collected the first powdered rock sample from its ultimate target. Perhaps the most striking announcement of the mission was made on 16 December 2014, when NASA scientists announced that Curiosity had definitively identified the first instance of organic molecules on Mars. These were found in a drilled sample of the Sheepbed mudstone in Gale Crater. While these could be from organisms, it is more

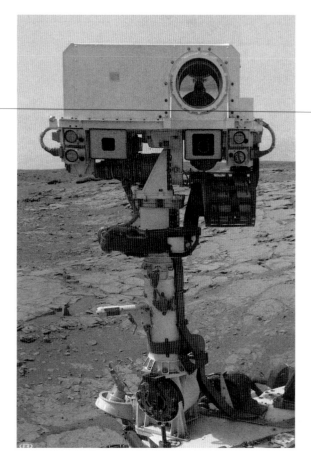

Visible at the bottom (the thick vertical rod) is the Rover Environmental Monitoring Station (REMS) built by the Centro de Astrobiologia (CAB) in Madrid, Spain. The pencil-like instrument sticking out to the left is Boom 1 which houses a suite of infrared sensors to measure the intensity of infrared radiation emitted by the ground. The large structure at the top includes the Mastcam (Mast Camera) and the white rectangular-shaped ChemCam (Chemistry & Camera) instruments. *Credit: NASA/JPL/MSSS/Ed Truthan*

a new, low-percussion-level drilling technique to collect sample powder from a rock target called Mojave 2. Preliminary results (based on analysis by the CheMin instrument) suggested more acidic qualities in the form of jarosite, than previous drilled samples. On 24 February, Curiosity used its drill for the sixth time, this time to collect sample powder from inside a rock known as Telegraph Peak that was resting on the upper portion of Pahrump Hills (also the site for the two previous drilling experiments). Three days later, Curiosity experienced a so-called "fault-protection action" that stopped the robot from transferring sample material between devices because of an "irregularity" in the electric current. After a series of tests running through early March, mission planners finally directed the robotic arm to resume its work on 11 March and have it deliver the sample to the CheMin analytic instrument. Later in March, scientists published results (in the *Proceedings of the National Academy of Science*) from the SAM instrument suite that indicated for the first time the existence of nitrogen on the surface of Mars released during the heating of Martian sediments. The nitrogen, which NASA called "biologically useful" was in the form of nitric oxide, perhaps released from the breakdown of nitrates. The discovery added more weight to the argument that ancient Mars might have been "habitable for life." On 16 April, Curiosity passed the 10-kilometer mark on its travels as it moved through a series of shallow valleys between Pahrump Hills and Logan Pass, its next science destination. Curiosity resumed full operations after a period of limited activity for most of June when Mars passed nearly behind the Sun (relative to Earth). In late July, the rover found unusual bedrock in a target named Elk, one with unexpectedly high level of silica, which suggests conditions suitable for preserving ancient organic material. On 12 August 2015, Curiosity finally finished its work at Marias Pass, where it had been since May, and where it had drilled a rock target named Buckskin and found rocks with high silica and hydrogen content. It then headed upward and

likely that the substance is from dust and meteorites that have landed on the planet. At the same time, new data showed that there was a recent tenfold spike and then decrease in the abundance of methane—still very tiny—in the Martian atmosphere. Curiosity was visible in a photograph taken on 13 December 2014 by the HiRISE camera on board Mars Reconnaissance Orbiter (MRO); the image clearly showed the rover as it was examining part of the basal layer of Mount Sharp inside Gale Crater. A later image, from 8 April 2015, also showed Curiosity very clearly on the lower slope of Mount Sharp. In late January 2015, Curiosity used southwest up Mount Sharp. Much later, in June 2016, scientists published results from an investigation of Buckskin noting that the silica mineral in question was tridymite, a material generally linked to silicic volcanism. Continuing its studies at Mount Sharp, on 29 September 2015 Curiosity drilled its eighth hole (and fifth at Mount Sharp), one that was 65-mm deep in a rock known as Big Sky as part of an experiment to analyze Martian rocks in both the CheMin and SAM instrument suites. New results from Curiosity, published in October 2015, confirmed that billions of years ago there were definitely water lakes on Mars. Scientists determined that water helped deposit sediment into Gale Crater, and the sediment was deposited as layers that formed the foundation for Mount Sharp, the mountain in the middle of Gale. At the end of the year, in December 2015, Curiosity began close examination (and returned spectacular images) of dark sand dunes up to two stories tall, located at Bagnold Dunes, a band along the northwestern side of Mount Sharp. Through the subsequent few weeks, the rover took several samples from the Samib Dune, that were sorted by grain size for closer studies. Moving on from the dunes in early March, Curiosity climbed onto the Naukluft Plateau on the lower side of Mount Sharp, ending up in a stretch of extremely rugged and difficult-to-navigate terrain, whose bedrock was shaped by long periods of wind erosion into ridges and knobs. Here, the rover continued to take drill samples (its 10th and 11th). On 2 July 2016, Curiosity suddenly entered into safe standby mode, but controllers were able to return it to normal operations a week later. The cause of the original switch to safe mode was a software "mismatch" in a particular mode, involving writing images from some cameras' memories into files on the rover's main computer. Among the many thousands of images returned by Curiosity, some of the most spectacular were those of the Murray Buttes region of lower Mount Sharp. Color images showed beautiful vistas not unlike "a bit of the American desert southwest," according to Curiosity Project

Scientist Ashwin Vasavada. Curiosity began a second two-year extended mission on 1 October 2016, continuing its explorations of lower Mount Sharp. At that point, the rover had returned more than 180,000 images to Earth; NASA declared that the mission "has already achieved its main goal of determining whether the landing region ever offered environmental conditions that would have been favorable for microbial life." Engineers put a halt on using the rover's drill soon after, while taking what would have been the seventh drill sample of the year. On 1 December, the Curiosity team discovered that the rover had not completed the commands for drilling; apparently, the rover had detected a fault in the "drill feed mechanism" which was supposed to extend the drill to touch the rock target. This problem remained unresolved until May 2018 when a new method of "percussive" drilling finally opened the path to further use of the instrument. Despite the problem with the drill, Curiosity was once again investigating active sand dunes (the so-called Bagnold Dunes). As of 23 February 2017, Curiosity had driven 15.63 kilometers since landing. By this time, it was clear to NASA engineers that the zig zag treads on Curiosity's wheels were suffering damage, jeopardizing the wheels' ability to carry the weight of the rover. Damage to only three treads (or "grousers") would indicate that the wheel had reached 60% of its lifetime. Lessons from damage to Curiosity's wheels will play a major role in the design of future Mars rovers. In March 2017, JPL controllers uploaded a software for traction control that helped the rover adjust wheel speed depending on the rocks it is climbing. The traction control algorithm uses real-time data to vary the wheel speed, thus reducing pressure from the rocks. "Armed" with the new software, in July, Curiosity began a campaign to study a ridge on lower Mount Sharp, informally named Vera Rubin Ridge after the recently departed astronomer, Vera Florence Cooper Rubin (1928–2016). The ridge was thought to be rich in an iron-oxide mineral known as hematite that can form under wet conditions. Two months later, the rover began making a steep ascent toward the ridge top. On 17 October 2017, for the first time in 10 months, the rover cautiously touched its sampler drill to a surface rock. While it was still several months away from resuming full-scale drilling operations on the Martian surface, this and subsequent tests allowed ground controllers to test techniques, including using the motion of the robotic arm directly to advance the extended bit into the rock, thus working around the mechanical problem that suspended drill work. NASA made some significant attempts to involve the public in the Curiosity mission. On the third anniversary of landing on Mars, in August 2015, the Agency made available two online tools for public engagement. Mars Trek was a free web-based application that provides high-quality visualizations of the planet derived from 50 years of NASA exploration of Mars, while Experience Curiosity was a platform to allow viewers to experience in 3D, movement along the surface of Mars based on data from both Curiosity and MRO. Around the time of Curiosity's third anniversary on Mars, in August 2016, NASA also released a social media game, Mars Rover, for use on mobile devices where users can drive a rover through Martian terrain while earning points.

2013

LADEE

Nation: USA (98)

Objective(s): lunar orbit

Spacecraft: LADEE

Spacecraft Mass: 383 kg

Mission Design and Management: NASA / ARC / GSFC

Launch Vehicle: Minotaur V (no. 1)

Launch Date and Time: 7 September 2013 / 03:27:00 UT

Launch Site: Mid Atlantic Regional Spaceport (MARS) / Pad 0B

Scientific Instruments:

1. ultraviolet and visible light spectrometer (UVS)
2. neutral mass spectrometer (NMS)
3. lunar dust experiment (LDEX)
4. lunar laser communications demonstration experiment (LLCD)

Results: The Lunar Atmosphere and Dust Environment Explorer (LADEE), the first mission in the Lunar Quest series, was designed to orbit the Moon and study its thin atmosphere and the lunar dust environment, specifically to collect data on the global density, composition, and time variability of the exosphere. By studying the Moon's exosphere—an atmosphere that is so thin that its molecules do not collide with each other—LADEE's instruments helped further the study of other planetary bodies with exospheres such as Mercury and some of Jupiter's moons. After insertion into low parking around Earth after launch from the Wallops Flight Facility—the first lunar launch from that location—Minotaur's fifth stage (with a Star 37FM solid motor) fired at 03:43 UT to boost the payload into a highly elliptical Earth orbit of 200 × 274,600 kilometers at 37.7° inclination. LADEE took a path to lunar orbit used by several other recent lunar spacecraft that involved flying increasingly larger Earth orbits (in this case, three orbits) over a period of a month, with the apogee increasing until it was at lunar distance. On the third orbit, on 6 October, as LADEE approached the Moon, it fired its own engine and entered into an initial elliptical lunar orbit with a 24-hour period. On 9 and 12 October, further burns brought LADEE down into a 235 × 250-kilometer orbit. These events occurred exactly during the period when the U.S. Government—and therefore NASA—shut down briefly, opening back up on 16 October. One of the early experiments was use of the LLCD system, carried out on 18 October 2013 when the spacecraft, using the optical laser system, transmitted good data to a ground station 385,000 kilometers away. Finally, on 20 November, LADEE successfully entered its planned equatorial orbit of 20 × 60 kilometers, allowing the probe to make frequent passes from lunar day to lunar night. When the Chinese Chang'e 3 lunar lander arrived at the Moon, LADEE (more specifically, it's NMS neutral mass spectrometer) was used to observe the specific masses of the substances (such as water, nitrogen, carbon monoxide, and hydrogen) that would be expected to be found given the operation of Chang'e's operation in near-lunar space. In the event, LADEE's data found no effects—no increase in dust, no propulsion products, etc.—that could be attributed to Chang'e 3. Another experiment that involved another spacecraft was NASA's Lunar Reconnaissance Orbiter (LRO) taking a photo of LADEE in orbit at 01:11 UT on 15 January 2014. Its 100-day science mission, during which LADEE collected an enormous amount of data, came formally to an end by early March 2014. The three science payloads worked fulltime during this period:

the UVS instrument acquired more than 700,000 spectra of the exosphere. The NMS instrument positively identified argon-40 in the atmosphere (first identified by an Apollo surface experiment 40 years before). Finally, the LDEX recorded more than 11,000 impacts from dust particles from a dust cloud engulfing the Moon. In early 2014, LADEE began to gradually lower its orbital altitude in anticipation of its final impact on the Moon. Controllers lowered LADEE's orbit to within 2 kilometers of the lunar surface to ensure impact. On its penultimate orbit, on 17 April, LADEE swooped down to as low as 300 meters of the lunar surface, and contact was lost at 04:30 UT on 18 April when it moved behind the Moon. Controllers estimated that the spacecraft probably struck the Moon on the eastern rim of Sundman V crater between 04:30 and 05:22 UT at a speed of 5,800 kilometers/hour. Later, on 28 October 2014, NASA announced that its LRO spacecraft had successfully imaged the impact location of LADEE on the far side of the Moon.

233

Mangalyaan / Mars Orbiter Mission (MOM)

Nation: India (2)
Objective(s): Mars orbit
Spacecraft: Mars Orbiter Spacecraft
Spacecraft Mass: 1,337 kg
Mission Design and Management: ISRO
Launch Vehicle: PSLV-XL (no. C25)
Launch Date and Time: 5 November 2013 / 09:08 UT
Launch Site: SHAR / PSLV pad
Scientific Instruments:

1. Mars color camera (MCC)
2. thermal infrared imaging spectrometer (TIS)
3. methane sensor for Mars (MSM)
4. Mars exospheric neutral composition analyzer (MENCA)
5. Lyman alpha photometer (LAP)

Results: India's first interplanetary spacecraft was designed and built in a relatively short period of time, with a total development time of 4 years and 2 months, although official government approval came as late as August 2012. The main spacecraft bus was a modified I-1 bus used on the lunar Chandrayaan-1 mission. The mission was essentially a technology demonstrator although it carried a set of modest scientific instruments to study the surface features, morphology, mineralogy, and Martian atmosphere. With the mission, India accomplished a remarkable feat, becoming only the fourth nation or agency to have a spacecraft orbit Mars, after the former Soviet Union, the United States, and the European Space Agency. Japan had tried and failed, while a Chinese rocket had yet to launch a probe to Mars (although its Yinghuo-1 orbiter launched by the Russians failed to leave Earth orbit due to the malfunctioning Russian upper stage). The name Mangalyaan was formally attached to the mission before launch while in the development period the spacecraft was known variously as the Mars Orbiter Spacecraft, the Mars Orbiter Satellite, or the Mars Orbiter Mission. Mangalyaan entered an initial orbit around Earth at 251 × 23,892 kilometers at 19.4° inclination. The mission profile to Mars involved six engine burns (the fourth on 10 November was partially successful) over a month in progressive larger Earth orbits to accumulate sufficient velocity to escape Earth's sphere of influence. A seventh engine burn lasting over 22 minutes, beginning at 19:19 UT on 30 November, inserted Mangalyaan into heliocentric orbit. On the way to Mars, the spacecraft conducted three mid-course corrections (on 11 December 2013, 11 June, and 22 September) before a successful burn (lasting over 23 minutes) of the main 44.9 kgf thrust engine put the probe into Mars orbit on 24 September 2014. This was a highly elliptical orbit at 421.7 × 76,993.6 kilometers with an orbital period of nearly 73 hours. Mangalyaan returned its first global image of Mars, a spectacular picture captured by the MCC instrument, on 28 September 2014 from an altitude of

74,580 kilometers with about a 4-kilometer resolution that showed various morphological features and thin clouds in the Martian atmosphere. The ground team maneuvered the spacecraft to avoid a possible encounter with Comet C/2013 A1 (Siding Spring) which passed by Mars on 19 October 2014, one of seven spacecraft on or around Mars that had to take measures to prevent damage. The comet passed by Mars at a range of about 132,000 kilometers shedding material around Mars, some at a velocity of 56 kilometers/second putting many of these spacecraft in danger. Mangalyaan's instruments remained fully operational through late 2014, and on 1 January 2015 ISRO scientists marked 100 days of successful operations around Mars. Mission planners were confident that the spacecraft would meet its planned lifetime of six months or 180 days in orbit around Mars. As with many other spacecraft in and around Mars, Mangalyaan was subjected to a communications blackout in June 2015 when Mars' orbit took it behind the Sun relative to Earth. Mangalyaan commemorated a successful operational year orbiting Mars in September 2015 although at the time, scientific results from its instruments had yet to be publicly shared. The 13 pictures released by that time were largely taken in September and October 2014 although, because of MOM's unique orbit, they show the kind of wide-angle images have rarely been seen from Mars orbit. Finally, in March 2016, scientists published the first results (in *Geophysical Research Letters*) of the MENCA instrument. Further results from other instruments were made available in the fall of 2016. Newly released images included some of the most spectacular views from orbit showing many surface patterns. One note of concern was a possible problem with the MSM methane sensor, data from which has yet to be released. There were some reports that the sensor itself had a design flaw. NASA Goddard Space Flight Center scientists had apparently briefed ISRO personnel on the problem in February 2016 and suggested that the instrument instead be repurposed for albedo mapping and measuring reflected sunlight. On 17 January

2017, ISRO controllers changed Mangalyaan's orbit as a strategy to avoid the eclipse season when the spacecraft would be in Mars' shadow for as much as 8 hours per day. The burn used about 20 kilograms of propellant, leaving 13 kilograms remaining for further maneuvers. It is hoped that the spacecraft can transmit data until 2020.

234

MAVEN

Nation: USA (99)

Objective(s): Mars orbit

Spacecraft: MAVEN

Spacecraft Mass: 2,454 kg

Mission Design and Management: NASA / GSFC / University of Colorado-Boulder

Launch Vehicle: Atlas V 401 (AV-038 + Centaur)

Launch Date and Time: 18 November 2013 / 18:28:00 UT

Launch Site: Cape Canaveral Air Force Station / SLC-41

Scientific Instruments:

1. P&F particle and fields package
   a. solar wind electron analyzer (SWEA)
   b. solar wind ion analyzer (SWIA)
   c. supra thermal and thermal ion composition (STATIC)
   d. solar energetic particle experiment (SEP)
   e. Langmuir probe and waves experiment (LPW)
   f. magnetometer (MAG)
2. RS remote sensing package
   a. imaging ultraviolet spectrometer (IUVS)
3. NGIMS neutral gas and ion mass spectrometer package

Results: The Mars Atmosphere and Volatile Evolution (MAVEN) mission was selected as part of NASA's now-cancelled Mars Scout Program to explore the atmosphere and ionosphere of the planet and their interaction with the Sun and solar wind. The goal is to use this data to determine how

Engineers and technicians test deploy the two solar panels on NASA's MAVEN spacecraft. The image was taken in the Payload Hazardous Servicing Facility at NASA's Kennedy Space Center before launch in November 2013. *Credit: NASA/ Kim Shiflett*

the loss of volatiles from the Martian atmosphere has affected the Martian climate over time, and thus contribute to a greater understanding of terrestrial climatology. The Mars Scout Program involved low cost spacecraft (less than $450 million) but was cancelled in 2010 after approval of MAVEN and the Phoenix lander. Further Mars missions would now be selected competitively under the Discovery Program. Because of a U.S. Government shutdown in the fall of 2013, MAVEN nearly didn't get off the ground. Fortunately, MAVEN was defined as part of "critical infrastructure," allowing the launch to proceed on time on 18 November 2013. The payload successfully reached a 167 × 315-kilometer parking orbit around Earth (at 26.7° inclination). Soon, the Centaur upper stage (with its RL-10A-4-2 engine), fired the spacecraft into a hyperbolic Earth orbit at 195 × 78,200 kilometers at 27.7° inclination. On 21 November, another

burn inserted the spacecraft on a Trans-Mars trajectory, with further mid-course corrections on 3 December 2013 and 27 February 2014. At 03:24 UT on 21 September 2014, MAVEN successfully entered orbit around Mars after a 10-month journey when its six main engines fired, two by two in succession, and burned for 33 minutes and 26 seconds to slow the craft. The spacecraft entered a six-week commissioning phase before beginning science operations. The initial orbital period was 35.02 hours. The primary mission, in an orbit with a period of 4.5 hours, included five "deep-dip" campaigns, during which MAVEN's periapsis was lowered from 150 kilometers to about 125 kilometers to collect data on the boundary between the upper and lower atmosphere. By mid-October all of its scientific instruments were turned on. Because of the close Martian flyby (about 139,500 kilometers) of Comet C/2013 A1 (Siding Spring) on 19 October, controllers took precautions to protect MAVEN from damage. In the event, MAVEN survived without any damage, and also returned valuable data on the comet's effects on the Martian atmosphere. MAVEN began its one-year primary science mission on 16 November, carrying out regular observations of the Martian upper atmosphere, ionosphere, and solar-wind interactions with its nine scientific instruments. MAVEN completed the first of its five "deep-dip" maneuvers between 10 and 18 February 2015. As with most of these dives, the first three days were used to lower the periapsis, with the remaining five days used for scientific investigations over roughly 20 orbits. Given that the planet rotates under the spacecraft, the 20 orbits allow the opportunity to explore different longitudes spaced around Mars, essentially giving it a global reach. The following month, based on data collected in December 2014, mission scientists announced that they had detected two unanticipated phenomena in the Martian atmosphere, one involving a high-altitude dust cloud (at about 150–300 kilometers altitude) and the other, a bright ultraviolet auroral glow in the northern hemisphere. Like all the other probes in orbit or on the surface of Mars, communications with MAVEN were put on hold during the Mars conjunction in June 2015. MAVEN commemorated a year in orbit in September 2015 by which time it had carried out four deep dive campaigns. By this time, NASA had approved an extended mission past November. In a major reveal in November 2015, scientists published the results (in the journals *Science* and *Geophysical Research Letters*) of their analysis of data from MAVEN, which identified the processes that contributed to the transition of the Martian climate from an early, warm, and wet environment that might have supported surface life to the cold and dry world at the present. More specifically, the information helped to determine the most accurate rate at which the Martian atmosphere was currently losing gas to space via "stripping" to the solar wind. In late November and early December 2015, MAVEN made a series of close flybys of Phobos,

coming within 500 kilometers of its surface, and collecting spectral images using the IUVS instrument. On 3 October 2016, MAVEN completed an entire Mars year of scientific observations. The following year, on 28 February 2017, MAVEN carried out a small orbital maneuver, the first of its kind, to avoid a possible impact with Phobos. During MAVEN's second Martian year in orbit, through 2017, research was being coordinated with simultaneous atmospheric observations by ESA's Trace Gas Orbiter.

235

Chang'e 3 and Yutu

Nation: China (4)
Objective(s): lunar landing and rover
Spacecraft: Chang'e sanhao
Spacecraft Mass: 3,780 kg (140 kg Yutu)
Mission Design and Management: CNSA
Launch Vehicle: CZ-3B (no. Y23)
Launch Date and Time: 1 December 2013 / 17:30 UT
Launch Site: Xichang / Launch Complex 2
Scientific Instruments:
Lander:
1. terrain camera (TCAM)
2. landing camera (LCAM)
3. Moon-based ultraviolet telescope (MUVT)
4. extreme ultraviolet camera (EUVC)
Yutu:
1. panoramic camera (PCAM)
2. lunar penetrating radar (LPR)
3. imaging spectrometer (VIS-NIR)
4. active particle induced x-ray spectrometer (APXS)

Results: Launched as part of the second phase of the Chinese Lunar Exploration Program, Chang'e 3 was the most sophisticated robotic probe launched by the Chinese to date. The spacecraft included a four-legged soft-lander and a small rover named Yutu ("Jade Rabbit") to traverse the lunar surface. The 765-kilogram lander was based around an

Image taken on 13 January 2014 by the panoramic camera (PCAM) on board the Chinese Yutu lunar rover as it looked back at the Chang'e 3 lander. *Credit: Chinese Academy of Sciences / China National Space Administration / The Science and Application Center for Moon and Deepspace Exploration / Emily Lakdawalla*

octagon-shaped body with four legs, a throttleable engine at the bottom, refoldable solar panels, a radioisotope heater unit to generate heat, and three engineering monitoring cameras at the top. The lander's mission lifetime was one year. Besides the technological objectives of the mission, the lander and rover also had some scientific objectives including investigating the lunar topography and geological structure of the Moon; studying the material composition of the lunar surface; surveying the space environment between the Moon, Earth, and Sun; and carrying out astronomical observations from the lunar surface. The launch was marred by hardware from the rocket that landed on villages downrange from the launch site, although no one was hurt. The payload was inserted directly into a lunar transfer orbit of 210 × 389,109 kilometers

at 28.5° inclination. After two mid-course corrections (on 2 and 3 December), the spacecraft fired its engine at 09:41 UT for 6 minutes and entered a 100-kilometer circular orbit around the Moon. On 10 December, Chang'e 3 lowered is orbit to a 15 × 100-kilometer orbit. When it reached perilune on 14 December, its engine fired at 13:00 UT for a powered descent to the lunar surface (apparently one orbit earlier than planned, probably to ensure a live broadcast on Chinese TV). The 11-minute landing sequence was punctuated by a "hazard avoidance maneuver" about 100 meters above the surface. The main engine cut out at 4 meters above the surface, and the lander settled down naturally the rest of the way, landing at 13:11 UT, thus becoming the first spacecraft to soft-land on the Moon since the Soviet Luna 24 in 1976. Landing

Image taken on 17 December 2013 by the Lander Terrain Camera (TCAM) on board the Chinese Chang'e 3 lander showing the Yutu rover on the surface of the Moon. *Credit: Chinese Academy of Sciences / China National Space Administration / The Science and Application Center for Moon and Deepspace Exploration / Emily Lakdawalla*

coordinates were announced as 19.51° W / 44.12° N, in the northwestern portion of Mare Imbrium which, as is turns out was different than what was announced earlier—Sinus Iridum. Immediately after landing, Chang'e 3 deployed its solar panels. Soon, a command was sent to Yutu to deploy its two solar panels and unlock its wheels. Yutu was a 6-wheeled mobile spacecraft capable of transmitting video in real time and able to perform relatively simple soil analyses. Like the lander, the rover had its own radioisotope heater units (using plutonium-238) for heating. It also had sensors that enabled it to automatically negotiate over obstacles on the surface. Its original planned lifetime was three months, during which it was expected to travel an area of 3 square kilometers. At 20:10 UT on 14 December, the rover began to slowly move

toward the rover transfer mechanism, which then slowly lowered itself until its tracks touched the lunar surface. At 20:35 UT, the rover wheeled down and touched the surface. The rover remained at Point A, about 10 meters north of the lander for about 13 hours until it turned around and cameras on the rover and lander took pictures of each other on 15 December. The rover then slowly moved to Point B and entered into sleep mode for four days. Woken up on 20 December, Yutu traveled about 21 meters from Point B to C and then to Point D, testing its robotic arm on 23 December. Both lander and rover entered into hibernation mode on 25 and 26 December, respectively, in anticipation of the lunar night. On 11–12 January the lander and Yutu were brought out of sleep to begin their second lunar day. The rover, however, ran into a major problem later in the month; on 25 January, the Chinese media announced that Yutu had suffered a "mechanical control anomaly" caused by the "complicated lunar surface environment." Because the problem occurred just before the rover was entering hibernation for its second lunar night, controllers had to wait until hibernation was over to confirm the problem. After failing to hear from the rover on 10 February, ground controllers established contact two days later but it was clear at this point that Yutu was no longer capable of moving across the lunar surface, apparently because of a control circuit malfunction in its driving unit. Instead, mission scientists reformulated a limited mission for Yutu as a stationary science platform. Total traverse distance was computed as 0.114 kilometers. Because the problem on the rover was in a unit that also controlled the mobility of the solar panels to face the Sun, controllers expected the rover to lose power, but the rover appears to have communicated with mission control through at least early September 2014. The lander meanwhile entered a "maintenance mode" at some point in mid-2014 but continued operating through the year. In January 2015, over a year after landing, the Chinese media announced that the lander was still returning data, having entered its 14th lunar night, thus making it the longest functioning spacecraft on the lunar surface. The lander was said to be in contact at least once a month well into fall 2016 and sending back scientific data from at least one instrument, its ultraviolet telescope. The Chinese also released entire datasets of imagery from both the lander and the rover in January 2016. As far as Yutu, in March 2015, Ye Peijian confirmed that the rover was immobile but still working. The rover was still maintaining communications to at least July 2016 but the following month, Yutu was officially declared dead after operating for 31 months. At least one instrument on the lander (the ultraviolet telescope) was still active as of June 2018.

236

Gaia

Nation: ESA (8)
Objective(s): Sun–Earth L2 Lagrange Point
Spacecraft: Gaia
Spacecraft Mass: 2,029 kg
Mission Design and Management: ESA
Launch Vehicle: Soyuz-ST-B / Fregat-MT (Soyuz-ST-B no. E15000-004/104, Fregat-MT no. 1039)
Launch Date and Time: 19 December 2013 / 09:12:18 UT
Launch Site: CSG / ELS pad
Scientific Instruments:

1. astrometric instrument (ASTRO)
2. photometric instrument
3. radial velocity spectrometer (RVS)

Results: Gaia is a European space observatory whose goal is to chart a three-dimensional map of the Milky Way galaxy in order to reveal the composition, formation, and evolution of the galaxy. More specifically, Gaia provides high-quality positional and radial velocity measurements to produce a stereoscopic and kinematic census of about one billion stars in our galaxy (about 1% of the total) and the Local Group. Launched by a Russian Soyuz, the Fregat upper stage pushed Gaia into a 175 × 175-kilometer Earth orbit, and then fired again for a long burn into a 344 × 962,690-kilometer orbit at 15.0° inclination. It performed a mid-course correction the day after launch. On 8 January 2014, Gaia entered its operational orbit around the Sun–Earth L2, about 1.5 million kilometers from Earth, when its engine fired to boost the spacecraft into a 263,000 × 707,000-kilometer halo orbit around L2 with a period of 180 days. After four further months of calibration, alignment, and proper focusing of the telescopes on board, Gaia began its five-year mission on 25 July 2014 to produce the

An artist's rendering of ESA's Gaia astrometry spacecraft. *Credit: ESA/ATG medialab*

most accurate 3D map of the Milky Way galaxy. In its observation mode, Gaia spins slowly (once every 6 hours), sweeping its two telescopes across the entire sky and focusing the received light simultaneously onto a single digital camera, the largest flown in space, with nearly a billion pixels (106 CCDs each with 4,500 × 1,996 pixels). It will observe each of its billion stars an average of 70 times over five years. In September 2014, ESA announced that Gaia had discovered its first supernova, Gaia14aaa, some 500 million light-years away from Earth. A minor anomaly, "a stray light problem" was detected shortly after launch that might degrade the quality of some of the results, especially for the faintest stars, but ESA scientists are confident that mitigation schemes will compensate for the problem. In August 2015, Gaia completed its first year of science observations, during which it had recorded 272 billion positional or astrometric measurements, 54.4 billion photometric data points, and 5.4 billion spectra. On 6 November 2015, the spacecraft was placed perfectly to see a lunar transit across the Sun, viewed from Gaia's vantage point as a small, dark circle crossing the face of the Sun. On 14 September 2016, ESA released its first dataset from Gaia that included positions and G magnitudes for about one billion stars based on observations from 25 July 2014 to 16 September 2015. A second data release is planned for April 2018 and will include positions, parallaxes, and proper motions for approximately one billion stars. Final Gaia results in the form of complete datasets are not expected to be publicly available until the early 2020s. Besides its primary goal of mapping stars, Gaia also carries out observations of known asteroids within our solar system, providing data on the orbits and physical properties of these bodies.

237

Chang'e 5-T1

Nation: China (5)

Objective(s): circumlunar flight, Earth–Moon L2 Lagrange Point, lunar orbit

Spacecraft: Zhongguo tanyue gongcheng san qi zai ru fanhui feixing shiyan qi

Spacecraft Mass: 3,300 kg

Mission Design and Management: CNSA

Launch Vehicle: Chang Zheng 3C/G2 (or Chang Zheng 3C/E)

Launch Date and Time: 23 October 2014 / 18:00:04 UT

Launch Site: Xichang

Scientific Instruments:

1. cameras
2. biological payloads

Results: This mission, called Chang'e 5 Flight Test Device (or CE 5-T1), was a precursor to the planned Chang'e 5 mission, which, at the time of writing, is slated to land on the Moon and return lunar samples (at least 2 kilograms) back to Earth in 2019. The spacecraft consists of a bus (DFH-3A) similar to the CE-1 and CE-2 lunar orbiters, but with a large decent vehicle, resembling a scaled-down version of the piloted Shen Zhou descent module. During this flight, the reentry module carried a complement of biological payloads. The goal of the test mission was to fly a full-scale circumlunar mission so as to flight-test the reentry vehicle—particularly the heatshield—in real conditions before the actual sample return mission. This was the first circumlunar mission and recovery of a spacecraft since the Soviet Zond 8 mission in 1970. The payload entered into a parking orbit around Earth after successful firing of the launch vehicle's first three stages. The third stage then reignited for about 3 minutes to send the payload on a translunar trajectory of 209 × 413,000 kilometers. Soon after TLI, the spacecraft deployed its solar arrays and implemented at least two mid-course corrections (on 24 and 25 October). Chang'e 5-T1 entered the Moon's gravitational influence at 13:30 UT, before flying around the far side of the Moon, with the closest distance of approximately 12,000 kilometers reached at 19:03 UT on 27 October. Throughout its voyage to the Moon, the spacecraft relayed back spectacular pictures of both Earth and the Moon. On the four-day trip home, the spacecraft performed one mid-course correction. Before entry into Earth's atmosphere, at 21:53 UT on 31 October, the headlight-shaped descent vehicle separated from the service module when about 5,000 kilometers from Earth; 5 minutes later, the service module fired its engine (for about 12 minutes) to go back into a translunar orbit. The descent vehicle, meanwhile, completed a "double-dip" reentry into Earth's atmosphere to reduce g-loads.

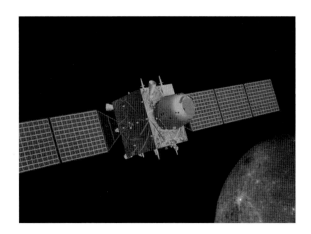

Artist's rendering of the Chinese Chang'e 5-T1 spacecraft. The headlight-shaped object is the descent module that was successfully recovered after its circumlunar mission. *Credit: chinaspacereport.com*

The second time it entered Earth's atmosphere, it deployed its parachute system at 10 kilometers altitude and safely touched down at 22:42 UT in China's Inner Mongolia Autonomous Region. The descent vehicle was located just 5 minutes after touchdown. Yang Mengfeil, the "commander" of the lunar exploration program at CASC noted that "[f]rom what we have seen, the capsule is in good condition... and will lay a solid foundation for our future space program." Meanwhile, the service module of CE 5-T1 circled back into a high Earth orbit (540,000 × 600 kilometers) that slung it back towards the Moon, which it passed by on 23 November after two mid-course corrections. On 23 November, after reaching perilune and with the help of gravity assist from the Moon, the spacecraft headed to the Earth–Moon L2 Lagrange Point where it arrived on 27 November and entered into a Lissajous-type orbit at 20,000 × 40,000 kilometers with a period of 14 days. A little over a month later, at 15:00 UT on 4 January 2015, the service module left L2 after three circuits around it, and headed back toward the Moon. Six days later, at around 19:00 UT, the spacecraft returned back into lunar orbit with initial parameters of 200 × 5,300 kilometers with a period of 8 hours. Over the following two days, CE 5-T1 made three orbital corrections, ending up, by 11 January, in a 200-kilometer orbit. Once at the Moon, the spacecraft conducted two "virtual target" rendezvous tests in February and March 2015 to rehearse for the future CE-5 mission. Later, between 30 August and 2 September 2015, CE 5-T1 carried out an intensive photography mission, some of it at altitudes as low as 30 kilometers. The campaign was dedicated to identifying potential landing sites for the Chang'e 5 sample return mission. By 1 March 2016, the spacecraft had completed over 4,100 orbits around the Moon. The Chinese launch vehicle's upper stage carried a secondary payload called the Manfred Memorial Moon Mission (LX0OHB-4M or 4M) prepared by the Germany company OHB System and managed by LuxSpace. The experiment, in honor of its OHB founder Manfred Fuchs (1938–2014) weighed 14 kilograms and contained two instruments, a radio beacon and a radiation dosimeter (provided by the Spanish company iC-Málaga). Last contact with the radio beacon was at 01:35 UT on 11 November 2014.

238

Hayabusa 2

Nation: Japan (10)
Objective(s): asteroid rendezvous and sample return
Spacecraft: Hayabusa 2, Rover1a, Rover1b, Rover2, MASCOT
Spacecraft Mass: 600 kg
Mission Design and Management: JAXA
Launch Vehicle: H-IIA (no. F26)
Launch Date and Time: 3 December 2014 / 04:22:04 UT
Launch Site: Tanegashima / Area Y1
Scientific Instruments:

1. near infrared spectrometer (NIRS3)
2. thermal infrared imager (TIR)
3. multiband imager (ONC-T)
4. laser altimeter (LIDAR)
5. separation camera (DCAM)

MASCOT:

1. MicrOmega infrared microscope
2. magnetometer (MAG)
3. radiometer (MARA)
4. wide-angle camera (CAM)

Results: Hayabusa 2 is a Japanese spacecraft on a six-year mission to rendezvous and land on a C-class asteroid, (162173) 199 JU3, dispatch a series of landers and a penetrator, collect samples from it, and then return to Earth. The probe is an evolutionary follow-on to the Hayabusa (1) mission, launched in 2003, which had similar goals. The new mission is designed to be "more reliable and robust" and has several improved systems from its predecessor. Hayabusa 2's flight profile involves an Earth gravity-assist flyby in December 2015 before a rendezvous with its target asteroid in July 2018. At the asteroid, Hayabusa 2 will deploy a

Artist's concept of Hayabusa 2 collecting samples from asteroid 1999 JU3. *Credit: JAXA*

target marker (somewhat like a little beanbag filled with granular particles) to establish an artificial landmark, and then land on the planetary body. Five of these target markers are carried aboard the spacecraft, as opposed to three on its predecessor. As it touches the surface multiple times, Hayabusa 2 will activate its SMP sampler mechanism which will release a small projectile to be shot into the ground so material is ejected that is collected by a "catcher." Additionally, Hayabusa 2 will deploy several passenger payloads, including the 10-kilogram MASCOT (Mobile Asteroid Surface Scout), a joint French-German lander that is capable of lifting off the asteroid and landing again to sample different areas with its suite of instruments. Hayabusa 2 will also carry the SCI (Small Carry-on Impactor), a small explosive penetrator that will use a 2-kilogram pure copper "lump" (or a "liner") 30 centimeters in diameter that will be dropped to the surface of the asteroid at a velocity of 2 kilometers/second to make an artificial crater. A palmsize deployable camera (DCAM3) will observe the explosive impact of the SCI while the mother ship is positioned on the opposite side of the asteroid during the impact to protect it. Hayabusa 2 also carries three 1-kilogram rovers installed on MINERVA-II1 (which has Rover1A and Rover1B) and MINERVA-II2 (which has Rover2). They will take pictures and collect temperature data at various points. Once its mission is over, Hayabusa 2

will leave orbit around the asteroid in December 2019 and return to Earth a year later, in December 2020, for a reentry and recovery of the samples collected. After a successful launch, Hayabusa 2 and its fellow travelers left the Earth–Moon system on 6 December 2014, entering heliocentric orbit at 0.915×1.089 AU at $6.8°$ inclination to the ecliptic. It completed an initial period of checkout by 2 March 2015 and then began to move into its "cruising phase" while heading to asteroid 1999 JU3. Less than a year later, on 3 December 2015, Hayabusa 2 carried out an Earth flyby at a range of 3,090 kilometers over the Hawaiian islands. The encounter increased the spacecraft's velocity by about 1.6 kilometers/second to about 31.9 kilometers/second (relative to the Sun). The spacecraft, remaining in good health, performed its first major firing of its ion engines from 22 March to 5 May 2016, followed by a shorter (3.5 hour) firing on 20 May to adjust its trajectory.

239

PROCYON

Nation: Japan (11)
Objective(s): asteroid flyby
Spacecraft: PROCYON
Spacecraft Mass: 65 kg
Mission Design and Management: JAXA / University of Tokyo
Launch Vehicle: H-IIA (no. F26)
Launch Date and Time: 3 December 2014 / 04:22:04 UT
Launch Site: Tanegashima / Area Y1
Scientific Instruments:
   1. optical telescope
   2. ion thrusters

Results: The PROCYON (Proximate Object Close Flyby with Optical Navigation) spacecraft, a low-cost "micro-spacecraft for deep space exploration" was designed to accomplish two goals: to demonstrate the successful use of a micro-spacecraft for deep space activity and to fly by an asteroid, 1999

JU3 (later renamed 162173 Ryugu). Launched at the same time as the Hayabusa 2 mission, PROCYON is accompanying the larger spacecraft to its own asteroid target, designated (185851) 2000 DP$_{107}$ (an asteroid that has its own moon). The plan for the mission was that it would pass within 50 kilometers of the target at a relative flyby velocity of 10 kilometers/second. The spacecraft, largely built from commercial off-the-shelf equipment is equipped with a micro-propulsion system named I-COUPS (Ion thruster and Cold-gas thruster developed for PROCYON) that uses ion engines and multiple cold gas thrusters fed from the same tank (containing 2.5 kilograms xenon at launch). After launch with Hayabusa 2 (see entry for that mission) into low Earth parking orbit, the H-IIA rocket's second stage fired again for 4 minutes, 1 second to accelerate the entire payload (Hayabusa 2, PROCYON, and ArtSat-2) into heliocentric orbit. While Hayabusa separated from the upper stage 3 minutes, 40 seconds after the burn was complete, PROCYON separated 15 minutes after Hayabusa and, like its compatriots, entered heliocentric orbit. University of Tokyo and JAXA announced that they had received the first signals from PROCYON at 00:51 UT on 3 December and that it was on its confirmed interplanetary trajectory. By 10 March 2015, the on-board ion engine had operated for 223 hours but then stopped functioning apparently due to a high voltage problem, thus potentially threatening both the planned Earth flyby on 3 December 2015 and the asteroid flyby on 12 May 2016. When it was clear that the engine could no longer be restarted, in May 2015, the asteroid flyby was abandoned. In the event, PROCYON flew past Earth on 3 December at a range of 2.7 million kilometers. Despite the loss of the asteroid flyby, PROCYON used its telescopic imager to return photos of Earth and the Earth–Moon system during its Earth flyby.

240

Shin'en 2

Nation: Japan (12)
Objective(s): heliocentric orbit
Spacecraft: Shin'en 2
Spacecraft Mass: 17 kg
Mission Design and Management: Kagoshima University
Launch Vehicle: H-IIA (no. F26)
Launch Date and Time: 3 December 2014 / 04:22:04 UT
Launch Site: Tanegashima / Area Y1
Scientific Instruments:

1. dosimeter

Results: Shin'en 2 is a small satellite amateur radio payload—also given the Oscar designation of Fuji-Oscar 82—that carried a radio operating in the amateur UHF and VHF bands. During its mission, it was to simply downlink basic telemetry parameters (voltage, current, temperature, etc.) as well as data from the NASA-supplied radiation dosimeter. Radio enthusiasts could also uplink messages to the satellite. Once the satellite and its sister payloads were sent out to heliocentric orbit after launch, Shin'en 2 separated from the H-IIA second stage at 06:16 UT about 6 minutes, 40 seconds after the separation of Hayabusa 2. Contact with Shin'en 2 was established soon after, although the last reported transmission was on 14 December from 20:13 UT to 23:00 UT when the spacecraft was between 4.67 and 4.72 million kilometers from Earth. Like its sister spacecraft, Shin'en 2 flew by Earth on 4 December at 10:27 UT, at a range of about 5.71 million kilometers. Attempts to detect signals from Shin'en 2 at the time proved to be unsuccessful.

241

DESPATCH / ArtSat-2

Nation: Japan (13)
Objective(s): heliocentric orbit
Spacecraft: DESPATCH / Fuji-Oscar 81
Spacecraft Mass: 30 kg
Mission Design and Management: Tama Art University
Launch Vehicle: H-IIA (no. F26)
Launch Date and Time: 3 December 2014 / 04:22:04 UT
Launch Site: Tanegashima / Y1
Scientific Instruments: [none]

Results: DESPATCH (Deep Space Amateur Troubadour's Challenge) or ArtSat-2 was a project developed at Tama Art University to build a 3D-printed sculpture and launch it into deep space. The sculpture, measuring 50 × 50 × 45 centimeters, contained a 7-watt CW radio transmitter that operated in an amateur UHF frequency (437.325 MHz) and transmitted automatically generated "poetic" messages (in English) from space using telemetry numbers derived from voltage and temperature readings on board. As a whole, the project was designed to combine art and technology into a singular whole, to create a "deep-space sculpture" while also producing "generative poetry." The on-board batteries were designed to work for only seven days until the spacecraft was about 3 million kilometers from Earth. After launch and insertion into heliocentric orbit, it separated from the H-IIA

A full-scale prototype (engineering model) of DESPATCH.
Credit: ArtSat

upper stage around 06:21 UT on 3 December 2014. ArtSat personnel received transmissions of its "cosmic poem" after launch, but the poem was transmitted at irregular intervals. By 14 December, ArtSat-2 was already 4.7 million kilometers from Earth and in a 0.7 × 1.3 AU orbit and still transmitting. On 3 January 2015, the Tama Art University team concluded attempts to receive further transmissions from ArtSat-2 since the on-board battery had a lifetime of 27 days after launch. Despite its death, it has fully achieved its goal of being "the most distant artwork in the world." Like its sister spacecraft, Hayabusa-2, PROCYON, and Shin'en 2, DESPATCH performed an Earth flyby on 4 December at a range of 5.8 million kilometers. DESPATCH was also known as Fuji-Oscar 81.

2015

DSCOVR

Nation: USA (100)

Objective(s): Sun–Earth L1 Lagrange Point

Spacecraft: Triana

Spacecraft Mass: 570 kg

Mission Design and Management: NASA / NOAA / USAF

Launch Vehicle: Falcon 9 v1.1

Launch Date and Time: 11 February 2015 / 23:03:02 UT

Launch Site: Cape Canaveral / SLC-40

Scientific Instruments:

1. PlasMag plasma-magnetometer (magnetometer, Faraday cup, electrostatic analyzer)
2. Earth Polychromatic Imaging Camera (EPIC)
3. National Institute of Standards and Technology Advanced Radiometer (NISTAR)

Results: Deep Space Climate Observatory (DSCOVR) is a joint mission between NASA, NOAA, and the USAF designed as a successor NASA's Advanced Composition Explorer (ACE), whose goal is to provide real-time solar wind observations from an L1 orbit. The roots of the project date back to Triana, originally conceived in 1998 by then-Vice President Albert A. Gore, Jr. (1948–), as a NASA Earth science mission to provide a (near) continuous view of Earth from space (and also use a radiometer to take direct measurements of sunlight reflected and emitted from Earth). Despite being originally slated for launch on STS-107 (the tragic mission of Space Shuttle Columbia in 2003), Triana was canceled in 2001 and the satellite put into storage. Seven years later, in 2008, the Committee on Space Environmental Sensor Mitigation Options (CSESMO) determined

that using that spacecraft would be "the optimal solution for meeting NOAA and USAF space weather requirements." The satellite was removed from storage in November 2008 and recertified for launch with some modifications. The satellite, designed on the basis of the Small Explorer Program (SMEX-Lite) bus, was launched into an initial orbit of 184 × 186 kilometers at 37° inclination. About 30 minutes following launch, Falcon 9's second stage re-ignited to boost DSCOVR into a 187 × 1,371,156-kilometer transfer orbit at 37° inclination. By 24 February, the spacecraft had reached the halfway mark to the L1 position, traveling nearly 0.8 million kilometers. By 8 June, 100 days after launch, DSCOVR finally reached the Sun–Earth L1 point, and entered a Lissajous orbit, about 1.5 million kilometers from Earth, where it has a continuous view of the Sun and the sunlit side of Earth. Its primary mission is to provide a suite of diverse data on variations in the solar wind, provide early warning on coronal mass ejections, and observe Earth climate changes in ozone, aerosols, dust and volcanic ash, cloud altitudes, and vegetation cover. DSCOVR also takes full Earth pictures about every 2 hours, returning its first views of the entire sunlit side of Earth from approximately 1.6 million kilometers (using the EPIC instrument). In October 2015, NASA launched a Web site that posted at least a dozen new color images every day from EPIC. Earlier, on 16–17 July, the spacecraft took striking images of the Moon moving over the Pacific Ocean near North America. Similar pictures, showing the farside of the Moon had been taken by Deep Impact in May 2008 but from a much further distance of 50 million kilometers. On 28 October 2015, NASA officially handed over control of DSCOVR to NOAA (more specifically, its Space Weather Prediction Center, SWPC) for the latter agency to begin optimizing the final settings for its space weather instruments. The spacecraft completed its first year in deep space on 11 February 2016, now serving as the U.S.'s primary warning system for solar magnetic storms and solar wind data. Real-time data from DSCOVR and space weather forecasts were available to the general public beginning July 2016 from the SWPC Web site; the center also began coordinating the work of DSCOVR with GOES-16 which was launched on 19 November 2016.

243

LISA Pathfinder

Nation: ESA (9)
Objective(s): Sun–Earth L1 Lagrange Point
Spacecraft: SMART-2
Spacecraft Mass: 1,910 kg
Mission Design and Management: ESA
Launch Vehicle: Vega (no. VV06)
Launch Date and Time: 3 December 2015 / 04:04:00 UT
Launch Site: Kourou / ELV
Scientific Instruments:

1. LISA Technology Package (LTP)
2. Disturbance Reduction System (DRS)

Results: LISA Pathfinder is a technology demonstrator for future spaceborne gravitational wave detectors, such as the proposed Evolved Laser Interferometer Space Antenna (eLISA), tentatively planned for 2034. The spacecraft is equipped to test one of the key concepts behind gravitational wave detectors, that free particles follow geodesics in space-time. It does this by tracking the relative motions of two test masses in nearly perfect gravitational free fall, using picometer resolution laser interferometry. The name "LISA" comes from "Laser Interferometer Space Antenna," an earlier abandoned concept for an observatory to study gravitational waves. The spacecraft has a main science spacecraft and a separable propulsion module, the latter used for raising LISA Pathfinder's

orbit after launch and sending it to its operational Lissajous orbit around the Sun–Earth L1 Lagrange Point. The science spacecraft carries two test packages. The first, LTP, contains two identical test masses (Gravitational Reference Sensors), each weighing 2 kilograms in the form of 46-mm gold-platinum cubes suspended in its own vacuum container. Contributors include teams from France, Germany, Italy, the Netherlands, Spain, Switzerland, and the UK. The second, DRS, is a NASA-built system, originally from the canceled Space Technology 7 mission, made up of two clusters of colloidal micro-propulsion thrusters and a computer. LISA Pathfinder was launched into an initial orbit of 208 × 1,165 kilometers. Vega's liquid propellant fourth stage (known as AVUM) then refired to put the spacecraft into a 209 × 1,521-kilometer orbit at 6.0° inclination. The propulsion module then slowly fired six times, thus raising LISA Pathfinder's apogee until it began a cruise to L1. After a six-week trip and a final 64-second firing, the spacecraft arrived in a Lissajous orbit around L1 on 22 January 2016. At 11:30 UT, the propulsion module separated from the science section. Its final orbit was a roughly 500,000 × 800,000-kilometer orbit around L1. About two weeks later, on 3 February, ESA controllers at the European Space Operations Center (ESOC) at Darmstadt, Germany, retracted eight locking "fingers" pressing on the two gold-platinum cubes, at the time held in position by two rods. These rods were retracted from the first test mass on 15 February, and from the second the following day. Finally, on 22 February, controllers set the two cubes completely free to move under the effect of gravity alone, with actively maneuvering spacecraft around them. In the subsequent few months, the LISA Pathfinder team applied a number of different forces to the cubes to study their reaction. For example, one experiment involved raising the temperature of the housing, thus heating the very few gas molecules remaining in there to measure the effect on the cubes. After several months of successful experiments, at 08:00 UT on 25 June 2016,

LISA Pathfinder operates from a vantage point in space about 1.5 million kilometers from Earth towards the Sun, orbiting the first Sun–Earth Lagrange point, L1. *Credit: ESA-C. Carreau*

the LTP completed its "nominal operations phase," thus transitioning to work on NASA's DRS experiment. Almost simultaneously, on 21–22 June, ESA approved a mission extension, beginning 1 November, for seven months, during which investigators will be working with LTP. The mission was finally concluded on 30 June 2017. Earlier, in June 2016, mission scientists published the first results of the LISA Pathfinder experiments, confirming that the two cubes were falling freely under the influence of gravity alone, to a precision level more than five times better than originally expected.

2016

ExoMars Trace Gas Orbiter / Schiaparelli EDM Lander

Nation: ESA / Russia (1)
Objective(s): Mars orbit and landing
Spacecraft: TGO / EDM
Spacecraft Mass: 4,332 kg total including 3,755 kg TGO and 577 kg Schiaparelli EDM
Mission Design and Management:
Launch Vehicle: Proton-M + Briz-M (8K82M no. 93560 + Briz-M no. 99560)
Launch Date and Time: 14 March 2016 / 09:31 UT
Launch Site: Baikonur Cosmodrome / Site 200/39
Scientific Instruments:

TGO:

1. Nadir and Occultation for Mars Discovery spectrometer (NOMAD)
2. Atmospheric Chemistry Suite spectrometers (ACS)
3. Color and Stereo Surface Imaging System (CaSSIS)
4. Fine-Resolution Epithermal Neutron Detector (FREND)

EDM:

1. Dust Characterization, Risk Assessment, and Environmental Analyzer on the Martian Surface (DREAMS)
2. Atmospheric Mars Entry and Landing Investigation and Analysis (AMELIA)
3. Descent Camera (DECA)
4. Combined Aerothermal Sensor Package (COMARS+)
5. Instrument for Landing – Roving Laser Retroreflector Investigations (INRRI)

Results: The first in a series of joint missions under the ExoMars program between ESA and Roskosmos, the Russian space agency, the ExoMars Trace Gas Orbiter (TGO) is designed to study methane and other atmospheric trace gases present in small concentrations in the Martian atmosphere. Despite their relative scarcity (less than 1% of the atmosphere), studying these gases could provide evidence for possible biological or geological activity. Organisms on Earth release methane when they digest nutrients, although geological processes (such as the oxidation of certain minerals) can also release methane. ExoMars TGO was originally a collaborative project with NASA but the latter's contribution was cut due to lack of support in 2012, leading to cooperation with the Russians. As currently envisioned, the two ExoMars missions involve the TGO and Schiaparelli (first mission) and a lander-rover (second mission) in 2020, both launched by the Russians, who also are contributing hardware to both missions. Besides its primary mission, ExoMars TGO also had two secondary missions: to deliver the Schiaparelli Entry, Descent, and Landing Demonstrator Module (EDM), an Italian-built lander designed to test technologies planned for use in future soft-landings on Mars; and to serve as a data relay to support communications for the ExoMars 2020 rover and surface science platform. The orbiter included instruments developed by Belgium (NOMAD), Russia (ACS and FREND), and Switzerland (CaSSIS). EDM's scientific mission was limited to measurements of several atmospheric parameters (such as wind speed and direction, humidity, pressure, etc.). Its small technical camera, weighing 0.6 kilograms, was a refurbished spare flight model of the Visual Monitoring Camera flown on ESA's Herschel and

Planck spacecraft. Teams from Italy, France, the Netherlands, Finland, Spain, and Belgium contributed to EDM. The Russian Proton-M + Briz-M combination inserted the payload stack into an initial 185 × 185-kilometer orbit at 51.5° inclination. A second Briz-M firing after one orbit raised the orbit to 250 × 5,800 kilometers, and a third burn raised it further to 696 × 21,086 kilometers. A final fourth firing sent the payload to escape velocity with spacecraft release from Briz-M occurring at 20:13 UT on the day of the launch. On 14 March 2016, ExoMars adjusted its trajectory with a 52-minute burn. Finally, on 19 October, TGO fired its engine for 139 minutes from 13:05 to 15:24 UT to enter its planned initial orbit around Mars of 346 × 95,228 kilometers at 9.7° inclination. Three days earlier, at 14:42 UT on 16 October, the orbiter had released the EDM that was programmed to autonomously perform an automated landing sequence. After it entered the atmosphere at 14:42 UT on 19 October, the sequence would include parachute deployment, front heat shield release (at between 11 and 7 kilometers altitude), retrorocket braking (starting at 1,100 meters altitude), and a final freefall from a height of 2 meters, cushioned by a crushable structure. During this phase, Schiaparelli was to have captured 15 images of the approaching surface. Unfortunately, the signal from the lander was lost a short time before the planned landing sequence initiated. Later analysis showed that the parachute deployed normally at an altitude of 12 kilometers and the heatshield was released as planned at 7.8 kilometers. However, at some point, an inertial measurement unit sent an incorrect reading of the vehicle's rotation, which in turn generated an incorrect estimated altitude (in fact, a negative altitude) which triggered premature release of the parachute and backshell, firing of the thrusters and activation of ground systems even though the lander was still at an altitude of 3.7 kilometers. As such, the Schiaparelli simply plummeted from that altitude to the ground at near terminal velocity and was destroyed. During atmospheric entry, some data was collected by the COMARS+ instrument. The day after the crash, NASA's Mars Reconnaissance Orbiter (MRO) photographed the crash site in Meridiani Planum. Higher resolution images taken on 25 October and 1 November (the latter in color) showed more detail. Impact coordinates were 6.11° W / 2.07° S. The orbiter, meanwhile, began several months of aerobraking to reach its nominal 400-kilometer circular orbit for its primary science mission slated to begin in late 2017. In November 2016, TGO was in a highly elliptical 230–310 × 98,0000-kilometer orbit with an orbital period of 4.2 days. Between 20 and 28 November, it tested and calibrated its four scientific instruments for the first time. In January 2017, the orbiter carried out three maneuvers by firing its main engine to adjust inclination to 74°, necessary for its primary science mission to begin later in the year. By March 2017, TGO was in a one-day 200 × 33,000-kilometer orbit and was using the atmosphere to adjust the orbit by a process of gradual aerobraking. The goal was to achieve a final operational circular orbit at 400 kilometers by March 2018 at which time full-scale science operations would begin. Earlier, in December 2016, ESA announced the formal approval of the second joint European-Russian ExoMars mission, tentatively slated for launch in 2020.

245

OSIRIS-REx

Nation: USA (101)

Objective(s): asteroid sample return

Spacecraft: OSIRIS-REx

Spacecraft Mass: 2,110 kg

Mission Design and Management: NASA GSFC / University of Arizona

Launch Vehicle: Atlas V 411 (no. AV-067)

Launch Date and Time: 8 September 2016 / 23:05 UT

Launch Site: Cape Canaveral / SLC-41

Scientific Instruments:

1. Camera Suite (PolyCam, MapCam, SamCam) (OCAMS OSIRIS-REx)
2. Laser Altimeter (OLA OSIRIS-REx)
3. Visible and IR Spectrometer (OVIRS OSIRIS-REx)
4. Thermal Emission Spectrometer (OTES OSIRIS-REx)
5. Regolith X-Ray Imaging Spectrometer (REXIS OSIRIS-REx)
6. Touch-And-Go Sample Acquisition Mechanism (TAGSAM)

Results: The Origins, Spectral Interpretation, Resource Identification, Security, Regolith Explorer (OSIRIS-REx) mission is the third major planetary science mission falling under NASA's New Frontiers Program (after New Horizons launched in 2006 and Juno launched in 2011). The goal of the mission is to reach a near-Earth asteroid 101955 Bennu (formerly known as 1999 RQ36), collect a 59.5-gram sample, and then return it to Earth. The science mission, developed by scientists at the University of Arizona, will open up the possibilities to glean more information on how planets formed and how life began and help scientists understand asteroids that could impact Earth in the future. About 55 minutes after launch, after a boost by the Centaur upper stage, OSIRIS-REx separated from the Atlas V and the solar arrays deployed. At 17:30 UT, on 9 September, the spacecraft crossed the orbital path of the Moon at a range of 386,500 kilometers. Three days later, it was in heliocentric orbit at 0.77 × 1.17 AU. Beginning 19 September, the mission team activated all of its scientific instruments. The larger Trajectory Correction Maneuver (TCM) thrusters were fired (for 12 seconds) for the first time on 7 October for a mid-course correction. The spacecraft also carries three other sets of thrusters—the Attitude Control System (ACS), a Main Engine (ME), and Low Thrust Reaction Engine Assembly (LTR) thrusters—thus providing significant redundancy for maneuvers. On 28 December 2016, the spacecraft conducted its first Deep Space Maneuver (DSM-1), firing the main engine to position it properly for an Earth gravity-assist

NASA's OSIRIS-REx spacecraft is shown here in an artist's impression. OSIRIS-REx is the third mission in NASA's New Frontiers Program. *Credit: NASA/University of Arizona*

encounter in late 2017. A second firing, the first to use the spacecraft's Attitude Control System (ACS) thrusters, on 25 August 2017, further sharpened its trajectory by changing the velocity by 47.9 centimeters/second. About a month later, on 22 September, OSIRIS-REx passed over Earth at a range of 17,237 kilometers as part of a gravity-assist maneuver that tilted its orbit to match Bennu. During the encounter, the spacecraft took several high-resolution pictures of both Earth and the Moon. The actual asteroid encounter is scheduled to begin in August 2018, culminating in a rendezvous with Bennu. OSIRIS-REx will survey the asteroid for about a year beginning October 2018 and select a final touchdown site. The actual sample collection will be carried out by the TAGSAM instrument which will release a burst of nitrogen gas to blow regolith particles into a sampler head at the end of a robotic arm. The spacecraft is capable of returning to the asteroid in case of a first failed attempt at sample collection. In March 2021, there will be a window to depart from Bennu, allowing OSIRIS-REx to begin its Earthward return trip. If all goes well, the spacecraft will return to Earth in September 2023, when the sample return capsule will separate from the main spacecraft and enter Earth's atmosphere, landing at the Utah Test and Training Range.

Tables

Table 1. Master Table of All Launch Attempts for Deep Space, Lunar, and Planetary Probes 1958–2016

	Official Name	Spacecraft / no.	Mass	Launch date / time	Launch place / pad¹	Launch vehicle / no.	Nation	Design and Operation	Objective	Outcome²
	1958									
1	[Pioneer 0]	Able 1	38.5 kg	08-17-58 / 12:18	CC / 17A	Thor Able I / 127	USA	ARPA / AFBMD	lunar orbit	F
2	[Luna]	Ye-1 / 1	c. 360 kg	09-23-58 / 07:03:23	NIIP-5 / 1	Luna / B1-3	USSR	OKB-1	lunar impact	F
3	Pioneer	Able 2	38.3 kg	10-11-58 / 08:42:13	CC / 17A	Thor Able I / 1	USA	NASA / AFBMD	lunar orbit	F
4	[Luna]	Ye-1 / 2	c. 360 kg	10-11-58 / 21:41:58	NIIP-5 / 1	Luna / B1-4	USSR	OKB-1	lunar impact	F
5	Pioneer II	Able 3	39.6 kg	11-08-58 / 07:30:20	CC / 17A	Thor-Able I / 2	USA	NASA / AFBMD	lunar orbit	F
6	[Luna]	Ye-1 / 3	c. 360 kg	12-04-58 / 18:18:44	NIIP-5 / 1	Luna / B1-5	USSR	OKB-1	lunar impact	F
7	Pioneer III	—	5.87 kg	12-06-58 / 05:44:52	CC / 5	Juno II / AM-11	USA	NASA / ABMA / JPL	lunar flyby	F
	1959									
8	Soviet Space Rocket [Luna 1]	Ye-1 / 4	361.3 kg	01-02-59 / 16:41:21	NIIP-5 / 1	Luna / B1-6	USSR	OKB-1	lunar impact	P
9	Pioneer IV	—	6.08 kg	03-03-59 / 05:10:56	CC / 5	Juno II / AM-14	USA	NASA / ABMA / JPL	lunar flyby	P
10	[Luna]	Ye-1A / 5	c. 390 kg	06-18-59 / 08:08	NIIP-5 / 1	Luna / 11-7	USSR	OKB-1	lunar impact	F
11	Second Soviet Space Rocket [Luna 2]	Ye-1A / 7	390.2 kg	09-12-59 / 06:39:42	NIIP-5 / 1	Luna / 11-7b	USSR	OKB-1	lunar impact	S
12	Automatic Interplanetary Station [Luna 3]	Ye-2A / 1	278.5 kg	10-04-59 / 00:43:40	NIIP-5 / 1	Luna / 11-8	USSR	OKB-1	lunar flyby	S
13	[Pioneer, P-3]	P-3 / Able IVB	168.7 kg	11-26-59 / 07:26	CC / 14	Atlas Able / 1	USA	NASA / AFBMD	lunar orbit	F
	1960									
14	Pioneer V	P-2 / Able 6	43.2 kg	03-11-60 / 13:00:07	CC / 17A	Thor Able IV / 4	USA	NASA / AFBMD	solar orbit	S
15	[Luna]	Ye-3 / 1	?	04-15-60 / 15:06:44	NIIP-5 / 1	Luna / 11-9	USSR	OKB-1	lunar flyby	F
16	[Luna]	Ye-3 / 2	?	04-19-60 / 16:07:43	NIIP-5 / 1	Luna / 11-9a	USSR	OKB-1	lunar flyby	F
17	[Pioneer, P-30]	P-30 / Able VA	175.5 kg	09-25-60 / 15:13	CC / 12	Atlas Able / 2	USA	NASA / AFBMD	lunar orbit	F
18	[Mars]	1M / 1	480 kg	10-10-60 / 14:27:49	NIIP-5 / 1	Molniya / L1-4M	USSR	OKB-1	Mars flyby	F

1 CC = Cape Canaveral, CK = Cape Kennedy, CSG = Le Centre Spatial Guyanais, GIK-5 = Baikonur, KSC = Kennedy Space Center, MARS = Mid-Atlantic Regional Spaceport, NIIP-5 = Baikonur, SHAR = Sriharikota, VAFB = Vandenberg Air Force Base

2 F = failure, IP = in progress, P = partial success, S = success

	Official Name	Spacecraft/no.	Mass	Launch date/time	Launch place/pad[1]	Launch vehicle/no.	Nation	Design and Operation	Objective	Outcome[2]
19	[Mars]	1M/2	480 kg	10-14-60/13:51:03	NIIP-5/1	Molniya/L1-5M	USSR	OKB-1	Mars flyby	F
20	[Pioneer, P-31]	P-31/Able VB	176 kg	12-15-60/09:10	CC/12	Atlas Able/3	USA	NASA/AFBMD	lunar orbit	F
1961										
21	Heavy Satellite [Venera]	1VA/1	c. 645 kg	02-04-61/01:18:04	NIIP-5/1	Molniya/L1-7V	USSR	OKB-1	Venus impact	F
22	Automatic Interplanetary Station [Venera 1]	1VA/2	643.5 kg	02-12-61/00:34:38	NIIP-5/1	Molniya/L1-6V	USSR	OKB-1	Venus impact	F
23	Ranger I	P-32	306.18 kg	08-23-61/10:04	CC/12	Atlas Agena B/1	USA	NASA/JPL	deep space orbit	F
24	Ranger II	P-33	306.18 kg	11-18-61/08:12	CC/12	Atlas Agena B/2	USA	NASA/JPL	deep space orbit	F
1962										
25	Ranger III	P-34	330 kg	01-26-62/20:30	CC/12	Atlas-Agena B/3	USA	NASA/JPL	lunar impact	F
26	Ranger IV	P-35	331.12 kg	04-23-62/20:50	CC/12	Atlas-Agena B/4	USA	NASA JPL	lunar impact	P
27	Mariner I	P-37	202.8 kg	07-22-62/09:21:23	CC/12	Atlas-Agena B/5	USA	NASA/JPL	Venus flyby	F
28	[Venera]	2MV-1/3	1,097 kg	08-25-62/02:56:06	NIIP-5/1	Molniya/T103-12	USSR	OKB-1	Venus impact	F
29	Mariner II	P-38	203.6 kg	08-27-62/06:53:14	CC/12	Atlas-Agena B/6	USA	NASA/JPL	Venus flyby	S
30	[Venera]	2MV-1/4	c. 1,100 kg	09-01-62/02:12:33	NIIP-5/1	Molniya/T103-13	USSR	OKB-1	Venus impact	F
31	[Venera]	2MV-2/1	?	09-12-62/00:59:13	NIIP-5/1	Molniya/T103-14	USSR	OKB-1	Venus flyby	F
32	Ranger V	P-36	342.46 kg	10-18-62/16:59:00	CC/12	Atlas-Agena B/7	USA	NASA/JPL	lunar landing	P
33	[Mars]	2MV-4/3	c. 900 kg	10-24-62/17:55:04	NIIP-5/1	Molniya/T103-15	USSR	OKB-1	Mars flyby	F
34	Mars 1	2MV-4/4	893.5 kg	11-01-62/16:14:06	NIIP-5/1	Molniya/T103-16	USSR	OKB-1	Mars flyby	P
35	[Mars]	2MV-3/1	?	11-04-62/15:35:14	NIIP-5/1	Molniya/T103-17	USSR	OKB-1	Mars impact	F
1963										
36	[Luna]	Ye-6/2	1,420 kg	01-04-63/08:48:58	NIIP-5/1	Molniya/T103-09	USSR	OKB-1	lunar landing	F
37	[Luna]	Ye-6/3	1,420 kg	02-03-63/09:29:14	NIIP-5/1	Molniya/T103-10	USSR	OKB-1	lunar landing	F
38	Luna 4	Ye-6/4	1,422 kg	04-02-63/08:16:38	NIIP-5/1	Molniya/T103-11	USSR	OKB-1	lunar landing	F
39	Kosmos 21 [Zond]	3MV-1A/2 or 1	c. 800 kg	11-11-63/06:23:34	NIIP-5/1	Molniya/G103-18	USSR	OKB-1	deep space and recovery	F

	Official Name	Spacecraft / no.	Mass	Launch date / time	Launch place / pad[1]	Launch vehicle / no.	Nation	Design and Operation	Objective	Outcome[2]
					1964					
40	Ranger VI	Ranger-A / P-53	364.69 kg	01-30-64 / 15:49:09	CK / 12	Atlas-Agena B / 8	USA	NASA / JPL	lunar impact	P
41	[Zond]	3MV-1A / 4A or 2	c. 800 kg	02-19-64 / 05:47:40	NIIP-5 / 1	Molniya / G15000-26	USSR	OKB-1	Venus flyby	F
42	[Luna]	Ye-6 / 6	c. 1,420 kg	03-21-64 / 08:14:33	NIIP-5 / 1	Molniya-M / T15000-20	USSR	OKB-1	lunar landing	F
43	Kosmos 27 [Venera]	3MV-1 / 5	948 kg	03-27-64 / 03:24:43	NIIP-5 / 1	Molniya / T15000-27	USSR	OKB-1	Venus impact	F
44	Zond 1 [Venera]	3MV-1 / 4	948 kg	04-02-64 / 02:42:40	NIIP-5 / 1	Molniya / T15000-28	USSR	OKB-1	Venus impact	F
45	[Luna]	Ye-6 / 5	c. 1,420 kg	04-20-64 / 08:08:28	NIIP-5 / 1	Molniya-M / T15000-21	USSR	OKB-1	lunar landing	F
46	Ranger VII	Ranger-B / P-54	365.6 kg	07-28-64 / 16:50:07	CK / 12	Atlas-Agena B / 9	USA	NASA / JPL	lunar impact	S
47	Mariner III	Mariner-64C	260.8 kg	11-05-64 / 19:22:05	CK / 13	Atlas-Agena D / 11	USA	NASA / JPL	Mars flyby	F
48	Mariner IV	Mariner-64D	260.8 kg	11-28-64 / 14:22:01	CK / 12	Atlas-Agena D / 12	USA	NASA / JPL	Mars flyby	S
49	Zond 2	3MV-4A / 2	996 kg	11-30-64 / 13:25	NIIP-5 / 1	Molniya / G15000-29	USSR	OKB-1	Mars flyby	P
					1965					
50	Ranger VIII	Ranger-C	366.87 kg	02-17-65 / 17:05:00	CK / 12	Atlas-Agena B / 13	USA	NASA / JPL	lunar impact	S
51	[Surveyor Model]	SD-1	951 kg	03-02-65 / 13:25	CK / 36A	Atlas Centaur / 5	USA	NASA / JPL	deep space orbit	F
52	Kosmos 60 [Luna]	Ye-6 / 9	c. 1,470 kg	03-12-65 / 09:25	NIIP-5 / 1	Molniya / G15000-24	USSR	OKB-1	lunar landing	F
53	Ranger IX	Ranger-D	366.87 kg	03-21-65 / 21:37:02	CK / 12	Atlas-Agena B / 14	USA	NASA / JPL	lunar impact	S
54	[Luna]	Ye-6 / 8	c. 1,470 kg	04-10-65 / 08:39	NIIP-5 / 1	Molniya-M / R103-26	USSR	OKB-1	lunar landing	F
55	Luna 5	Ye-6 / 10	1,476 kg	05-09-65 / 07:49:37	NIIP-5 / 1	Molniya-M / U103-30	USSR	OKB-1	lunar landing	F
56	Luna 6	Ye-6 / 7	1,442 kg	06-08-65 / 07:40	NIIP-5 / 1	Molniya-M / U103-31	USSR	OKB-1	lunar landing	F
57	Zond 3	3MV-4 / 3	950 kg	07-18-65 / 14:32	NIIP-5 / 1	Molniya / U103-32	USSR	OKB-1	lunar flyby	S
58	Surveyor Model 1	SD-2	950 kg	08-11-65 / 14:31:04	CK / 36B	Atlas Centaur / 6	USA	NASA / JPL	deep space orbit	S
59	Luna 7	Ye-6 / 11	1,506 kg	10-04-65 / 07:56:40	NIIP-5 / 1	Molniya-M / U103-27	USSR	OKB-1	lunar landing	F
60	Venera 2	3MV-4 / 4	963 kg	11-12-65 / 04:46	NIIP-5 / 31	Molniya-M / U103-42	USSR	OKB-1	Venus flyby	S
61	Venera 3	3MV-3 / 1	960 kg	11-16-65 / 04:13	NIIP-5 / 31	Molniya / U103-31	USSR	OKB-1	Venus impact	S
62	Kosmos 96 [Venera]	3MV-4 / 6	c. 950 kg	11-23-65 / 03:14	NIIP-5 / 31	Molniya / U103-30	USSR	OKB-1	Venus flyby	F
63	Luna 8	Ye-6 / 12	1,552 kg	12-03-65 / 10:46:14	NIIP-5 / 31	Molniya-M / U103-28	USSR	OKB-1	lunar landing	F
64	Pioneer VI	Pioneer-A	62.14 kg	12-16-65 / 07:31:21	CK / 17A	Thor Delta E / 35	USA	NASA / ARC	solar orbit	S

	Official Name	Spacecraft / no.	Mass	Launch date / time	Launch place / pad[1]	Launch vehicle / no.	Nation	Design and Operation	Objective	Outcome[2]
1966										
65	Luna 9	Ye-6M / 202	1,583.7 kg	01-31-66 / 11:41:37	NIIP-5 / 31	Molniya-M / U103-32	USSR	Lavochkin	lunar landing	S
66	Kosmos 111 [Luna]	Ye-6S / 204	c. 1,580 kg	03-01-66 / 11:03:49	NIIP-5 / 31	Molniya-M / N103-41	USSR	Lavochkin	lunar orbit	F
67	Luna 10	Ye-6S / 206	1,583.7 kg	03-31-66 / 10:46:59	NIIP-5 / 31	Molniya-M / N103-42	USSR	Lavochkin	lunar orbit	S
68	Surveyor Model 2	SD-3	784 kg	04-08-66 / 01:00:02	CK / 36B	Atlas Centaur / 8	USA	NASA / JPL	deep space orbit	F
69	Surveyor I	Surveyor-A	995.2 kg	05-30-66 / 14:41:01	CK / 36A	Atlas Centaur / 10	USA	NASA / JPL	lunar landing	S
70	Explorer XXXIII	AIMP-D	93.4 kg	07-01-66 / 16:02:25	CK / 17A	Thor Delta E1 / 39	USA	NASA / GSFC	lunar orbit	P
71	Lunar Orbiter I	LO-A	386.9 kg	08-10-66 / 19:26:00	CK / 13	Atlas Agena D / 17	USA	NASA / LaRC	lunar orbit	S
72	Pioneer VII	Pioneer-B	62.75 kg	08-17-66 / 15:20:17	CK / 17A	Thor Delta E1 / 40	USA	NASA / ARC	solar orbit	S
73	Luna 11	Ye-6LF / 101	1,640 kg	08-24-66 / 08:03:21	NIIP-5 / 31	Molniya-M / N103-43	USSR	Lavochkin	lunar orbit	S
74	Surveyor II	Surveyor-B	995.2 kg	09-20-66 / 12:32:00	CK / 36A	Atlas Centaur / 7	USA	NASA / JPL	lunar landing	F
75	Luna 12	Ye-6LF / 102	1,640 kg	10-22-66 / 08:42:26	NIIP-5 / 31	Molniya-M / N103-44	USSR	Lavochkin	lunar orbit	S
76	Lunar Orbiter II	S/C 5	385.6 kg	11-06-66 / 23:21:00	CK / 13	Atlas Agena D / 18	USA	NASA / LaRC	lunar orbit	S
77	Luna 13	Ye-6M / 205	1,620 kg	12-21-66 / 10:17:08	NIIP-5 / 1	Molniya-M / N103-45	USSR	Lavochkin	lunar landing	S
1967										
78	Lunar Orbiter III	S/C 6	385.6 kg	02-05-67 / 01:17:01	CK / 13	Atlas Agena D / 20	USA	NASA / LaRC	lunar orbit	S
79	Surveyor III	Surveyor-C	997.9 kg	04-17-67 / 07:05:01	CK / 36B	Atlas Centaur / 12	USA	NASA / JPL	lunar landing	S
80	Lunar Orbiter IV	S/C 7	385.6 kg	05-04-67 / 22:25:00	CK / 13	Atlas Agena D / 22	USA	NASA / LaRC	lunar orbit	S
81	Kosmos 159 [Luna]	Ye-6LS / 111	1,640 kg	05-16-67 / 21:43:57	NIIP-5 / 1	Molniya-M / Ya716-56	USSR	Lavochkin	lunar orbit	P
82	Venera 4	1V / 310	1,106 kg	06-12-67 / 02:39:45	NIIP-5 / 1	Molniya-M / Ya716-70	USSR	Lavochkin	Venus impact	S
83	Mariner V	Mariner-67E	244.9 kg	06-14-67 / 06:01:00	CK / 12	Atlas Agena D / 23	USA	NASA / JPL	Venus flyby	S
84	Kosmos 167 [Venera]	1V / 311	c. 1,100 kg	06-17-67 / 02:36:38	NIIP-5 / 1	Molniya-M / Y7 6-71	USSR	Lavochkin	Venus impact	F
85	Surveyor IV	Surveyor-D	1,037.4 kg	07-14-67 / 11:53:29	CK / 36A	Atlas Centaur / 11	USA	NASA / JPL	lunar landing	F
86	Explorer XXXV	AIMP-E	104.3 kg	07-19-67 / 14:19:02	CK / 17B	Thor Delta E1 / 50	USA	NASA / GSFC	lunar orbit	S
87	Lunar Orbiter V	S/C 3	385.6 kg	08-01-67 / 22:33:00	CK / 13	Atlas Agena D / 24	USA	NASA / LaRC	lunar orbit	S
88	Surveyor V	Surveyor-E	1,006 kg	09-08-67 / 07:57:01	CK / 36B	Atlas Centaur / 13	USA	NASA / JPL	lunar landing	S

	Official Name	Spacecraft / no.	Mass	Launch date / time	Launch place / pad[1]	Launch vehicle / no.	Nation	Design and Operation	Objective	Outcome[2]
89	[Zond]	7K-L1 / 4L	c. 5,375 kg	09-27-67 / 22:11:54	NIIP-5 / 81L	Proton-K / 229-01	USSR	TsKBEM	circumlunar	F
90	Surveyor VI	Surveyor-F	1,008.3 kg	11-07-67 / 07:39:01	CK / 36B	Atlas Centaur / 14	USA	NASA / JPL	lunar landing	S
91	[Zond]	7K-L1 / 5L	c. 5,375 kg	11-22-67 / 19:07:59	NIIP-5 / 81P	Proton-K / 230-01	USSR	TsKBEM	circumlunar	F
92	Pioneer VIII	Pioneer-C	65.36 kg	12-13-67 / 14:08:00	CK / 17B	Thor Delta E-1 / 55	USA	NASA / ARC	solar orbit	S
				1968						
93	Surveyor VII	Surveyor-G	1,040.1 kg	01-07-68 / 06:30:00	CK / 36A	Atlas Centaur / 15	USA	NASA / JPL	lunar landing	S
94	[Luna]	Ye-6LS / 112	1,640 kg	02-07-68 / 10:43:54	NIIP-5 / 1	Molniya-M / Ya716-57	USSR	Lavochkin	lunar orbit	F
95	Zond 4	7K-L1 / 6L	c. 5,375 kg	03-02-68 / 18:29:23	NIIP-5 / 81L	Proton-K / 231-01	USSR	TsKBEM	deep space	P
96	Luna 14	Ye-6LS / 113	1,640 kg	04-07-68 / 10:09:32	NIIP-5 / 1	Molniya-M / Ya716-58	USSR	Lavochkin	lunar orbit	S
97	[Zond]	7K-L1 / 7L	c. 5,375 kg	04-22-68 / 23:01:27	NIIP-5 / 81P	Proton-K / 232-01	USSR	TsKBEM	circumlunar	F
98	Zond 5	7K-L1 / 9L	c. 5,375 kg	09-14-68 / 21:42:11	NIIP-5 / 81L	Proton-K / 234-01	USSR	TsKBEM	circumlunar	S
99	Pioneer IX	Pioneer-D	65.36	11-08-68 / 09:46:29	CK / 17B	Thor Delta E1 / 60	USA	NASA / ARC	solar orbit	S
100	Zond 6	7K-L1 / 12L	c. 5,375 kg	11-10-68 / 19:11:31	NIIP-5 / 81L	Proton-K / 235-01	USSR	TsKBEM	circumlunar	S
				1969						
101	Venera 5	2V / 330	1,130 kg	01-05-69 / 06:28:08	NIIP-5 / 1	Molniya-M / V716-72	USSR	Lavochkin	Venus landing	S
102	Venera 6	2V / 331	1,130 kg	01-10-69 / 05:51:52	NIIP-5 / 1	Molniya-M / V716-73	USSR	Lavochkin	Venus landing	S
103	[Zond]	7K-L1 / 13L	c. 5,375 kg	01-20-69 / 04:14:36	NIIP-5 / 81L	Proton-K / 237-01	USSR	TsKBEM	circumlunar	F
104	[Luna/Lunokhod]	Ye-8 / 201	c. 5,700 kg	02-19-69 / 06:48:48	NIIP-5 / 81P	Proton-K / 239-01	USSR	Lavochkin	lunar rover	F
105	[N1]	7K-L1S / 2	6,900 kg	02-21-69 / 09:18:07	NIIP-5 / 110P	N1 / 3L	USSR	TsKBEM	lunar orbit	F
106	Mariner VI	Mariner-69F	381 kg	02-25-69 / 01:29:02	CK / 36B	Atlas Centaur / 20	USA	NASA / JPL	Mars flyby	S
107	[Mars]	M-69 / 521	c. 4,850 kg	03-27-69 / 10:40:45	NIIP-5 / 81L	Proton-K / 240-01	USSR	Lavochkin	Mars orbit	F
108	Mariner VII	Mariner-69G	381 kg	03-27-69 / 22:22:01	CK / 36A	Atlas Centaur / 19	USA	NASA / JPL	Mars flyby	S
109	[Mars]	M-69 / 522	c. 4,850 kg	04-02-69 / 10:33:00	NIIP-5 / 81P	Proton-K / 233-01	USSR	Lavochkin	Mars orbit	F
110	[Luna]	Ye-8-5 / 402	c. 5,700 kg	06-14-69 / 04:00:48	NIIP-5 / 81P	Proton-K / 238-01	USSR	Lavochkin	lunar sample	F
111	[N1]	7K-L1S / 5	c. 6,900 kg	07-03-69 / 20:18:32	NIIP-5 / 110P	N1 / 15005	USSR	TsKBEM	lunar orbit	F
112	Luna 15	Ye-8-5 / 401	5,667 kg	07-13-69 / 02:54:42	NIIP-5 / 81P	Proton-K / 242-01	USSR	Lavochkin	lunar sample	F

	Official Name	Spacecraft / no.	Mass	Launch date / time	Launch place / pad[1]	Launch vehicle / no.	Nation	Design and Operation	Objective	Outcome[2]
113	Zond 7	7K-L1 / 11	c. 5,375 kg	08-07-69 / 23:48:06	NIIP-5 / 81L	Proton-K / 243-01	USSR	TsKBEM	circumlunar	S
114	[Pioneer]	Pioneer-E	65.4 kg	08-27-69 / 21:59:00	CK / 17A	Thor Delta L / D73	USA	NASA / ARC	solar orbit	F
115	Kosmos 300 [Luna]	Ye-8-5 / 403	c. 5,700 kg	09-23-69 / 14:07:37	NIIP-5 / 81P	Proton-K / 244-01	USSR	Lavochkin	lunar sample	F
116	Kosmos 305 [Luna]	Ye-8-5 / 404	c. 5,700 kg	10-22-69 / 14:09:59	NIIP-5 / 81P	Proton-K / 241-01	USSR	Lavochkin	lunar sample	F
				1970						
117	[Luna]	Ye-8-5 / 405	c. 5,700 kg	02-06-70 / 04:16:05	NIIP-5 / 81L	Proton-K / 247-01	USSR	Lavochkin	lunar sample	F
118	Venera 7	3V / 630	1,180 kg	08-17-70 / 05:38:22	NIIP-5 / 31	Molniya-M / Kh15000-62	USSR	Lavochkin	Venus landing	S
119	Kosmos 359 [Venera]	3V / 631	c. 1,200 kg	08-22-70 / 05:06:08	NIIP-5 / 31	Molniya-M / Kh15000-61	USSR	Lavochkin	Venus landing	F
120	Luna 16	Ye-8-5 / 406	5,725 kg	09-12-70 / 13:25:52	NIIP-5 / 81L	Proton-K / 248-01	USSR	Lavochkin	lunar sample	S
121	Zond 8	7K-L1 / 14	c. 5,375 kg	10-20-70 / 19:55:39	NIIP-5 / 81L	Proton-K / 250-01	USSR	TsKBEM	circumlunar	S
122	Luna 17 Lunokhod 1	Ye-8 / 203	5,700 kg	11-10-70 / 14:44:01	NIIP-5 / 81L	Proton-K / 251-01	USSR	Lavochkin	lunar rover	S
				1971						
123	Mariner 8	Mariner-71H	997.9 kg	05-09-71 / 01:11:02	CK / 36A	Atlas Centaur / 24	USA	NASA / JPL	Mars orbit	F
124	Kosmos 419 [Mars]	3MS / 170	4,549 kg	05-10-71 / 16:58:42	NIIP-5 / 81L	Proton-K / 253-01	USSR	Lavochkin	Mars orbit	F
125	Mars 2	4M / 171	4,625 kg	05-19-71 / 16:22:44	NIIP-5 / 81P	Proton-K / 255-01	USSR	Lavochkin	Mars orbit / landing	P
126	Mars 3	4M / 172	4,625 kg	05-28-71 / 15:26:30	NIIP-5 / 81L	Proton-K / 249-01	USSR	Lavochkin	Mars orbit / landing	P
127	Mariner 9	Mariner-71I	997.9 kg	05-30-71 / 22:23:04	CK / 36B	Atlas Centaur / 23	USA	NASA / JPL	Mars orbit	S
128	Apollo 15 Particle & Fields SubSat		35.6 kg	08-04-71 / 20:13:19	Apollo 15 CSM 112	Saturn V / 510	USA	NASA / MSC	lunar orbit	S
129	Luna 18	Ye-8-5 / 407	5,725 kg	09-02-71 / 13:40:40	NIIP-5 / 81P	Proton-K / 256-01	USSR	Lavochkin	lunar sample	F
130	Luna 19	Ye-8LS / 202	5,330 kg	09-28-71 / 10:00:22	NIIP-5 / 81P	Proton-K / 257-01	USSR	Lavochkin	lunar orbit	P
				1972						
131	Luna 20	Ye-8-5 / 408	5,725 kg	02-14-72 / 03:27:58	NIIP-5 / 81P	Proton-K / 258-01	USSR	Lavochkin	lunar sample	S
123	Pioneer 10	Pioneer-F	258 kg	03-02-72 / 01:49:04	CK / 36A	Atlas Centaur / 27	USA	NASA / ARC	Jupiter flyby	S

	Official Name	Spacecraft / no.	Mass	Launch date / time	Launch place / pad[1]	Launch vehicle / no.[1]	Nation	Design and Operation	Objective	Outcome[2]
133	Venera 8	3V / 670	1,184 kg	03-27-72 / 04:15:06	NIIP-5 / 31	Molniya-M / S1500-63	USSR	Lavochkin	Venus landing	S
134	Kosmos 482 [Venera]	3V / 671	c. 1,180 kg	03-31-72 / 04:02:33	NIIP-5 / 31	Molniya-M / S1500-64	USSR	Lavochkin	Venus landing	F
135	Apollo 16 Particle & Fields SubSat		42 kg	04-24-72 / 09:56:09	Apollo 16 CSM 113	Saturn V / 511	USA	NASA / MSC	lunar orbit	S
136	[N1]	7K-L0K / 6A	c. 9,500 kg	11-23-72 / 06:11:55	NIIP-5 / 110L	N1 / 15007	USSR	TsKBEM	lunar orbit	F
1973										
137	Luna 21 Lunokhod 2	Ye-8 / 204	5,700 kg	01-08-73 / 06:55:38	NIIP-5 / 81L	Proton-K / 259-01	USSR	Lavochkin	lunar rover	S
138	Pioneer 11	Pioneer-G	258.5 kg	04-06-73 / 02:11:00	CK / 36B	Atlas Centaur / 30	USA	NASA / ARC	Jupiter, Saturn flyby	S
139	Explorer 49	RAE-B	330.2 kg	06-10-73 / 14:13:00	CC / 17B	Delta 1913 / 95	USA	NASA / GSFC	lunar orbit	S
140	Mars 4	3MS / 52S	4,000 kg	07-21-73 / 19:30:59	NIIP-5 / 81L	Proton-K / 261-01	USSR	Lavochkin	Mars orbit	F
141	Mars 5	3MS / 53S	4,000 kg	07-25-73 / 18:55:48	NIIP-5 / 81P	Proton-K / 262-01	USSR	Lavochkin	Mars orbit	S
142	Mars 6	3MP / 50P	3,880 kg	08-05-73 / 17:45:48	NIIP-5 / 81L	Proton-K / 281-01	USSR	Lavochkin	Mars landing	P
143	Mars 7	3MP / 51P	3,880 kg	08-09-73 / 17:00:17	NIIP-5 / 81P	Proton-K / 281-02	USSR	Lavochkin	Mars landing	F
144	Mariner 10	Mariner-73J	502.9 kg	11-03-73 / 05:45:00	CC / 36B	Atlas Centaur / 34	USA	NASA / JPL	Mercury, Venus flyby	S
1974										
145	Luna 22	Ye-8LS / 206	5,700 kg	05-29-74 / 08:56:51	NIIP-5 / 81P	Proton-K / 282-02	USSR	Lavochkin	lunar orbit	S
146	Luna 23	Ye-8-5M / 410	5,795 kg	10-28-74 / 14:30:32	NIIP-5 / 81P	Proton-K / 285-01	USSR	Lavochkin	lunar sample	F
147	Helios 1	Helios-A	370 kg	11-10-74 / 07:11:02	CC / 41	Titan IIIE-Centaur / 2	FRG	DFVLR / NASA	solar orbit	S
1975										
148	Venera 9	4V-1 / 660	4,936 kg	06-08-75 / 02:38:00	NIIP-5 / 81P	Proton-K / 286-01	USSR	Lavochkin	Venus orbit / landing	S
149	Venera 10	4V-1 / 661	5,033 kg	06-14-75 / 03:00:31	NIIP-5 / 81P	Proton-K / 285-02	USSR	Lavochkin	Venus orbit / landing	S
150	Viking 1	Viking-B	3,527 kg	08-20-75 / 21:22:00	CC / 41	Titan IIIE-Centaur / 4	USA	NASA / JPL	Mars orbit / landing	S
151	Viking 2	Viking-A	3,527 kg	09-09-75 / 18:39:00	CC / 41	Titan IIIE-Centaur / 3	USA	NASA / JPL	Mars orbit / landing	S

	Official Name	Spacecraft / no.	Mass	Launch date / time	Launch place / pad[1]	Launch vehicle / no.	Nation	Design and Operation	Objective	Outcome[2]
152	[Luna]	Ye-8-5M / 412	5,795 kg	10-16-75 / 04:04:56	NIIP-5 / 81L	Proton-K / 287-02	USSR	Lavochkin	lunar sample	F
1976										
153	Helios 2	Helios-B	370 kg	01-15-76 / 05:34:00	CC / 41	Titan IIIE-Centaur / 5	FRG	DFVLR / NASA	solar orbit	S
154	Luna 24	Ye-8-5M / 413	c. 5,800 kg	08-09-76 / 15:04:12	NIIP-5 / 81L	Proton-K / 288-02	USSR	Lavochkin	lunar sample	S
1977										
155	Voyager 2	Voyager-2	721.9 kg	08-20-77 / 14:29:44	CC / 41	Titan IIIE-Centaur / 7	USA	NASA / JPL	Jupiter, Saturn, Uranus, Neptune flyby	S
156	Voyager 1	Voyager-1	721.9 kg	09-05-77 / 12:56:01	CC / 41	Titan IIIE-Centaur / 6	USA	NASA / JPL	Jupiter, Saturn flyby	S
1978										
157	Pioneer Venus 1	Pioneer Venus Orbiter	582 kg	05-20-78 / 13:13:00	CC / 36A	Atlas Centaur / 50	USA	NASA / ARC	Venus orbiter	S
158	Pioneer Venus 2	Pioneer Venus Multiprobe	904 kg	08-08-78 / 07:33	CC / 36A	Atlas Centaur / 51	USA	NASA / ARC	Venus impacts	S
159	ISEE-C	ISEE-3	479 kg	08-12-78 / 15:12	CC / 17B	Delta 2914 / 144	USA	NASA / GSFC	Sun–Earth L1	S
160	Venera 11	4V-1 / 360	4,447.3 kg	09-09-78 / 03:25:39	NIIP-5 / 81L	Proton-K / 296-01	USSR	Lavochkin	Venus landing	S
161	Venera 12	4V-1 / 361	4,457.9 kg	09-14-78 / 02:25:13	NIIP-5 / 81P	Proton-K / 296-02	USSR	Lavochkin	Venus landing	S
1981										
162	Venera 13	4V-1M / 760	4,397.8 kg	10-30-81 / 06:04	NIIP-5 / 200P	Proton-K / 311-01	USSR	Lavochkin	Venus landing	S
163	Venera 14	4V-1M / 761	4,394.5 kg	11-04-81 / 05:31	NIIP-5 / 200L	Proton-K / 311-02	USSR	Lavochkin	Venus landing	S
1983										
164	Venera 15	4V-2 / 860	5,250 kg	06-02-83 / 02:38:39	NIIP-5 / 200L	Proton-K / 321-01	USSR	Lavochkin	Venus orbit	S
165	Venera-16	4V-2 / 861	5,300 kg	06-07-83 / 02:32	NIIP-5 / 200P	Proton-K / 321-02	USSR	Lavochkin	Venus orbit	S
1984										
166	Vega 1	5VK / 901	c. 4,840 kg	12-15-84 / 09:16:24	NIIP-5 / 200L	Proton-K / 329-01	USSR	Lavochkin	Venus landing, Halley flyby	S
167	Vega 2	5VK / 902	c. 4,840 kg	12-21-84 / 09:13:52	NIIP-5 / 200P	Proton-K / 325-02	USSR	Lavochkin	Venus landing, Halley flyby	S

#	Official Name	Spacecraft/no.	Mass	Launch date/time	Launch place/pad[1]	Launch vehicle/no.	Nation	Design and Operation	Objective	Outcome[2]
	1985									
168	Sakigake	MS-T5	138.1 kg	01-07-85/19:26	Kagoshima/M1	Mu-3S-II/1	Japan	ISAS	Halley flyby	S
169	Giotto	Giotto	960 kg	07-02-85/11:23:16	CSG/ELA 1	Ariane 1/V14	ESA	ESA	Halley flyby	S
170	Suisei	Planet-A	139.5 kg	08-18-85/23:33	Kagoshima/M1	Mu-3S-II/2	Japan	ISAS	Halley flyby	S
	1988									
171	Fobos 1	1F/101	6,220 kg	07-07-88/17:38:04	NIIP-5/200L	Proton-K/356-02	USSR	Lavochkin	Mars orbit/Phobos flyby/landings	F
172	Fobos 2	1F/102	6,220 kg	07-12-88/17:01:43	NIIP-5/200P	Proton-K/356-01	USSR	Lavochkin	Mars orbit/Phobos flyby/landings	P
	1989									
173	Magellan	Magellan	3,445 kg	05-04-89/18:47:00	KSC/39B	STS-30R/IUS	USA	NASA/JPL	Venus orbit	S
174	Galileo	Galileo	2,380 kg	10-18-89/16:53:40	KSC/39B	STS-34R/IUS	USA	NASA/JPL	Jupiter orbit/probe entry	S
	1990									
175	Hiten/Hagoromo	MUSES-A/MUSES-A subsat	185 kg/12 kg	01-24-90/11:46:00	Kagoshima/M1	Mu-3S-II/5	Japan	ISAS	lunar encounter/orbit	S
176	Ulysses	Ulysses	371 kg	10-06-90/11:47:16	KSC/39B	STS-41/IUS	ESA/USA	ESA/NASA/JPL	solar orbit	S
	1992									
177	Geotail	Geotail	1,009 kg	07-24-92/14:26	CC/17A	Delta 6925/D212	Japan/USA	ISAS/NASA	high Earth orbit	S
178	Mars Observer	Mars Observer	1,018 kg	09-25-92/17:05:01	CC/40	Titan III/4	USA	NASA/JPL	Mars orbit	F
	1994									
179	Clementine	Clementine	424 kg	01-25-94/16:34	VAFB/SLC-4W	Titan IIG/11	USA	BMDO/NASA	lunar orbit	S
180	Wind	Wind	1,250 kg	11-01-94/09:31:00	CC/17B	Delta 7925-10/D227	USA	NASA/GSFC	Sun–Earth L1	S
	1995									
181	SOHO	SOHO	1,864 kg	12-02-95/08:08:01	CC/36B	Atlas Centaur IIAS/121	ESA/USA	ESA/NASA	Sun–Earth L1	S

	Official Name	Spacecraft/no.	Mass	Launch date/time	Launch place/pad[1]	Launch vehicle/no.	Nation	Design and Operation	Objective	Outcome[2]
	1996									
182	NEAR Shoemaker	NEAR	805 kg	02-17-96 / 20:43:27	CC / 17B	Delta 7925-8 / D232	USA	NASA / GSFC / APL	asteroid orbit, landing	S
183	Mars Global Surveyor	MGS	1,030.5 kg	11-07-96 / 17:00:49	CC / 17A	Delta 7925 / D239	USA	NASA / JPL	Mars orbit	S
184	Mars 8	M1 / 520	6,795 kg	11-16-96 / 20:48:53	GIK-5 / 200L	Proton-K / 392-02	Russia	Lavochkin	Mars landing	F
185	Mars Pathfinder	Mars Pathfinder	870 kg	12-04-96 / 06:58:07	CC / 17B	Delta 7925 / D240	USA	NASA / JPL	Mars landing	S
	1997									
186	ACE	ACE	752 kg	08-25-97 / 14:39	CC / 17A	Delta 7920-8 / D247	USA	NASA / GSFC	Sun–Earth L1	S
187	Cassini-Huygens	Cassini / Huygens	5,655 kg	10-15-97 / 08:43	CC / 40	Titan 401B-Centaur / 21	USA / ESA	NASA / JPL / ESA	Saturn orbit, Titan landing	S
188	Asiasat 3	HGS 1	3,465 kg	12-24-97 / 23:19	GIK-5 / 81L	Proton-K / 394-01	Asiasat	Asiasat / Hughes	circumlunar	S
	1998									
189	Lunar Prospector	Lunar Prospector	300 kg	01-07-98 / 02:28:44	CC / 46	Athena 2 / LM-004	USA	NASA / ARC	lunar orbit	S
190	Nozomi	Planet-B	536 kg	07-03-98 / 18:12	Kagoshima / M-5	M-V / 3	Japan	ISAS	Mars orbit	F
191	Deep Space 1	DS1	486 kg	10-24-98 / 12:08:00	CC / 17A	Delta 7326-9.5 / D261	USA	NASA / JPL	asteroid flyby, comet flyby	S
192	Mars Climate Orbiter	MCO	638 kg	12-11-98 / 18:45:51	CC / 17A	Delta 7425-9.5 / D264	USA	NASA / JPL	Mars orbit	F
	1999									
193	Mars Polar Lander / Deep Space 2	MPL / DS-2	576 kg	01-03-99 / 20:21:10	CC / 17B	Delta 7425-9.5 / D265	USA	NASA / JPL	Mars landing	F
194	Stardust	Stardust	385 kg	02-07-99 / 21:04:15	CC / 17A	Delta 7426-9.5 / D266	USA	NASA / JPL	comet sample return, comet flybys	S
	2001									
195	2001 Mars Odyssey	2001 Mars Odyssey	1,608.7 kg	04-07-01 / 15:02:22	CC / 17A	Delta 7925-9.5 / D284	USA	NASA / JPL	Mars orbit	S
196	WMAP	Explorer 80	840 kg	06-30-01 / 19:46:46	CC / 17B	Delta 7425-10 / D286	USA	NASA / GSFC	Sun–Earth L2	S
197	Genesis	Genesis	636 kg	08-08-01 / 16:13:40	CC / 17A	Delta 7326-9.5 / D287	USA	NASA / JPL	solar wind sample return, Sun–Earth L1	P

	Official Name	Spacecraft / no.	Mass	Launch date / time	Launch place / pad[1]	Launch vehicle / no.	Nation	Design and Operation	Objective	Outcome[2]
					2002					
198	CONTOUR	CONTOUR	970 kg	07-03-02 / 06:47:41	CC / 17A	Delta 7425-9.5 / D292	USA	NASA / APL	comet encounters	F
					2003					
199	Hayabusa	MUSES-C, MINERVA	510 kg	05-09-03 / 04:29:25	Kagoshima / M-V	M-V / 5	Japan	ISAS / JAXA	asteroid sample return	P
200	Mars Express	Mars Express, Beagle 2	1,186 kg	06-02-03 / 17:45:26	GIK-5 / 31	Soyuz-FG / E15000-005	Europe	ESA	Mars orbit, landing	P
201	Spirit	MER-2	1,062 kg	06-10-03 / 17:58:47	CC / 17A	Delta 7925-9.5 / D298	USA	NASA / JPL	Mars surface exploration	S
202	Opportunity	MER-1	1,062 kg	07-08-03 / 03:18:15	CC / 17B	Delta 7925H / D299	USA	NASA / JPL	Mars surface exploration	S
203	Spitzer Space Telescope	SIRTF	950 kg	08-25-03 / 05:35:39	CC / 17B	Delta 7920H / D300	USA	NASA / JPL / Caltech	solar orbit	S
204	SMART-1	SMART-1	367 kg	09-27-03 / 23:14:46	CSG / ELA 3	Ariane 5G / V162	Europe	ESA	lunar orbit	S
					2004					
205	Rosetta, Philae	Rosetta, Philae	3,000 kg	03-02-04 / 07:17:44	CSG / ELA 3	Ariane 5G+ / V158	Europe	ESA	comet orbit and landing	S
206	MESSENGER	MESSENGER	1,107.9 kg	08-03-04 / 06:15:57	CC / 17B	Delta 7925H-9.5 / D307	USA	NASA / APL	Mercury orbit	S
					2005					
207	Deep Impact	DIF, DI	650 kg	01-12-05 / 18:47:08	CC / 17B	Delta 7925-9.5 / D311	USA	NASA / JPL	comet impact, flyby	S
208	Mars Reconnaissance Orbiter	MRO	2,180 kg	08-12-05 / 11:43:00	CC / 41	Atlas V 401 / 007	USA	NASA / JPL	Mars orbit	S
209	Venus Express	VEX	1,270 kg	11-09-05 / 03:33:34	Baikonur / 31	Soyuz-FG / Zh15000-010	Europe	ESA	Venus orbit	S
					2006					
210	New Horizons	New Horizons	478 kg	01-19-06 / 19:00:00	CC / 41	Atlas V 551 / 010	USA	NASA / APL	Pluto flyby	S
211	STEREO Ahead, STEREO Behind	Stereo A, Stereo B	623 kg 658 kg	10-26-06 / 00:52:00	CC / 17B	Delta 7925-10L (D319)	USA	NASA / GSFC / APL	solar orbit	S
					2007					
212	Artemis P1, Artemis P2	THEMIS B, THEMIS C	2×126 kg	02-17-07 / 23:01:00	CC / 17B	Delta 7925-10C / D323	USA	NASA / UC-Berkeley	Earth–Moon L1, Earth–Moon L2, lunar orbit	S

	Official Name	Spacecraft/no.	Mass	Launch date/time	Launch place/pad[1]	Launch vehicle/no.	Nation	Design and Operation	Objective	Outcome[2]
213	Phoenix	Phoenix Lander	680 kg	08-04-07/09:26:34	CC/17A	Delta 7925-9.5/D325	USA	NASA/JPL/UofA	Mars landing	S
214	Kaguya	SELENE	2,900 kg	09-14-07/01:31:01	Tanegashima/Y1	H-IIA 2022/13	Japan	JAXA	lunar orbit	S
215	Dawn	Dawn	1,218 kg	09-27-07/11:34:00	CC/17B	Delta 7925H-9.5/D327	USA	NASA/JPL	Vesta, Ceres orbit	S
216	Chang'e 1	Chang'e 1	2,350 kg	10-24-07/10:05:04	Xichang/LC3	CZ-3A/Y14	China	CNSA	lunar orbit	S
				2008						
217	Chandrayaan-1	Chandrayaan-1, MIP	1,380 kg	10-22-08/00:52:11	SHAR/SLP	PSLV-XL/C11	India	ISRO	lunar orbit, lunar impact	S
				2009						
218	Kepler	Kepler	1,039 kg	03-07-09/03:49:57	CC/17B	Delta 7925-10L/D339	USA	NASA/ARC/JPL	solar orbit	S
219	Herschel	Herschel	3,400 kg	05-14-09/13:12	CSG/ELA 3	Ariane 5ECA/V188	ESA	ESA	Sun–Earth L2	S
220	Planck	Planck	1,950 kg	05-14-09/13:12	CSG/ELA 3	Ariane 5ECA/V188	ESA	ESA	Sun–Earth L2	S
221	Lunar Reconnaissance Orbiter	LRO	1,850 kg	06-18-09/21:32:00	CC/41	Atlas V 401/020	USA	NASA/GSFC	lunar orbit	S
222	Lunar Crater Observation and Sensing Satellite	LCROSS	621 kg	06-18-09/21:32:00	CC/41	Atlas V 401/020	USA	NASA/ARC	lunar orbit	S
				2010						
223	Akatsuki	VCO	517.6 kg	05-20-10/21:58:22	Tanegashima/Y1	H-IIA 202/17	Japan	JAXA	Venus orbit	S
224	Shin'en	UNITEC 1	20 kg	05-20-10/21:58:22	Tanegashima/Y1	H-IIA 202/17	Japan	JAXA	Venus encounter	P
225	IKAROS	IKAROS	310 kg	05-20-10/21:58:22	Tanegashima/Y1	H-IIA 202/17	Japan	JAXA	Venus flyby	S
226	Chang'e 2	Chang'e 2	2,480 kg	10-01-10/10:59:57	Xichang/LC2	CZ-3C	China	CNSA	lunar orbit, Sun–Earth L2, asteroid flyby	S
				2011						
227	Juno	Juno	3,625 kg	08-05-11/16:25:00	CC/41	Atlas V 551/29	USA	NASA/JPL	Jupiter orbit	S
228	Ebb, Flow	GRAIL-A, GRAIL-B	2 × 202.4 kg	09-10-11/13:08:53	CC/17B	Delta 7920H-10/D356	USA	NASA/JPL	lunar orbit	S
229	Fobos-Grunt		13,505 kg	11-08-11/20:16:03	Baikonur/45	Zenit-2SB41.1	Russia	Lavochkin/IKI/RAN	Mars orbit, Phobos flyby, sample return	F

	Official Name	Spacecraft / no.	Mass	Launch date / time	Launch place / pad[1]	Launch vehicle / no.	Nation	Design and Operation	Objective	Outcome[2]
230	Yinghuo-1		113 kg	11-08-11 / 20:16:03	Baikonur / 45	Zenit-2SB41.1	China	CNSA	Mars orbit	F
231	Curiosity	MSL	3,893 kg	11-26-11 / 15:02:00	CC / 41	Atlas V 541 / 28	USA	NASA / JPL	Mars landing and roving	S
						2013				
232	LADEE	LADEE	383 kg	09-07-13 / 03:27:00	MARS / 0B	Minotaur V / 1	USA	NASA / ARC / GSFC	lunar orbit	S
233	Mangalyaan	MOM	1,337 kg	11-05-13 / 09:08	SHAR / PSLV pad	PSLV-XL / C25	India	ISRO	Mars orbit	S
234	MAVEN	MAVEN	2,454 kg	11-18-13 / 18:28:00	CC / 41	Atlas V 401 / 38	USA	NASA / GSFC / UofC	Mars orbit	S
235	Chang'e 3, Yutu	Chang'e 3, Yutu	3,780 kg	12-01-13 / 17:30	Xichang / LC2	CZ-3B / Y23	China	CNSA	lunar landing, surface exploration	S
236	Gaia	Gaia	2,029 kg	12-19-13 / 09:12:18	CSG / ELS	Soyuz-ST-B / E15000-004	ESA	ESA	Sun–Earth L2	S
						2014				
237	Chang'e 5-T1		3,300 kg	10-23-14 / 18:00:04	Xichang	CZ-3C/G2	China	CNSA	circumlunar, Earth–Moon L2, lunar orbit	S
238	Hayabusa 2		600 kg	12-03-14 / 04:22:04	Tanegashima / Y1	H-IIA / F26	Japan	JAXA	asteroid rendezvous, sample return	IP
239	PROCYON		65 kg	12-03-14 / 04:22:04	Tanegashima / Y1	H-IIA / F26	Japan	JAXA	asteroid flyby	P
240	Shin'en 2		17 kg	12-03-14 / 04:22:04	Tanegashima / Y1	H-IIA / F26	Japan	JAXA	solar orbit	F
241	ArtSat-2	DESPATCH	30 kg	12-03-14 / 04:22:04	Tanegashima / Y1	H-IIA / F26	Japan	JAXA	solar orbit	S
						2015				
242	DSCOVR	Triana	570 kg	02-11-15 / 23:03:02	CC / 40	Falcon 9 v.1.1	USA	NASA / NOAA / USAF	Sun–Earth L1	S
243	LISA Pathfinder	SMART-2	1,910 kg	12-03-15 / 04:04:00	CSG / ELV	Vega / VV06	ESA	ESA	Sun–Earth L1	S
						2016				
244	ExoMars TGO / Schiaparelli	TGO / EDM Lander	4,332 kg	03-14-16 / 09:31	Baikonur / 200L	Proton-M / 93560	ESA / Russia	ESA / Roskosmos	Mars orbit and landing	S/P
245	OSIRIS-REx	OSIRIS-REx	2,110 kg	09-08-16 / 23:05	CC / 41	Atlas V 411 / AV-067	USA	NASA	asteroid sample return	IP

Table 2. Programs

UNITED STATES

PIONEER

Official Name	Spacecraft / no.	Mass	Launch date / time	Launch place / pad[1]	Launch vehicle / no.	Nation	Design & Operation	Objective	Out-come[2]
[Pioneer 0]	Able 1	38.5 kg	08-17-58 / 12:18	CC / 17A	Thor Able I / 127	USA	ARPA / AFBMD	lunar orbit	F
Pioneer	Able 2	38.3 kg	10-11-58 / 08:42:13	CC / 17A	Thor Able I / 1	USA	NASA / AFBMD	lunar orbit	F
Pioneer II	Able 3	39.6 kg	11-08-58 / 07:30:20	CC / 17A	Thor-Able I / 2	USA	NASA / AFBMD	lunar orbit	F
Pioneer III	—	5.87 kg	12-06-58 / 05:44:52	CC / 5	Juno II / AM-11	USA	NASA / ABMA / JPL	lunar flyby	F
Pioneer IV	—	6.08 kg	03-03-59 / 05:10:56	CC / 5	Juno II / AM-14	USA	NASA / ABMA / JPL	lunar flyby	P
[Pioneer, P-3]	P-3 / Able IVB	168.7 kg	11-26-59 / 07:26	CC / 14	Atlas Able / 1	USA	NASA / AFBMD	lunar orbit	F
Pioneer V	P-2 / Able 6	43.2 kg	03-11-60 / 13:00:07	CC / 17A	Thor Able IV / 4	USA	NASA / AFBMD	solar orbit	S
[Pioneer, P-30]	P-30 / Able VA	175.5 kg	09-25-60 / 15:13	CC / 12	Atlas Able / 2	USA	NASA / AFBMD	lunar orbit	F
[Pioneer, P-31]	P-31 / Able VB	176 kg	12-15-60 / 09:10	CC / 12	Atlas Able / 3	USA	NASA / AFBMD	lunar orbit	F
Pioneer VI	Pioneer-A	62.14 kg	12-16-65 / 07:31:21	CK / 17A	Thor Delta E / 35	USA	NASA / ARC	solar orbit	S
Pioneer VII	Pioneer-B	62.75 kg	08-17-66 / 15:20:17	CK / 17A	Thor Delta E1 / 40	USA	NASA / ARC	solar orbit	S
Pioneer VIII	Pioneer-C	65.36 kg	12-13-67 / 14:08	CK / 17B	Thor Delta E-1 / 55	USA	NASA / ARC	solar orbit	S
Pioneer IX	Pioneer-D	65.36	11-08-68 / 09:46:29	CK / 17B	Thor Delta E1 / 60	USA	NASA / ARC	solar orbit	S
[Pioneer]	Pioneer-E	65.4 kg	08-27-69 / 21:59	CC / 17A	Thor Delta L / 73	USA	NASA ARC	solar orbit	F
Pioneer 10	Pioneer-F	258 kg	03-02-72 / 01:49:04	CK / 36A	Atlas Centaur / 27	USA	NASA / ARC	Jupiter flyby	S
Pioneer 11	Pioneer-G	258.5 kg	04-06-73 / 02:11	CK / 36B	Atlas Centaur / 30	USA	NASA / ARC	Jupiter, Saturn flyby	S
Pioneer Venus Orbiter		582 kg	05-20-78 / 13:13:00	CC / 36A	Atlas Centaur / 50	USA	NASA / ARC	Venus orbiter	S
Pioneer Venus Multiprobe		904 kg	08-08-78 / 07:33	CC / 36A	Atlas Centaur / 51	USA	NASA / ARC	Venus impacts	S

1   CC = Cape Canaveral, CK = Cape Kennedy, CSG = Le Centre Spatial Guyanais, GIK-5 = Baikonur, KSC = Kennedy Space Center, MARS = Mid-Atlantic Regional Spaceport, NIIP-5 = Baikonur, SHAR = Sriharikota, VAFB = Vandenberg Air Force Base

2   F = failure, IP = in progress, P = partial success, S = success

Official Name	Spacecraft/no.	Mass	Launch date/time	Launch place/pad[1]	Launch vehicle/no.	Nation	Design & Operation	Objective	Out-come[2]
RANGER									
Ranger I	P-32	306.18 kg	08-23-61/10:04	CC/12	Atlas Agena B/1	USA	NASA/JPL	deep space orbit	F
Ranger II	P-33	306.18 kg	11-18-61/08:12	CC/12	Atlas Agena B/2	USA	NASA/JPL	deep space orbit	F
Ranger III	P-34	330 kg	01-26-62/20:30	CC/12	Atlas-Agena B/3	USA	NASA/JPL	lunar impact	F
Ranger IV	P-35	331.12 kg	04-23-62/20:50	CC/12	Atlas-Agena B/4	USA	NASA JPL	lunar impact	P
Ranger V	P-36	342.46 kg	10-18-62/16:59:00	CC/12	Atlas-Agena B/7	USA	NASA/JPL	lunar landing	P
Ranger VI	Ranger-A/P-53	364.69 kg	01-30-64/15:49:09	CK/12	Atlas-Agena B/8	USA	NASA/JPL	lunar impact	P
Ranger VII	Ranger-B/P-54	365.6 kg	07-28-64/16:50:07	CK/12	Atlas-Agena B/9	USA	NASA/JPL	lunar impact	S
Ranger VIII	Ranger-C	366.87 kg	02-17-65/17:05:00	CK/12	Atlas-Agena B/13	USA	NASA/JPL	lunar impact	S
Ranger IX	Ranger-D	366.87 kg	03-21-65/21:37:02	CK/12	Atlas-Agena B/14	USA	NASA/JPL	lunar impact	S
MARINER									
Mariner I	P-37	202.8 kg	07-22-62/09:21:23	CC/12	Atlas-Agena B/5	USA	NASA/JPL	Venus flyby	F
Mariner II	P-38	203.6 kg	08-27-62/06:53:14	CC/12	Atlas-Agena B/6	USA	NASA/JPL	Venus flyby	S
Mariner III	Mariner-64C	260.8 kg	11-05-64/19:22:05	CK/13	Atlas-Agena D/11	USA	NASA/JPL	Mars flyby	F
Mariner IV	Mariner-64D	260.8 kg	11-28-64/14:22:01	CK/12	Atlas-Agena D/12	USA	NASA/JPL	Mars flyby	S
Mariner V	Mariner-67E	244.9 kg	06-14-67/06:01:00	CK/12	Atlas Agena D/23	USA	NASA/JPL	Venus flyby	S
Mariner VI	Mariner-69F	381 kg	02-25-69/01:29:02	CK/36B	Atlas Centaur/20	USA	NASA/JPL	Mars flyby	S
Mariner VII	Mariner-69G	381 kg	03-27-69/22:22:01	CK/36A	Atlas Centaur/19	USA	NASA/JPL	Mars flyby	S
Mariner 8	Mariner-71H	997.9 kg	05-09-71/01:11:01	CK/36A	Atlas Centaur/24	USA	NASA/JPL	Mars orbit	F
Mariner 9	Mariner-71I	997.9 kg	05-30-71/22:23:04	CK/36B	Atlas Centaur/23	USA	NASA/JPL	Mars orbit	S
Mariner 10	Mariner-73J	502.9 kg	11-03-73/05:45:00	CC/36B	Atlas Centaur/34	USA	NASA/JPL	Mercury, Venus flyby	S
SURVEYOR									
[Surveyor Model]	SD-1	951 kg	03-02-65/13:25	CK/36A	Atlas Centaur/5	USA	NASA/JPL	deep space orbit	F
Surveyor Model 1	SD-2	950 kg	08-11-65/14:31:04	CK/36B	Atlas Centaur/6	USA	NASA/JPL	deep space orbit	S
Surveyor Model 2	SD-3	784 kg	04-08-66/01:00:02	CK/36B	Atlas Centaur/8	USA	NASA/JPL	deep space orbit	F
Surveyor 1	Surveyor-A	995.2 kg	05-30-66/14:41:01	CK/36A	Atlas Centaur/10	USA	NASA/JPL	lunar landing	S

Official Name	Spacecraft / no.	Mass	Launch date / time	Launch place / pad[1]	Launch vehicle / no.	Nation	Design & Operation	Objective	Outcome[2]
Surveyor II	Surveyor-B	995.2 kg	09-20-66 / 12:32:00	CK / 36A	Atlas Centaur / 7	USA	NASA / JPL	lunar landing	F
Surveyor III	Surveyor-C	997.9 kg	04-17-67 / 07:05:01	CK / 36B	Atlas Centaur / 12	USA	NASA / JPL	lunar landing	S
Surveyor IV	Surveyor-D	1,037.4 kg	07-14-67 / 11:53:29	CK / 36A	Atlas Centaur / 11	USA	NASA / JPL	lunar landing	F
Surveyor V	Surveyor-E	1,006 kg	09-08-67 / 07:57:01	CK / 36B	Atlas Centaur / 13	USA	NASA / JPL	lunar landing	S
Surveyor VI	Surveyor-F	1,008.3 kg	11-07-67 / 07:39:01	CK / 36B	Atlas Centaur / 14	USA	NASA / JPL	lunar landing	S
Surveyor VII	Surveyor-G	1,040.1 kg	01-07-68 / 06:30:00	CK / 36A	Atlas Centaur / 15	USA	NASA / JPL	lunar landing	S
EXPLORER									
Explorer XXXIII	AIMP-D	93.4 kg	07-01-66 / 16:02:25	CK / 17A	Thor Delta E1 / 39	USA	NASA / GSFC	lunar orbit	P
Explorer XXXV	AIMP-E	104.3 kg	07-19-67 / 14:19:02	CK / 17B	Thor Delta E1 / 50	USA	NASA / GSFC	lunar orbit	S
Explorer 49	RAE-B	330.2 kg	06-10-73 / 14:13:00	CC / 17B	Delta 1913 / 95	USA	NASA / GSFC	lunar orbit	S
LUNAR ORBITER									
Lunar Orbiter I	LO-A	386.9 kg	08-10-66 / 19:26:00	CK / 13	Atlas Agena D / 17	USA	NASA / LaRC	lunar orbit	S
Lunar Orbiter II	S/C 5	385.6 kg	11-06-66 / 23:21:00	CK / 13	Atlas Agena D / 18	USA	NASA / LaRC	lunar orbit	S
Lunar Orbiter III	S/C 6	385.6 kg	02-05-67 / 01:17:01	CK / 13	Atlas Agena D / 20	USA	NASA / LaRC	lunar orbit	S
Lunar Orbiter IV	S/C 7	385.6 kg	05-04-67 / 22:25:00	CK / 13	Atlas Agena D / 22	USA	NASA / LaRC	lunar orbit	S
Lunar Orbiter V	S/C 3	385.6 kg	08-01-67 / 22:33:00	CK / 13	Atlas Agena D / 24	USA	NASA / LaRC	lunar orbit	S
APOLLO PARTICLE AND FIELDS SATELLITE									
Apollo 15 Particle & Fields Sat		35.6 kg	08-04-71 / 20:13:19	Apollo 15 CSM 112	Saturn V / 510	USA	NASA / MSC	lunar orbit	S
Apollo 16 Particle & Fields Sat		42 kg	04-24-72 / 09:56:09	Apollo 16 CSM 113	Saturn V / 511	USA	NASA / MSC	lunar orbit	S
VIKING									
Viking 1	Viking-B	3,527 kg	08-20-75 / 21:22:00	CC / 41	Titan IIIE-Centaur / 4	USA	NASA / JPL	Mars orbit / landing	S
Viking 2	Viking-A	3,527 kg	09-09-75 / 18:39:00	CC / 41	Titan IIIE-Centaur / 3	USA	NASA / JPL	Mars orbit / landing	S
VOYAGER									
Voyager 2	Voyager-2	721.9 kg	08-20-77 / 14:29:44	CC / 41	Titan IIIE-Centaur / 7	USA	NASA / JPL	Jupiter, Saturn, Uranus, Neptune flyby	S
Voyager 1	Voyager-1	721.9 kg	09-05-77 / 12:56:01	CC / 41	Titan IIIE-Centaur / 6	USA	NASA / JPL	Jupiter, Saturn flyby	S

Official Name	Spacecraft/no.	Mass	Launch date/time	Launch place/pad[1]	Launch vehicle/no.	Nation	Design & Operation	Objective	Outcome[2]
ISEE									
ISEE-3	ISEE-C	479 kg	08-12-78/15:12	CC/17B	Delta 2914/144	USA	NASA GSFC	Sun–Earth L1	S
MAGELLAN									
Magellan	Magellan	3,445 kg	05-04-89/18:47:00	KSC/39B	STS-30R/IUS	USA	NASA/JPL	Venus orbit	S
GALILEO									
Galileo	Galileo	2,380 kg	10-18-89/16:53:40	KSC/39B	STS-34R/IUS	USA	NASA/JPL	Jupiter orbit/probe entry	S
MARS OBSERVER									
Mars Observer	Mars Observer	1,018 kg	09-25-92/17:05:01	CC/40	Titan III/4	USA	NASA/JPL	Mars orbit	F
CLEMENTINE									
Clementine	Clementine	424 kg	01-25-94/16:34	VAFB/SLC-4W	Titan IIG/11	USA	BMDO/NASA	lunar orbit	S
WIND									
Wind	Wind	1,250 kg	11-01-94/09:31:00	CC/17B	Delta 7925-10/D227	USA	NASA/GSFC	Sun–Earth L1	S
NEAR									
NEAR Shoemaker	NEAR	805 kg	02-17-96/20:43:27	CC/17B	Delta 7925-8/D232	USA	NASA/GSFC/APL	asteroid orbit, landing	S
MARS GLOBAL SURVEYOR									
Mars Global Surveyor	MGS	1,030.5 kg	11-07-96/17:00:49	CC/17A	Delta 7925/D239	USA	NASA/JPL	Mars orbit	S
MARS PATHFINDER									
Mars Pathfinder	Mars Pathfinder	870 kg	12-04-96/06:58:07	CC/17B	Delta 7925/D240	USA	NASA/JPL	Mars landing	S
ACE									
ACE	ACE	752 kg	08-25-97/14:39	CC/17A	Delta 7920-8/D247	USA	NASA/GSFC	Sun–Earth L1	S
LUNAR PROSPECTOR									
Lunar Prospector	Lunar Prospector	300 kg	01-07-98/02:28:44	CC/46	Athena 2/LM-004	USA	NASA/ARC	lunar orbit	S
DEEP SPACE									
Deep Space 1	DS1	486 kg	10-24-98/12:08:00	CC/17A	Delta 7326-9.5/D261	USA	NASA/JPL	asteroid flyby, comet flyby	S
MARS CLIMATE ORBITER									
Mars Climate Orbiter	MCO	638 kg	12-11-98/18:45:51	CC/17A	Delta 7425-9.5/D264	USA	NASA/JPL	Mars orbit	F

Official Name	Spacecraft / no.	Mass	Launch date / time	Launch place / pad[1]	Launch vehicle / no.	Nation	Design & Operation	Objective	Out-come[2]
MARS POLAR LANDER									
Mars Polar Lander / Deep Space 2	MPL / DS-2	576 kg	01-03-99 / 20:21:10	CC / 17B	Delta 7425-9.5 / D265	USA	NASA / JPL	Mars landing	F
STARDUST									
Stardust	Stardust	385 kg	02-07-99 / 21:04:15	CC / 17A	Delta 7426-9.5 / D266	USA	NASA / JPL	comet sample return, comet flybys	S
2001 MARS ODYSSEY									
2001 Mars Odyssey	2001 Mars Odyssey	1,608.7 kg	04-07-01 / 15:02:22	CC / 17A	Delta 7925-9.5 / D284	USA	NASA / JPL	Mars orbit	S
MAP									
MAP	Explorer 80	840 kg	06-30-01 / 19:46:46	CC / 17B	Delta 7425-10 / D286	USA	NASA / GSFC	Sun–Earth L2	S
GENESIS									
Genesis	Genesis	636 kg	08-08-01 / 16:13:40	CC / 17A	Delta 7326-9.5 / D287	USA	NASA / JPL	solar wind sample return, Sun–Earth L1	P
CONTOUR									
CONTOUR	CONTOUR	970 kg	07-03-02 / 06:47:41	CC / 17A	Delta 7425-9.5 / D292	USA	NASA / APL	comet flyby	F
MER									
Spirit	MER-2	1,062 kg	06-10-03 / 17:58:47	CC / 17A	Delta 7925-9.5 / D298	USA	NASA / JPL	Mars surface exploration	S
Opportunity	MER-1	1,062 kg	07-08-03 / 03:18:15	CC / 17B	Delta 7925H / D299	USA	NASA / JPL	Mars surface exploration	S
SPITZER									
Spitzer Space Telescope	SIRTF	950 kg	08-25-03 / 05:35:39	CC / 17B	Delta 7920H / D300	USA	NASA / JPL / Caltech	Solar orbit	S
MESSENGER									
MESSENGER	MESSENGER	1,107.9 kg	08-03-04 / 06:15:57	CC / 17B	Delta 7925H-9.5 / D307	USA	NASA / APL	Mercury orbit	S
DEEP IMPACT									
Deep Impact	DIF, DI	650 kg	01-12-05 / 18:47:08	CC / 17B	Delta 7925-9.5 / D311	USA	NASA / JPL	comet impact, flyby	S
MRO									
Mars Reconnaissance Orbiter	MRO	2,180 kg	08-12-05 / 11:43:00	CC / 41	Atlas V 401 / 007	USA	NASA / JPL	Mars orbit	S

Official Name	Spacecraft / no.	Mass	Launch date / time	Launch place / pad[1]	Launch vehicle / no.	Nation	Design & Operation	Objective	Outcome[2]
NEW HORIZONS									
New Horizons	New Horizons	478 kg	01-19-06 / 19:00:00	CC / 41	Atlas V 551 / 010	USA	NASA / APL	Pluto flyby	S
STEREO									
STEREO Ahead, STEREO Behind	Stereo A, Stereo B	623 kg 658 kg	10-26-06 / 00:52:00	CC / 17B	Delta 7925-10L (D319)	USA	NASA / GSFC / APL	solar orbit	S
ARTEMIS									
Artemis P1, Artemis P2	THEMIS B, THEMIS C	2 × 126 kg	02-17-07 / 23:01:00	CC / 17B	Delta 7925-10C / D323	USA	NASA / UC-Berkeley	Earth–Moon L1, Earth–Moon L2, lunar orbit	S
PHOENIX									
Phoenix	Phoenix Lander	664 kg	08-04-07 / 09:26:34	CC / 17A	Delta 7925-9.5 / D325	USA	NASA / JPL	Mars landing	S
DAWN									
Dawn	Dawn	1,218 kg	09-27-07 / 11:34:00	CC / 17B	Delta 7925H / D327	USA	NASA / JPL	Vesta, Ceres orbit	S
KEPLER									
Kepler	Kepler	1,039 kg	03-07-09 / 03:49:57	CC / 17B	Delta 7925-10L / D339	USA	NASA / ARC / JPL	solar orbit	S
LRO / LCROSS									
Lunar Reconnaissance Orbiter	LRO	1,850 kg	06-18-09 / 21:32:00	CC / 41	Atlas V 401 / 020	USA	NASA / GSFC	lunar orbit	S
Lunar Crater Observation and Sensing Satellite	LCROSS	621 kg	06-18-09 / 21:32:00	CC / 41	Atlas V 401 / 020	USA	NASA / GSFC	lunar orbit	S
JUNO									
Juno	Juno	3,625 kg	08-05-11 / 16:25:00	CC / 41	Atlas V 551 / 29	USA	NASA / JPL	Jupiter orbit	IP
GRAIL									
Ebb, Flow	GRAIL-A, GRAIL-B	2 × 202.4 kg	09-10-11 / 13:08:52	CC / 17B	Delta 7920H-10 / D356	USA	NASA / JPL	lunar orbit	S
MSL									
Curiosity	MSL	3,893 kg	11-26-11 / 15:02:00	CC / 41	Atlas V 541 / 28	USA	NASA / JPL	Mars landing and surface exploration	S
LADEE									
LADEE	LADEE	383 kg	09-07-13 / 03:27:00	MARS / 0B	Minotaur V / 1	USA	NASA / ARC / GSFC	lunar orbit	S

Official Name	Spacecraft / no.	Mass	Launch date / time	Launch place / pad[1]	Launch vehicle / no.	Nation	Design & Operation	Objective	Outcome[2]
MAVEN									
MAVEN	MAVEN	2,454 kg	11-18-13 / 18:28:00	CC / 41	Atlas V 401 / 38	USA	NASA / GSFC / UofC	Mars orbit	S
DSCOVR									
DSCOVR	Triana	570 kg	02-11-15 / 23:03:02	CC / 40	Falcon 9 v.1.1	USA	NASA / NOAA / USAF	Sun–Earth L1	S

SOVIET UNION / RUSSIA

Official Name	Spacecraft / no.	Mass	Launch date / time	Launch place / pad	Launch vehicle / no.	Nation	Design & Operation	Objective	Outcome
LUNA									
[Luna]	Ye-1 / 1	c. 360 kg	09-23-58 / 07:03:23	NIIP-5 / 1	Luna / B1-3	USSR	OKB-1	lunar impact	F
[Luna]	Ye-1 / 2	c. 360 kg	10-11-58 / 21:41:58	NIIP-5 / 1	Luna / B1-4	USSR	OKB-1	lunar impact	F
[Luna]	Ye-1 / 3	c. 360 kg	12-04-58 / 18:18:44	NIIP-5 / 1	Luna / B1-5	USSR	OKB-1	lunar impact	F
Soviet Space Rocket [Luna 1]	Ye-1 / 4	361.3 kg	01-02-59 / 16:41:21	NIIP-5 / 1	Luna / B1-6	USSR	OKB-1	lunar impact	P
Second Soviet Space Rocket [Luna 2]	Ye-1A / 7	390.2 kg	09-12-59 / 06:39:42	NIIP-5 / 1	Luna / I1-7b	USSR	OKB-1	lunar impact	S
Automatic Interplanetary Station [Luna 3]	Ye-2A / 1	278.5 kg	10-04-59 / 00:43:40	NIIP-5 / 1	Luna / I1-8	USSR	OKB-1	lunar flyby	S
[Luna]	Ye-3 / 1	?	04-15-60 / 15:06:44	NIIP-5 / 1	Luna / I1-9	USSR	OKB-1	lunar flyby	F
[Luna]	Ye-3 / 2	?	04-19-60 / 16:07:43	NIIP-5 / 1	Luna / I1-9a	USSR	OKB-1	lunar flyby	F
[Luna]	Ye-6 / 2	1,420 kg	01-04-63 / 08:48:58	NIIP-5 / 1	Molniya / T103-09	USSR	OKB-1	lunar landing	F
[Luna]	Ye-6 / 3	1,420 kg	02-03-63 / 09:29:14	NIIP-5 / 1	Molniya / T103-10	USSR	OKB-1	lunar landing	F
Luna 4	Ye-6 / 4	1,422 kg	04-02-63 / 08:16:38	NIIP-5 / 1	Molniya / T103-11	USSR	OKB-1	lunar landing	F
[Luna]	Ye-6 / 6	c. 1,420 kg	03-21-64 / 08:14:33	NIIP-5 / 1	Molniya / T15000-20	USSR	OKB-1	lunar landing	F
[Luna]	Ye-6 / 5	c. 1,420 kg	04-20-64 / 08:08:28	NIIP-5 / 1	Molniya / T15000-21	USSR	OKB-1	lunar landing	F
Kosmos 60 [Luna]	Ye-6 / 9	c. 1,470 kg	03-12-65 / 09:25	NIIP-5 / 1	Molniya / G15000-24	USSR	OKB-1	lunar landing	F

Official Name	Spacecraft / no.	Mass	Launch date / time	Launch place / pad	Launch vehicle / no.	Nation	Design & Operation	Objective	Out-come
[Luna]	Ye-6 / 8	c. 1,470 kg	04-10-65 / 08:39	NIIP-5 / 1	Molniya-M / R103-26	USSR	OKB-1	lunar landing	F
Luna 5	Ye-6 / 10	1,476 kg	05-09-65 / 07:49:37	NIIP-5 / 1	Molniya-M / U103-30	USSR	OKB-1	lunar landing	F
Luna 6	Ye-6 / 7	1,442 kg	06-08-65 / 07:40	NIIP-5 / 1	Molniya-M / U103-31	USSR	OKB-1	lunar landing	F
Luna 7	Ye-6 / 11	1,506 kg	10-04-65 / 07:56:40	NIIP-5 / 1	Molniya / U103-27	USSR	OKB-1	lunar landing	F
Luna 8	Ye-6 / 12	1,552 kg	12-03-65 / 10:46:14	NIIP-5 / 1	Molniya-M / U103-28	USSR	OKB-1	lunar landing	F
Luna 9	Ye-6M / 202	1,583.7 kg	01-31-66 / 11:41:37	NIIP-5 / 31	Molniya-M / U103-32	USSR	Lavochkin	lunar landing	S
Kosmos 111 [Luna]	Ye-6S / 204	c. 1,580 kg	03-01-66 / 11:03:49	NIIP-5 / 31	Molniya-M / N103-41	USSR	Lavochkin	lunar orbit	F
Luna 10	Ye-6S / 206	1,583.7 kg	03-31-66 / 10:46:59	NIIP-5 / 31	Molniya-M / N103-42	USSR	Lavochkin	lunar orbit	S
Luna 11	Ye-6LF / 101	1,640 kg	08-24-66 / 08:03:21	NIIP-5 / 31	Molniya-M / N103-43	USSR	Lavochkin	lunar orbit	S
Luna 12	Ye-6LF / 102	1,640 kg	10-22-66 / 08:42:26	NIIP-5 / 31	Molniya-M / N103-44	USSR	Lavochkin	lunar orbit	S
Luna 13	Ye-6M / 205	1,620 kg	12-21-66 / 10:17:08	NIIP-5 / 1	Molniya-M / N103-45	USSR	Lavochkin	lunar landing	S
Kosmos 159 [Luna]	Ye-6LS / 111	1.640 kg	05-16-67 / 21:43:57	NIIP-5 / 1	Molniya-M / Ya716-56	USSR	Lavochkin	lunar orbit	P
[Luna]	Ye-6LS / 112	1,640 kg	02-07-68 / 10:43:54	NIIP-5 / 1	Molniya-M / Ya716-57	USSR	Lavochkin	lunar orbit	F
Luna 14	Ye-6LS / 113	1,640 kg	04-07-68 / 10:09:32	NIIP-5 / 1	Molniya-M / Ya716-58	USSR	Lavochkin	lunar orbit	S
[Luna/Lunokhod]	Ye-8 / 201	c. 5,700 kg	02-19-69 / 06:48:48	NIIP-5 / 81P	Proton-K / 239-01	USSR	Lavochkin	lunar rover	F
[Luna]	Ye-8-5 / 402	c. 5,700 kg	06-14-69 / 040:0:48	NIIP-5 / 81P	Proton-K / 238-01	USSR	Lavochkin	lunar sample	F
Luna 15	Ye-8-5 / 401	5,667 kg	07-13-69 / 02:54:42	NIIP-5 / 81P	Proton-K / 242-01	USSR	Lavochkin	lunar sample	F
Kosmos 300 [Luna]	Ye-8-5 / 403	c. 5,700 kg	09-23-69 / 14:07:37	NIIP-5 / 81P	Proton-K / 244-01	USSR	Lavochkin	lunar sample	F
Kosmos 305 [Luna]	Ye-8-5 / 404	c. 5,700 kg	10-22-69 / 14:09:59	NIIP-5 / 81P	Proton-K / 241-01	USSR	Lavochkin	lunar sample	F
[Luna]	Ye-8-5 / 405	c. 5,700 kg	02-06-70 / 04:16:05	NIIP-5 / 81	Proton-K / 247-01	USSR	Lavochkin	lunar sample	F
Luna 16	Ye-8-5 / 406	5,725 kg	09-12-70 / 13:25:52	NIIP-5 / 81L	Proton-K / 248-01	USSR	Lavochkin	lunar sample	S
Luna 17 Lunokhod 1	Ye-8 / 203	5,700 kg	11-10-70 / 14:44:01	NIIP-5 / 81L	Proton-K / 251-01	USSR	Lavochkin	lunar rover	S
Luna 18	Ye-8-5 / 407	5,725 kg	09-02-71 / 13:40:40	NIIP-5 / 81P	Proton-K / 256-01	USSR	Lavochkin	lunar sample	F
Luna 19	Ye-8LS / 202	5,330 kg	09-28-71 / 10:00:22	NIIP-5 / 81P	Proton-K / 257-01	USSR	Lavochkin	lunar orbit	P
Luna 20	Ye-8-5 / 408	5,725 kg	02-14-72 / 03:27:58	NIIP-5 / 81P	Proton-K / 258-01	USSR	Lavochkin	lunar sample	S

Official Name	Spacecraft / no.	Mass	Launch date / time	Launch place / pad	Launch vehicle / no.	Nation	Design & Operation	Objective	Outcome
Luna 21 Lunokhod 2	Ye-8 / 204	5,700 kg	01-08-73 / 06:55:38	NIIP-5 / 81L	Proton-K / 259-01	USSR	Lavochkin	lunar rover	S
Luna 22	Ye-8LS / 206	5,700 kg	05-29-74 / 08:56:51	NIIP-5 / 81P	Proton-K / 282-02	USSR	Lavochkin	lunar orbit	S
Luna 23	Ye-8-5M / 410	5,795 kg	10-28-74 / 14:30:32	NIIP-5 / 81P	Proton-K / 285-01	USSR	Lavochkin	lunar sample	F
[Luna]	Ye-8-5M / 412	5,795 kg	10-16-75 / 04:04:56	NIIP-5 / 81L	Proton-K / 287-02	USSR	Lavochkin	lunar sample	F
Luna 24	Ye-8-5M / 413	c. 5,800 kg	08-09-76 / 15:04:12	NIIP-5 / 81L	Proton-K / 288-02	USSR	Lavochkin	lunar sample	S
MARS									
[Mars]	1M / 1	480 kg	10-10-60 / 14:27:49	NIIP-5 / 1	Molniya / L1-4M	USSR	OKB-1	Mars flyby	F
[Mars]	1M / 2	480 kg	10-14-60 / 13:51:03	NIIP-5 / 1	Molniya / L1-5M	USSR	OKB-1	Mars flyby	F
[Mars]	2MV-4 / 3	c. 900 kg	10-24-62 / 17:55:04	NIIP-5 / 1	Molniya / T103-15	USSR	OKB-1	Mars flyby	F
Mars 1	2MV-4 / 4	893.5 kg	11-01-62 / 16:14:06	NIIP-5 / 1	Molniya / T103-16	USSR	OKB-1	Mars flyby	P
[Mars]	2MV-3 / 1	?	11-04-62 / 15:35:15	NIIP-5 / 1	Molniya / T103-17	USSR	OKB-1	Mars impact	F
[Mars]	M-69 / 521	c. 4,850 kg	03-27-69 / 10:40:45	NIIP-5 / 81L	Proton-K / 240-01	USSR	Lavochkin	Mars orbit	F
[Mars]	M-69 / 522	c. 4,850 kg	04-02-69 / 10:33:00	NIIP-5 / 81P	Proton-K / 233-01	USSR	Lavochkin	Mars orbit	F
Kosmos 419 [Mars]	3MS / 170	4,549 kg	05-10-71 / 16:58:42	NIIP-5 / 81L	Proton-K / 253-01	USSR	Lavochkin	Mars orbit	F
Mars 2	4M / 171	4,625 kg	05-19-71 / 16:22:44	NIIP-5 / 81P	Proton-K / 255-01	USSR	Lavochkin	Mars orbit / landing	P
Mars 3	4M / 172	4,625 kg	05-28-71 / 15:26:30	NIIP-5 / 81L	Proton-K / 249-01	USSR	Lavochkin	Mars orbit / landing	P
Mars 4	3MS / 52S	4,000 kg	07-21-73 / 19:30:59	NIIP-5 / 81L	Proton-K / 261-01	USSR	Lavochkin	Mars orbit	F
Mars 5	3MS / 53S	4,000 kg	07-25-73 / 18:55:48	NIIP-5 / 81P	Proton-K / 262-01	USSR	Lavochkin	Mars orbit	S
Mars 6	3MP / 50P	3,880 kg	08-05-73 / 17:45:48	NIIP-5 / 81L	Proton-K / 281-01	USSR	Lavochkin	Mars landing	P
Mars 7	3MP / 51P	3,880 kg	08-09-73 / 17:00:17	NIIP-5 / 81P	Proton-K / 281-02	USSR	Lavochkin	Mars landing	F
Fobos 1	1F / 101	6,220 kg	07-07-88 / 17:38:04	NIIP-5 / 200L	Proton-K / 356-02	USSR	Lavochkin	Mars orbit / Phobos flyby / landings	F
Fobos 2	1F / 102	6,220 kg	07-12-88 / 17:01:43	NIIP-5 / 200P	Proton-K / 356-01	USSR	Lavochkin	Mars orbit / Phobos flyby / landings	P
Mars 8	M1 / 520	6,795 kg	11-16-96 / 20:48:53	GIK-5 / 200L	Proton-K / 392-02	Russia	Lavochkin	Mars landing	F

Official Name	Spacecraft/no.	Mass	Launch date/time	Launch place/pad	Launch vehicle/no.	Nation	Design & Operation	Objective	Outcome
Fobos-Grunt		13,505 kg	11-08-11 / 20:16:03	Baikonur / 45	Zenit-2SB41.1	Russia	Lavochkin / IKI	Mars orbit, Phobos flyby, sample return	F
VENERA									
Heavy Satellite [Venera]	1VA / 1	c. 645 kg	02-04-61 / 01:18:04	NIIP-5 / 1	Molniya / L1-7V	USSR	OKB-1	Venus impact	F
Automatic Interplanetary Station [Venera 1]	1VA / 2	643.5 kg	02-12-61 / 00:34:37	NIIP-5 / 1	Molniya / L1-6V	USSR	OKB-1	Venus impact	F
[Venera]	2MV-1 / 3	1,097 kg	08-25-62 / 02:56:06	NIIP-5 / 1	Molniya / T103-12	USSR	OKB-1	Venus impact	F
[Venera]	2MV-1 / 4	c. 1,100 kg	09-01-62 / 02:12:30	NIIP-5 / 1	Molniya / T103-13	USSR	OKB-1	Venus impact	F
[Venera]	2MV-2 / 1	?	09-12-62 / 00:59:13	NIIP-5 / 1	Molniya / T103-14	USSR	OKB-1	Venus flyby	F
Kosmos 27 [Venera]	3MV-1 / 5	948 kg	03-27-64 / 03:24:43	NIIP-5 / 1	Molniya / T15000-27	USSR	OKB-1	Venus impact	F
Zond 1 [Venera]	3MV-1 / 4	948 kg	04-02-64 / 02:42:40	NIIP-5 / 1	Molniya / T15000-28	USSR	OKB-1	Venus impact	F
Venera 2	3MV-4 / 4	963 kg	11-12-65 / 04:46	NIIP-5 / 31	Molniya-M / U103-42	USSR	OKB-1	Venus flyby	S
Venera 3	3MV-3 / 1	960 kg	11-16-65 / 04:13	NIIP-5 / 31	Molniya / U103-31	USSR	OKB-1	Venus impact	S
Kosmos 96 [Venera]	3MV-4 / 6	c. 950 kg	11-23-65 / 03:14	NIIP-5 / 31	Molniya / U103-30	USSR	OKB-1	Venus flyby	F
Venera 4	1V / 310	1,106 kg	06-12-67 / 02:39:45	NIIP-5 / 1	Molniya-M / Ya716-70	USSR	Lavochkin	Venus impact	S
Kosmos 167 [Venera]	1V / 311	c. 1,100 kg	06-17-67 / 02:36:38	NIIP-5 / 1	Molniya-M / Y7'6-71	USSR	Lavochkin	Venus impact	F
Venera 5	2V / 330	1,130 kg	01-05-69 / 06:28:08	NIIP-5 / 1	Molniya-M / V716-72	USSR	Lavochkin	Venus landing	S
Venera 6	2V / 331	1,130 kg	01-10-69 / 05:51:52	NIIP-5 / 1	Molniya-M / V716-73	USSR	Lavochkin	Venus landing	S
Venera 7	3V / 630	1,180 kg	08-17-70 / 05:38:22	NIIP-5 / 31	Molniya-M / Kh15000-62	USSR	Lavochkin	Venus landing	S
Kosmos 359 [Venera]	3V / 631	c. 1,200 kg	08-22-70 / 05:06:08	NIIP-5 / 31	Molniya-M / Kh15000-61	USSR	Lavochkin	Venus landing	F
Venera 8	3V / 670	1,184 kg	03-27-72 / 04:15:06	NIIP-5 / 31	Molniya-M / S1500-63	USSR	Lavochkin	Venus landing	S
Kosmos 482 [Venera]	3V / 671	c. 1,180 kg	03-31-72 / 04:02:33	NIIP-5 / 31	Molniya-M / S1500-64	USSR	Lavochkin	Venus landing	F
Venera 9	4V-1 / 660	4,936 kg	06-08-75 / 02:38:00	NIIP-5 / 81P	Proton-K / 286-01	USSR	Lavochkin	Venus orbit / landing	S
Venera 10	4V-1 / 661	5,033 kg	06-14-75 / 03:00:31	NIIP-5 / 81P	Proton-K / 285-02	USSR	Lavochkin	Venus orbit / landing	S
Venera 11	4V-1 / 360	4,447.3 kg	09-09-78 / 03:25:39	NIIP-5 / 81L	Proton-K / 296-01	USSR	Lavochkin	Venus landing	S

Official Name	Spacecraft / no.	Mass	Launch date / time	Launch place / pad	Launch vehicle / no.	Nation	Design & Operation	Objective	Out-come
Venera 12	4V-1 / 361	4,457.9 kg	09-14-78 / 02:25:13	NIIP-5 / 81P	Proton-K / 296-02	USSR	Lavochkin	Venus landing	S
Venera 13	4V-1M / 760	4,397.8 kg	10-30-81 / 06:04	NIIP-5 / 200P	Proton-K / 311-01	USSR	Lavochkin	Venus landing	S
Venera 14	4V-1M / 761	4,394.5 kg	11-04-81 / 05:31	NIIP-5 / 200L	Proton-K / 311-02	USSR	Lavochkin	Venus landing	S
Venera 15	4V-2 / 860	5,250 kg	06-02-83 / 02:38:39	NIIP-5 / 200L	Proton-K / 321-01	USSR	Lavochkin	Venus orbit	S
Venera 16	4V-2 / 861	5,300 kg	06-07-83 / 02:32	NIIP-5 / 200P	Proton-K / 321-02	USSR	Lavochkin	Venus orbit	S
Vega 1	5VK / 901	c. 4,840 kg	12-15-84 / 09:16:24	NIIP-5 / 200L	Proton-K / 329-01	USSR	Lavochkin	Venus landing, Halley flyby	S
Vega 2	5VK / 902	c. 4,840 kg	12-21-84 / 09:13:52	NIIP-5 / 200P	Proton-K / 325-02	USSR	Lavochkin	Venus landing, Halley flyby	S
OBJECT-PROBE (EARLY ZOND)									
Kosmos 21 [Zond]	3MV-1A / 2	c. 800 kg	11-11-63 / 06:23:34	NIIP-5 / 1	Molniya / G103-18	USSR	OKB-1	deep space and recovery	F
[Zond]	3MV-1A / 4A	c. 800 kg	02-19-64 / 05:47:40	NIIP-5 / 1	Molniya / G15000-26	USSR	OKB-1	Venus flyby	F
Zond 2	3MV-4A / 2	996 kg	11-30-64 / 13:25	NIIP-5 / 1	Molniya / G15000-29	USSR	OKB-1	Mars flyby	P
Zond 3	3MV-4A / 3	950 kg	07-18-65 / 14:32	NIIP-5 / 1	Molniya / U103-32	USSR	OKB-1	lunar flyby	S
ZOND (LATER)									
[Zond]	7K-L1 / 4L	c. 5,375 kg	09-27-67 / 22:11:54	NIIP-5 / 81L	Proton-K / 229-01	USSR	TsKBEM	circumlunar	F
[Zond]	7K-L1 / 5L	c. 5,375 kg	11-22-67 / 19:07:59	NIIP-5 / 81P	Proton-K / 230-01	USSR	TsKBEM	circumlunar	F
Zond 4	7K-L1 / 6L	c. 5,375 kg	03-02-68 / 18:29:23	NIIP-5 / 81L	Proton-K / 231-01	USSR	TsKBEM	deep space	P
[Zond]	7K-L1 / 7L	c. 5,375 kg	04-22-68 / 23:01:27	NIIP-5 / 81P	Proton-K / 232-01	USSR	TsKBEM	circumlunar	F
Zond 5	7K-L1 / 9L	c. 5,375 kg	09-14-68 / 21:42:11	NIIP-5 / 81L	Proton-K / 234-01	USSR	TsKBEM	circumlunar	S
Zond 6	7K-L1 / 12L	c. 5,375 kg	11-10-68 / 19:11:31	NIIP-5 / 81L	Proton-K / 235-01	USSR	TsKBEM	circumlunar	S
[Zond]	7K-L1 / 13L	c. 5,375 kg	01-20-69 / 04:14:36	NIIP-5 / 81L	Proton-K / 237-01	USSR	TsKBEM	circumlunar	F
Zond 7	7K-L1 / 11	c. 5,375 kg	08-07-69 / 23:48:06	NIIP-5 / 81L	Proton-K / 243-01	USSR	TsKBEM	circumlunar	S
Zond 8	7K-L1 / 14	c. 5,375 kg	10-20-70 / 19:55:39	NIIP-5 / 81L	Proton-K / 250-01	USSR	TsKBEM	circumlunar	S
N1 DEEP SPACE TESTS									
[N1]	7K-L1S / 2	6,900 kg	02-21-69 / 09:18:07	NIIP-5 / 110P	N1 / 3L	USSR	TsKBEM	lunar orbit	F
[N1]	7K-L1S / 5	c. 6,900 kg	07-03-69 / 20:18:32	NIIP-5 / 110P	N1 / 15005	USSR	TsKBEM	lunar orbit	F
[N1]	7K-LOK / 6A	c. 9,500 kg	11-23-72 / 06:11:55	NIIP-5 / 110L	N1 / 15007	USSR	TsKBEM	lunar orbit	F

GERMANY

Official Name	Spacecraft / no.	Mass	Launch date / time	Launch place / pad	Launch vehicle / no.	Nation	Design & Operation	Objective	Out-come
					HELIOS				
Helios 1	Helios-A	370 kg	11-10-74 / 07:11:02	CC / 41	Titan IIIE-Centaur / 2	FRG	DFVLR / NASA	solar orbit	S
Helios 2	Helios-B	370 kg	01-15-76 / 05:34:00	CC / 41	Titan IIIE-Centaur / 5	FRG	DFVLR / NASA	solar orbit	S

JAPAN

Official Name	Spacecraft / no.	Mass	Launch date / time	Launch place / pad	Launch vehicle / no.	Nation	Design & Operation	Objective	Out-come
					PLANET				
Sakigake	MS-T5	138.1 kg	01-07-85 / 19:26	Kagoshima / M1	Mu-3S-II / 1	Japan	ISAS	Halley flyby	S
Suisei	Planet-A	139.5 kg	08-18-85 / 23:33	Kagoshima / M1	Mu-3S-II / 2	Japan	ISAS	Halley flyby	S
Nozomi	Planet-B	536 kg	07-03-98 / 18:12	Kagoshima / M-5	M-V / 3	Japan	ISAS	Mars orbit	F
					MUSES				
Hiten / Hagoromo	MUSES-A / MUSES-A subsat	185 kg / 12 kg	01-24-90 / 11:46:00	Kagoshima / M1	Mu-3S-II / 5	Japan	ISAS	lunar flyby / orbit	S
Hayabusa	MUSES-C, MINERVA	510 kg	05-09-03 / 04:29:25	Kagoshima / M-V	M-V / 5	Japan	ISAS / JAXA	asteroid sample return	P
Hayabusa 2		600 kg	12-03-14 / 04:22:04	Tanegashima / Y1	H-IIA / F26	Japan	JAXA	asteroid rendezvous, sample return	IP
					SELENE				
Kaguya	SELENE	2,900 kg	09-14-07 / 01:31:01	Tanegashima / Y1	H-IIA 2022 / 13	Japan	JAXA	lunar orbit	S
					VCO				
Akatsuki	VCO	517.6 kg	05-20-10 / 21:58:22	Tanegashima / Y	H-IIA 202 / 17	Japan	JAXA	Venus orbit	IP
					SHIN'EN				
Shin'en	UNITEC 1	20 kg	05-20-10 / 21:58:22	Tanegashima / Y	H-IIA 202 / 17	Japan	JAXA	Venus flyby	P
Shin'en 2		17 kg	12-03-14 / 04:22:04	Tanegashima / Y1	H-IIA / F26	Japan	JAXA	solar orbit	F

Official Name	Spacecraft / no.	Mass	Launch date / time	Launch place / pad	Launch vehicle / no.	Nation	Design & Operation	Objective	Out-come
				IKAROS					
IKAROS	IKAROS	310 kg	05-20-10 / 21:58:22	Tanegashima / Y	H-IIA 202 / 17	Japan	JAXA	Venus flyby	S
				PROCYON					
PROCYON		65 kg	12-03-14 / 04:22:04	Tanegashima / Y1	H-IIA / F26	Japan	JAXA	asteroid flyby	IP
				DESPATCH					
ArtSat-2	DESPATCH	30 kg	12-03-14 / 04:22:04	Tanegashima / Y1	H-IIA / F26	Japan	JAXA	solar orbit	S

EUROPEAN SPACE AGENCY

Official Name	Spacecraft / no.	Mass	Launch date / time	Launch place / pad	Launch vehicle / no.	Nation	Design & Operation	Objective	Out-come
				GIOTTO					
Giotto	Giotto	960 kg	07-02-85 / 11:23:16	CSG / ELA 1	Ariane 1 / V14	ESA	ESA	Halley flyby	S
				MARS EXPRESS					
Mars Express	Mars Express, Beagle 2	1,186 kg	06-02-03 / 17:45:26	GIK-5 / 31	Soyuz-FG / E15000-005	Europe	ESA	Mars orbit, landing	P
				SMART					
SMART-1	SMART-1	367 kg	09-27-03 / 23:14:46	CSG / ELA 3	Ariane 5G / V162	Europe	ESA	lunar orbit	S
				ROSETTA					
Rosetta, Philae	Rosetta, Philae	3,000 kg	03-02-04 / 07:17:44	CSG / ELA 3	Ariane 5G+ / V158	Europe	ESA	comet orbit and landing	S
				VENUS EXPRESS					
Venus Express	VEX	1,270 kg	11-09-04 / 03:33:34	GIK-5 / 31	Soyuz-FG / Zh15000-010	Europe	ESA	Venus orbit	S
				HERSCHEL / PLANCK					
Herschel	Herschel	3,400 kg	05-14-09 / 13:12	CSG / ELA 3	Ariane 5ECA / V188	ESA	ESA	Sun–Earth L2	S
Planck	Planck	1,950 kg	05-14-09 / 13:12	CSG / ELA 3	Ariane 5ECA / V188	ESA	ESA	Sun–Earth L2	S
				GAIA					
Gaia	Gaia	2,029 kg	12-19-13 / 09:12:18	CSG / ELS	Soyuz-ST-B / E15000-004	ESA	ESA	Sun–Earth L2	S
				LISA					
LISA Pathfinder	SMART-2	1,910 kg	12-03-15 / 04:04:00	CSG / ELV	Vega / VV06	ESA	ESA	Sun–Earth L1	S

EUROPEAN SPACE AGENCY / UNITED STATES

Official Name	Spacecraft / no.	Mass	Launch date / time	Launch place / pad	Launch vehicle / no.	Nation	Design & Operation	Objective	Outcome
ULYSSES									
Ulysses	Ulysses	371 kg	10-06-90 / 11:47:16	KSC / 39B	STS-41 / IUS	ESA / USA	ESA / NASA / JPL	solar orbit	S
SOHO									
SOHO	SOHO	1,864 kg	12-02-95 / 08:08:01	CC / 36B	Atlas Centaur IIAS / 121	ESA / USA	ESA / NASA	Sun–Earth L1	S
CASSINI-HYUGENS									
Cassini-Huygens	Cassini / Huygens	5,655 kg	10-15-97 / 08:43	CC / 40	Titan 401B-Centaur / 21	USA / ESA	NASA / JPL / ESA	Saturn orbit, Titan landing	S

JAPAN / UNITED STATES

Official Name	Spacecraft / no.	Mass	Launch date / time	Launch place / pad	Launch vehicle / no.	Nation	Design & Operation	Objective	Outcome
GEOTAIL									
Geotail	Geotail	1,009 kg	07-24-92 / 14:26	CC / 17A	Delta 6925 / D212	Japan / USA	ISAS / NASA	high Earth orbit	S

ASIASAT

Official Name	Spacecraft / no.	Mass	Launch date / time	Launch place / pad	Launch vehicle / no.	Nation	Design & Operation	Objective	Outcome
ASIASAT									
Asiasat 3	HGS 1	3,465 kg	12-24-97 / 23:19	GIK-5 / 81L	Proton-K / 394-01	Asiasat	Asiasat / Hughes	circumlunar	S

CHINA

Official Name	Spacecraft / no.	Mass	Launch date / time	Launch place / pad	Launch vehicle / no.	Nation	Design & Operation	Objective	Out-come
CHANG'E									
Chang'e 1	Chang'e 1	2,350 kg	10-24-07 / 10:05:04	Xichang / LC 3	CZ-3A / Y14	China	CNSA	lunar orbit	S
Chang'e 2	Chang'e 2	2,480 kg	10-01-10 / 10:59:57	Xichang / LC2	CZ-3C	China	CNSA	lunar orbit, Sun–Earth L2, asteroid flyby	S
Yinghuo-1		113 kg	11-08-11 / 20:16:03	Baikonur / 45	Zenit-2SB41.1	China	CNSA	Mars orbit	F
Chang'e 3, Yutu	Chang'e 3, Yutu	3,780 kg	12-01-13 / 17:30	Xichang / LC2	CZ-3B / Y23	China	CNSA	lunar landing, roving	S
Chang'e 5-T1		3,300 kg	10-23-14 / 18:00:04	Xichang	CZ-3C / G2	China	CNSA	circumlunar, Earth–Moon L2, lunar orbit	S

INDIA

Official Name	Spacecraft / no.	Mass	Launch date / time	Launch place / pad	Launch vehicle / no.	Nation	Design & Operation	Objective	Out-come
CHANDRYAAN									
Chandrayaan-1	Chandrayaan-1, MIP	1,380 kg	10-22-08 / 00:42:11	SHAR / SLP	PSLV-XL / C11	India	ISRO	lunar orbit, lunar impact	S
MANGALYAAN									
Mangalyaan	MOM	1,337 kg	11-05-13 / 09:08	SHAR / PSLV pad	PSLV-XL / C11	India	ISRO	Mars orbit	S

EUROPEAN SPACE AGENCY / RUSSIA

Official Name	Spacecraft / no.	Mass	Launch date / time	Launch place / pad	Launch vehicle / no.	Nation	Design & Operation	Objective	Out-come
EXOMARS									
ExoMars TGO / Schiaparelli	TGO / EDM Lander	4,332 kg	03-14-16 / 09:31	Baikonur / 200L	Proton-M / 93560	ESA / Russia	ESA / Roskosmos	Mars orbit and landing	S/P

Table 3. Total Lunar Spacecraft Attempts by Nation/Agency 1958–2016

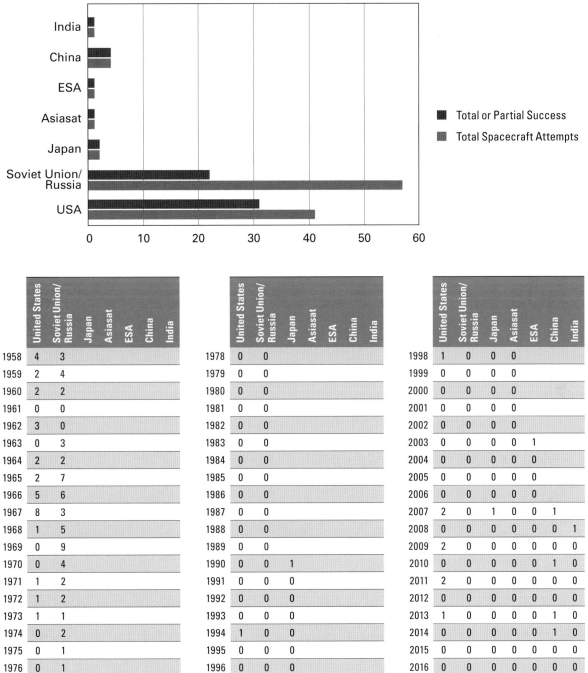

Legend: ■ Total or Partial Success ■ Total Spacecraft Attempts

Year	United States	Soviet Union/Russia	Japan	Asiasat	ESA	China	India
1958	4	3					
1959	2	4					
1960	2	2					
1961	0	0					
1962	3	0					
1963	0	3					
1964	2	2					
1965	2	7					
1966	5	6					
1967	8	3					
1968	1	5					
1969	0	9					
1970	0	4					
1971	1	2					
1972	1	2					
1973	1	1					
1974	0	2					
1975	0	1					
1976	0	1					
1977	0	0					
1978	0	0					
1979	0	0					
1980	0	0					
1981	0	0					
1982	0	0					
1983	0	0					
1984	0	0					
1985	0	0					
1986	0	0					
1987	0	0					
1988	0	0					
1989	0	0					
1990	0	0	1				
1991	0	0	0				
1992	0	0	0				
1993	0	0	0				
1994	1	0	0				
1995	0	0	0				
1996	0	0	0				
1997	0	0	0	1			
1998	1	0	0	0			
1999	0	0	0	0			
2000	0	0	0	0			
2001	0	0	0	0			
2002	0	0	0	0			
2003	0	0	0	0	1		
2004	0	0	0	0	0		
2005	0	0	0	0	0		
2006	0	0	0	0	0		
2007	2	0	1	0	0	1	
2008	0	0	0	0	0	0	1
2009	2	0	0	0	0	0	0
2010	0	0	0	0	0	1	0
2011	2	0	0	0	0	0	0
2012	0	0	0	0	0	0	0
2013	1	0	0	0	0	1	0
2014	0	0	0	0	0	1	0
2015	0	0	0	0	0	0	0
2016	0	0	0	0	0	0	0
Total Spacecraft Attempts	41	57	2	1	1	4	1
Total or Partial Success	31	22	2	1	1	4	1

Table 4. Total Mars Spacecraft Attempts by Nation/Agency 1958–2016

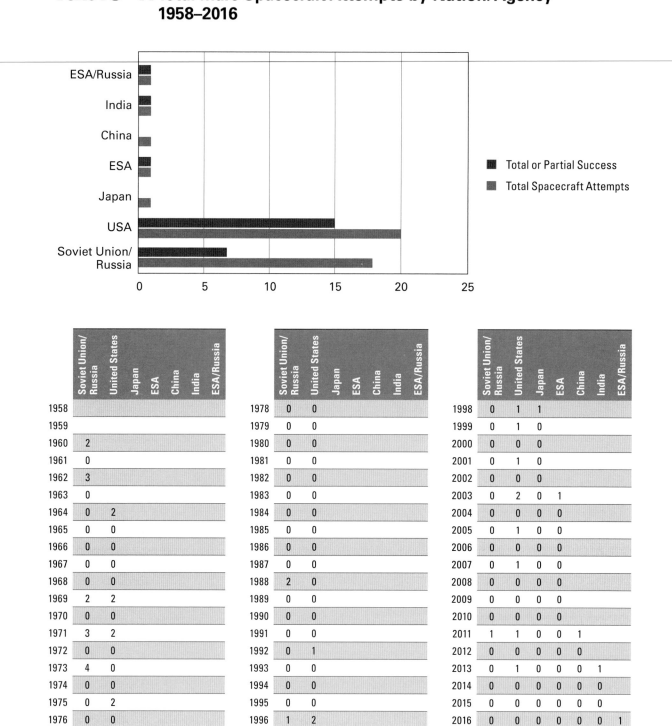

Legend:
- Total or Partial Success
- Total Spacecraft Attempts

Year	Soviet Union/Russia	United States	Japan	ESA	China	India	ESA/Russia
1958							
1959							
1960	2						
1961	0						
1962	3						
1963	0						
1964	0	2					
1965	0	0					
1966	0	0					
1967	0	0					
1968	0	0					
1969	2	2					
1970	0	0					
1971	3	2					
1972	0	0					
1973	4	0					
1974	0	0					
1975	0	2					
1976	0	0					
1977	0	0					
1978	0	0					
1979	0	0					
1980	0	0					
1981	0	0					
1982	0	0					
1983	0	0					
1984	0	0					
1985	0	0					
1986	0	0					
1987	0	0					
1988	2	0					
1989	0	0					
1990	0	0					
1991	0	0					
1992	0	1					
1993	0	0					
1994	0	0					
1995	0	0					
1996	1	2					
1997	0	0					
1998	0	1	1				
1999	0	1	0				
2000	0	0	0				
2001	0	1	0				
2002	0	0	0				
2003	0	2	0	1			
2004	0	0	0	0			
2005	0	1	0	0			
2006	0	0	0	0			
2007	0	1	0	0			
2008	0	0	0	0			
2009	0	0	0	0			
2010	0	0	0	0			
2011	1	1	0	0	1		
2012	0	0	0	0	0		
2013	0	1	0	0	0	1	
2014	0	0	0	0	0	0	
2015	0	0	0	0	0	0	
2016	0	0	0	0	0	0	1
Total Spacecraft Attempts	18	20	1	1	1	1	1
Total or Partial Success	7	15	0	1	0	1	1

Table 5. Total Venus Spacecraft Attempts by Nation/Agency 1958–2016

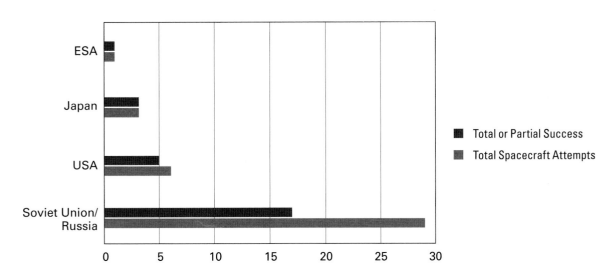

	Soviet Union/ Russia	United States	ESA	Japan			Soviet Union/ Russia	United States	ESA	Japan			Soviet Union/ Russia	United States	ESA	Japan
1958						1978	2	2				1998	0	0		
1959						1979	0	0				1999	0	0		
1960						1980	0	0				2000	0	0		
1961	2					1981	2	0				2001	0	0		
1962	3	2				1982	0	0				2002	0	0		
1963	0	0				1983	2	0				2003	0	0		
1964	3	0				1984	2	0				2004	0	0	1	
1965	3	0				1985	0	0				2005	0	0	0	
1966	0	0				1986	0	0				2006	0	0	0	
1967	2	1				1987	0	0				2007	0	0	0	
1968	0	0				1988	0	0				2008	0	0	0	
1969	2	0				1989	0	1				2009	0	0	0	
1970	2	0				1990	0	0				2010	0	0	0	3
1971	0	0				1991	0	0				2011	0	0	0	0
1972	2	0				1992	0	0				2012	0	0	0	0
1973	0	0				1993	0	0				2013	0	0	0	0
1974	0	0				1994	0	0				2014	0	0	0	0
1975	2	0				1995	0	0				2015	0	0	0	0
1976	0	0				1996	0	0				2016	0	0	0	0
1977	0	0				1997	0	0								

	Soviet Union/ Russia	United States	ESA	Japan
Total Spacecraft Attempts	**29**	**6**	**1**	**3**
Total or Partial Success	17	5	1	3

Abbreviations

ABMA	Army Ballistic Missile Agency
ACE	Advanced Composition Explorer
AFBMD	Air Force Ballistic Missile Division
AIMP	Advanced International Monitoring Platform
ALS	Automatic Lunar Station
AMS	Automatic Interplanetary Station
APL	Applied Physics Laboratory
ARC	Ames Research Center
ARPA	Advanced Projects Research Agency
ARTEMIS	Acceleration, Reconnection, Turbulence and Electrodynamics of the Moon's Interaction with the Sun
AU	astronomical unit
BMDO	Ballistic Missile Defense Organization
CIA	Central Intelligence Agency
CMB	Cosmic Microwave Background
CNSA	Chinese National Space Administration
COBE	Cosmic Background Explorer
CONTOUR	Comet Nucleus Tour
CTA	Cryogenic Telescope Assembly
DAS	Long-Term Autonomous Station
DESPATCH	Deep Space Amateur Troubadour's Challenge
DFVLR	German Test and Research Institute for Aviation and Space Flight
DIXI	Deep Impact Extended Investigation
DLR	German Aerospace Center
DOK	Engine Orientation Complex
DRK	Long-Range Radio Complex
DSCOVR	Deep Space Climate Observatory
DSN	Deep Space Network
DSPSE	Deep Space Program Science Experiment
EDM	Entry, Descent and Landing Demonstrator Module
ELA	Launch Assembly
EPOCh	Extrasolar Planet Observations
ESA	European Space Agency
ESC	Cryogenic Upper Stage
ESMD	Exploration Systems Mission Directorate
ESOC	European Space Operations Center
EVA	Extra-Vehicular Activity
GEM	Galileo Europa Mission
GIK	State Testing Cosmodrome
GOODS	Great Observatories Origins Deep Survey
GRAIL	Gravity Recovery and Interior Laboratory
GSFC	Goddard Space Flight Center
GSMZ	State Union Machine Building Factory
IASTP	International Solar-Terrestrial Physics
ICE	International Cometary Explorer
IGY	International Geophysical Year
IKAROS	Interplanetary Kite-Craft Accelerated by Radiation of the Sun
IKI	Institute of Space Research
IMP	Interplanetary Monitoring Platform
ISAS	Institute of Space and Astronautical Sciences
ISEE	International Sun-Earth Explorers
ISRO	Indian Space Research Organization
ISTP	International Solar Terrestrial Physics
IUS	Inertial Upper Stage
JAXA	Japan Aerospace Exploration Agency
JPL	Jet Propulsion Laboratory
LADEE	Lunar Atmosphere and Dust Environment Explorer
LaRC	Langley Research Center
LC	Launch Complex
LCROSS	Lunar Crater Observation and Sensing Satellite
LK	Lunar Ship
LOK	Lunar Orbital Ship

LRO	Lunar Reconnaissance Orbiter	**RIFMA**	X-Ray Isotopic Fluorescence Analysis Method
MAP	Microwave Anisotropy Probe	**RTG**	radioisotope thermal generator
MASCOT	Mobile Asteroid Surface Scout	**SAR**	synthetic aperture radar
MAVEN	Mars Atmosphere and Volatile Evolution	**SDI**	Strategic Defense Initiative
MCO	Mars Climate Orbiter	**SELENE**	Selenological and Engineering Explorer
MDU	Main Propulsion Unit	**SETI**	Search for Extraterrestrial Intelligence
MER	Mars Exploration Rover	**SIM**	Spectral Imaging Module
MESSENGER	Mercury Surface, Space Environment, Geochemistry, and Ranging	**SIRTF**	Space Infrared Telescope Facility
MGS	Mars Global Surveyor	**SLC**	Space Launch Complex
MINERVA	Micro/Nano Experimental Robot Vehicle for Asteroid	**SMART**	Small Missions for Advanced Research in Technology
MIP	Moon Impact Probe	**SMD**	Science Mission Directorate
MOM	Mars Orbiter Mission	**SOHO**	Solar and Heliospheric Observatory
MPL	Mars Polar Lander	**STEREO**	Solar Terrestrial Relations Observatory
MRO	Mars Reconnaissance Orbiter	**STL**	Space Technology Laboratories
MSC	Manned Spacecraft Center	**STSP**	Solar Terrestrial Science Program
MSL	Mars Science Laboratory	**TETR**	Test and Training Satellite
MUSES	Mu Space Engineering Spacecraft	**TCM**	trajectory correction maneuver
NASA	National Aeronautics and Space Administration	**TGO**	Trace Gas Orbiter
NEAR	Near Earth Asteroid Rendezvous	**THEMIS**	Time History of Events and Macro-scale Interactions during Substorms
NExT	New Exploration of Tempel	**TsKBEM**	Central Design Bureau of Experimental Machine Building
NGSLR	Next Generation Satellite Laser Ranging	**TsKBM**	Central Design Bureau of Machine Building
NIIP	Scientific-Research and Testing Range	**UCLA**	University of California Los Angeles
NOAA	National Oceanic and Atmospheric Administration	**UMVL**	Unified Mars Venus Moon
NPO	Scientific-Production Association	**UNISEC**	University Space Engineering Consortium
NRO	National Reconnaissance Office	**UNITEC**	UNISEC Technology Experiment Carrier
OKB	Experimental-Design Bureau	**USAF**	United States Air Force
OSIRIS-REx	Origins, Spectral Interpretation, Resource Identification, Security, Regolith Explorer	**USSR**	Union of Soviet Socialist Republics
PAM	Payload Assist Module	**UT**	Universal Time
PROCYON	Proximate Object Close Flyby with Optical Navigation	**VCO**	Venus Climate Orbiter
PrOP	Device to Evaluate Mobility	**VIM**	Voyager Interstellar Mission
RAE	Radio Astronomy Explorer	**WMAP**	Wilkinson Microwave Anisotropy Probe
RAN	Russian Academy of Sciences		

Bibliography of Secondary Sources

English Language

Andrew J. Ball et al., *Planetary Landers and Entry Probes* (Cambridge, UK: Cambridge University Press, 2007).

M. H. Carr et al., *Viking Orbiter Views of Mars* (Washington, DC: NASA SP-441, 1980).

Boris Chertok, *Rockets and People, Vol. II: Creating a Rocket Industry*, ed. Asif Siddiqi (Washington, DC: NASA SP-2006-4110, 2006).

Boris Chertok, *Rockets and People, Vol. III: Hot Days of the Cold War*, ed., Asif Siddiqi (Washington, DC: NASA SP-2009-4110, 2009).

Boris Chertok, *Rockets and People, Vol. IV: The Moon Race*, ed. Asif Siddiqi (Washington, DC: NASA SP-2011-4110, 2011).

Erik M. Conway, *Exploration and Engineering: The Jet Propulsion Laboratory and the Quest for Mars* (Baltimore: Johns Hopkins University Press, 2015).

James A. Dunne and Eric Burgess, *The Voyage of Mariner 10: Mission to Venus and Mercury* (Washington, DC: NASA SP-424, 1978).

Edward Clinton Ezell and Linda Neuman Ezell, *On Mars: Exploration of the Red Planet, 1958–1978* (Washington, DC: NASA SP-4212, 1984).

Linda Neumann Ezell, *NASA Historical Data Book, Volume II: Programs and Projects 1958–1968* (Washington, DC: NASA SP-4012, 1988).

Linda Neumann Ezell, *NASA Historical Data Book, Volume III: Programs and Projects 1969–1978* (Washington, DC: NASA SP-4012, 1988).

Richard O. Fimmel et al., *Pioneer Odyssey* (Washington, DC: NASA SP-396, 1977).

Jay Gallentine, *Ambassadors From Earth: Pioneering Explorations with Unmanned Spacecraft* (Lincoln, Nebraska: University of Nebraska Press, 2009).

Kenneth Gatland, *Robot Explorers* (London, UK: Blanford Press, 1972).

R. Cargill Hall, *Lunar Impact: A History of Project Ranger* (Washington, DC: NASA SP-4210, 1977).

Wesley T. Huntress, Jr. and Mikhail Ya. Marov, *Soviet Robots in the Solar System: Mission Technologies and Discoveries* (Chichester, UK: Springer-Praxis, 2011).

Nicholas L. Johnson, *Handbook of Soviet Lunar and Planetary Exploration* (San Diego: Univelt, 1979).

Vadym Kaydash et al., "Landing of the probes Luna 23 and Luna 24 remains an enigma," *Planetary and Space Science* 89 (2013): 172–182.

Alexander Kemurdjian [Aleksandr Kemurdzhian], "Experience in Creating Self-Propelled Undercarriages for Planet Rovers," *Journal of Aerospace Engineering* 4, no. 4 (October 1991): 317–329.

Robert S. Kraemer, *Beyond the Moon: A Golden Age of Planetary Exploration, 1971–1978* (Washington, DC: Smithsonian Institution Press, 2000).

Christian Lardier and Stefan Barensky, *The Soyuz Launch Vehicle: The Two Lives of an Engineering Triumph* (Chichester, UK: Springer-Praxis, 2013).

Roger D. Launius, ed., *Exploring the Solar System: The History and Science of Planetary Exploration* (Basingstoke: Palgrave Macmillan, 2013).

Major NASA Launches, Total Major ETR and WTR Launches, NASA Information Summaries, PMS 031 (KSC), December 1989.

G. Yu. Maksimov, "Construction and Testing of the First Soviet Automatic Interplanetary Stations" in *History of Rocketry and Astronautics, Vol. 20*, ed. J. D. Hunley, 233–246 (San Diego: American Astronautical Society, 1997).

M. Ya. Marov and G. I. Petrov, "Investigations of Mars from the Soviet Automatic Station Mars 2 and 3," *Icarus* 19 (1973): 163–179.

Michael Meltzer, *Mission to Jupiter: A History of the Galileo Project* (Washington, DC: NASA SP-2007-4231, 2007).

P. M. Muller and W. L. Sjogren, "Lunar Mass Concentrations," *Science* 161, no. 3842 (August 16, 1968): 680–684.

V. G. Perminov, *A Difficult Road to Mars: A Brief History of Mars Exploration in the Soviet Union* (NASA: Washington, DC, 1999).

Joel W. Powell, "Thor-Able and Atlas-Able," *Journal of the British Interplanetary Society* 37 (1984): 219–225.

Marc Rayman, "The Successful Conclusion of the Deep Space 1 Mission: Important Results without a Flashy Title," *Space Technology* 23, nos. 2-3 (2003): 185–198.

M. S. Robinson et al., "Soviet Lunar Sample Return Missions: Landing Site Identification and Geologic Context," *Planetary and Space Science* 69 (2012): 76–88.

Judy A. Rumerman, *NASA Historical Data Book, Volume V: NASA Launch Systems, Space Transportation, Human Spaceflight, and Space Science 1979–1988* (Washington, DC: NASA SP-4012, 1999).

R. Z. Sagdeev and A. V. Zakharov, "Brief History of the Phobos Mission," *Nature* 341 (October 10, 1989): 581–585.

Asif Siddiqi, "First to the Moon," *Journal of the British Interplanetary Society* 51 (1998): 231–238.

Asif Siddiqi, Bart Hendrickx and Timothy Varfolomeyev, "The Tough Road Travelled: A New Look at the Second Generation Luna Probes," *Journal of the British Interplanetary Society* 52, nos. 9/10 (September/October 2000): 319–356.

Phil Stooke, *Atlas of Lunar Landing Sites: Luna 9 to Chang'E 3* (Philip J. Stooke, 2018).

A Summary of Major NASA Launches, October 1, 1958–December 31, 1989, KSC Historical Report No. (KHR-1), June 1992.

G. A. Soffen and C. W. Snyder, "The First Viking Mission to Mars," *Science* 193, no. 4255 (August 27, 1976): 759–766.

Paolo Ulivi and David M. Harland, *Robotic Exploration of the Solar System, Part 1: The Golden Age 1957–1982* (Chichester, UK: Springer-Praxis, 2007).

Paolo Ulivi and David M. Harland, *Robotic Exploration of the Solar System, Part 2: Hiatus and Renewal 1983–1996* (Chichester, UK: Springer-Praxis, 2007).

Andrew Wilson, *The Eagle Has Wings: The Story of American Space Exploration, 1945–1975* (London, UK: British Interplanetary Society, 1982).

Andrew Wilson, *Solar System Log* (London, UK: Jane's Publishing Company Limited, 1987)

Russian Language

N. G. Babakin, A. N. Banketov, and V. N. Smorkalov, *G. N. Babakin: zhizn' i deyatel'nost'* [G. N. Babakin: Life and Activities] (Moscow: Adamant, 1996).

V. P. Glushko, ed., *Kosmonavtika: entsiklopediya* [Cosmonautics: Encyclopedia] (Moscow: Sovetskaya entsiklopediya, 1985).

M. V. Keldysh, ed., *Tvorcheskoye naslediye akademika Sergeya Pavlovicha Koroleva: izbrannyye trudy i dokumenty* [The Creative Legacy of Sergey Pavlovich Korolev: Selected Works and Documents] (Moscow: Nauka, 1980).

S. S. Kryukov, *Izbrannyye raboti: iz lichnogo arkhiva* [Selected Works: From the Personal Archive] (Moscow: MGTU im. N. E. Baumana, 2010).

K. Lantratov, " 'Opozdavshiy' lunnyy grunt" ["The 'Late' Lunar Soil"] *Novosti kosmonavtiki* no. 15 (1994): 41–43.

K. Lantratov, "25 let Lunokhodu-1" ["25 Years for Lunokhod-1"] *Novosti kosmonavtiki* no. 23 (1995): 79-83 and no. 24 (1995): 70–79.

K. Lantratov, "Na Mars!" ["On Mars!"] *Novosti kosmonavtiki* no. 20 (1996): 53–72 and no. 21 (1996): 41–51.

I. L. Shevalev, *Shest' let i vsya zhizn' konstruktora G. N. Babakina: vospominaniya otechestvennykh issledovateley kosmosa* [Six Years and the Full Life of Designer G. N. Babakin: Recollections of a Native Space Researcher] (Khimki: NPO im. S. A. Lavochkina, 2004).

S. S. Sokolov [S. S. Kryukov], "Sovetskiye avtomaticheskiye mezhplanetnyye stantsii issleduyut Mars" [Soviet automatic interplanetary stations research Mars], *Vestnik akademicheskii nauk sssr* 10 (1974): 21–38.

Unofficial Websites

"Pioneer 10 Position and Data," *https://theskylive.com/pioneer10-tracker*

"Pioneer 11 Position and Data," *https://theskylive.com/pioneer11-tracker*

"STL On-line Archive," *http://www.sdfo.org/stl/pioneer02.php.*

"Venera: The Soviet Exploration of Venus," *http://mentallandscape.com/V_Venus.htm.*

Online Fora

Novosti kosmonavtiki forum
NASAspaceflight.com forum

About the Author

Asif A. Siddiqi is a Professor of History at Fordham University in New York. He specializes in the history of science and technology and modern Russian history, having received his Ph.D. from Carnegie Mellon University in 2004. He has written extensively on a variety of topics related to the history of science and technology, especially on the social and cultural dimensions of the Russian fascination with the cosmos. His most recent book was *The Rockets' Red Glare: Spaceflight and the Soviet Imagination, 1857–1957* (Cambridge University Press, 2010), exploring the co-production of technology and culture in Russian society in the late 19th and early 20th centuries that led to the launch of Sputnik. His earlier book, *Challenge to Apollo: The Soviet Union and the Space Race, 1945–1974* (NASA SP-2000-4408, 2000) was the first major scholarly work on Cold War–era Soviet space achievements. He is currently working on several book projects including one on the Stalinist Gulag and another on the early history of the Indian space program. In 2016, he received the Guggenheim Fellowship from the John Simon Guggenheim Memorial Foundation for a projected work on a global history of space exploration entitled "Departure Gates: Histories of Spaceflight on Earth."

The NASA History Series

Reference Works, NASA SP-4000

Grimwood, James M. *Project Mercury: A Chronology.* NASA SP-4001, 1963.

Grimwood, James M., and Barton C. Hacker, with Peter J. Vorzimmer. *Project Gemini Technology and Operations: A Chronology.* NASA SP-4002, 1969.

Link, Mae Mills. *Space Medicine in Project Mercury.* NASA SP-4003, 1965.

Astronautics and Aeronautics, 1963: Chronology of Science, Technology, and Policy. NASA SP-4004, 1964.

Astronautics and Aeronautics, 1964: Chronology of Science, Technology, and Policy. NASA SP-4005, 1965.

Astronautics and Aeronautics, 1965: Chronology of Science, Technology, and Policy. NASA SP-4006, 1966.

Astronautics and Aeronautics, 1966: Chronology of Science, Technology, and Policy. NASA SP-4007, 1967.

Astronautics and Aeronautics, 1967: Chronology of Science, Technology, and Policy. NASA SP-4008, 1968.

Ertel, Ivan D., and Mary Louise Morse. *The Apollo Spacecraft: A Chronology, Volume I, Through November 7, 1962.* NASA SP-4009, 1969.

Morse, Mary Louise, and Jean Kernahan Bays. *The Apollo Spacecraft: A Chronology, Volume II, November 8, 1962–September 30, 1964.* NASA SP-4009, 1973.

Brooks, Courtney G., and Ivan D. Ertel. *The Apollo Spacecraft: A Chronology, Volume III, October 1, 1964–January 20, 1966.* NASA SP-4009, 1973.

Ertel, Ivan D., and Roland W. Newkirk, with Courtney G. Brooks. *The Apollo Spacecraft: A Chronology, Volume IV, January 21, 1966–July 13, 1974.* NASA SP-4009, 1978.

Astronautics and Aeronautics, 1968: Chronology of Science, Technology, and Policy. NASA SP-4010, 1969.

Newkirk, Roland W., and Ivan D. Ertel, with Courtney G. Brooks. *Skylab: A Chronology.* NASA SP-4011, 1977.

Van Nimmen, Jane, and Leonard C. Bruno, with Robert L. Rosholt. *NASA Historical Data Book, Volume I: NASA Resources, 1958–1968.* NASA SP-4012, 1976; rep. ed. 1988.

Ezell, Linda Neuman. *NASA Historical Data Book, Volume II: Programs and Projects, 1958–1968.* NASA SP-4012, 1988.

Ezell, Linda Neuman. *NASA Historical Data Book, Volume III: Programs and Projects, 1969–1978.* NASA SP-4012, 1988.

Gawdiak, Ihor, with Helen Fedor. *NASA Historical Data Book, Volume IV: NASA Resources, 1969–1978.* NASA SP-4012, 1994.

Rumerman, Judy A. *NASA Historical Data Book, Volume V: NASA Launch Systems, Space Transportation, Human Spaceflight, and Space Science, 1979–1988.* NASA SP-4012, 1999.

Rumerman, Judy A. *NASA Historical Data Book, Volume VI: NASA Space Applications, Aeronautics and Space Research and Technology, Tracking and Data Acquisition/Support Operations, Commercial Programs, and Resources, 1979–1988.* NASA SP-4012, 1999.

Rumerman, Judy A. *NASA Historical Data Book, Volume VII: NASA Launch Systems, Space Transportation, Human Spaceflight, and Space Science, 1989–1998.* NASA SP-2009-4012, 2009.

Rumerman, Judy A. *NASA Historical Data Book, Volume VIII: NASA Earth Science and Space*

Applications, Aeronautics, Technology, and Exploration, Tracking and Data Acquisition/ Space Operations, Facilities and Resources, 1989–1998. NASA SP-2012-4012, 2012.

No SP-4013.

Astronautics and Aeronautics, 1969: Chronology of Science, Technology, and Policy. NASA SP-4014, 1970.

Astronautics and Aeronautics, 1970: Chronology of Science, Technology, and Policy. NASA SP-4015, 1972.

Astronautics and Aeronautics, 1971: Chronology of Science, Technology, and Policy. NASA SP-4016, 1972.

Astronautics and Aeronautics, 1972: Chronology of Science, Technology, and Policy. NASA SP-4017, 1974.

Astronautics and Aeronautics, 1973: Chronology of Science, Technology, and Policy. NASA SP-4018, 1975.

Astronautics and Aeronautics, 1974: Chronology of Science, Technology, and Policy. NASA SP-4019, 1977.

Astronautics and Aeronautics, 1975: Chronology of Science, Technology, and Policy. NASA SP-4020, 1979.

Astronautics and Aeronautics, 1976: Chronology of Science, Technology, and Policy. NASA SP-4021, 1984.

Astronautics and Aeronautics, 1977: Chronology of Science, Technology, and Policy. NASA SP-4022, 1986.

Astronautics and Aeronautics, 1978: Chronology of Science, Technology, and Policy. NASA SP-4023, 1986.

Astronautics and Aeronautics, 1979–1984: Chronology of Science, Technology, and Policy. NASA SP-4024, 1988.

Astronautics and Aeronautics, 1985: Chronology of Science, Technology, and Policy. NASA SP-4025, 1990.

Noordung, Hermann. *The Problem of Space Travel: The Rocket Motor.* Edited by Ernst Stuhlinger and J. D. Hunley, with Jennifer Garland. NASA SP-4026, 1995.

Gawdiak, Ihor Y., Ramon J. Miro, and Sam Stueland. *Astronautics and Aeronautics, 1986– 1990: A Chronology.* NASA SP-4027, 1997.

Gawdiak, Ihor Y., and Charles Shetland. *Astronautics and Aeronautics, 1991–1995: A Chronology.* NASA SP-2000-4028, 2000.

Orloff, Richard W. *Apollo by the Numbers: A Statistical Reference.* NASA SP-2000-4029, 2000.

Lewis, Marieke, and Ryan Swanson. *Astronautics and Aeronautics: A Chronology, 1996–2000.* NASA SP-2009-4030, 2009.

Ivey, William Noel, and Marieke Lewis. *Astronautics and Aeronautics: A Chronology, 2001–2005.* NASA SP-2010-4031, 2010.

Buchalter, Alice R., and William Noel Ivey. *Astronautics and Aeronautics: A Chronology, 2006.* NASA SP-2011-4032, 2010.

Lewis, Marieke. *Astronautics and Aeronautics: A Chronology, 2007.* NASA SP-2011-4033, 2011.

Lewis, Marieke. *Astronautics and Aeronautics: A Chronology, 2008.* NASA SP-2012-4034, 2012.

Lewis, Marieke. *Astronautics and Aeronautics: A Chronology, 2009.* NASA SP-2012-4035, 2012.

Flattery, Meaghan. *Astronautics and Aeronautics: A Chronology, 2010.* NASA SP-2013-4037, 2014.

Siddiqi, Asif A. *Beyond Earth: A Chronicle of Deep Space Exploration, 1958–2016.* NASA SP-2018-4041, 2018.

Management Histories, NASA SP-4100

Rosholt, Robert L. *An Administrative History of NASA, 1958–1963.* NASA SP-4101, 1966.

Levine, Arnold S. *Managing NASA in the Apollo Era.* NASA SP-4102, 1982.

Roland, Alex. *Model Research: The National Advisory Committee for Aeronautics, 1915–1958.* NASA SP-4103, 1985.

Fries, Sylvia D. *NASA Engineers and the Age of Apollo.* NASA SP-4104, 1992.

Glennan, T. Keith. *The Birth of NASA: The Diary of T. Keith Glennan.* Edited by J. D. Hunley. NASA SP-4105, 1993.

Seamans, Robert C. *Aiming at Targets: The Autobiography of Robert C. Seamans.* NASA SP-4106, 1996.

Garber, Stephen J., ed. *Looking Backward, Looking Forward: Forty Years of Human Spaceflight Symposium.* NASA SP-2002-4107, 2002.

Mallick, Donald L., with Peter W. Merlin. *The Smell of Kerosene: A Test Pilot's Odyssey.* NASA SP-4108, 2003.

Iliff, Kenneth W., and Curtis L. Peebles. *From Runway to Orbit: Reflections of a NASA Engineer.* NASA SP-2004-4109, 2004.

Chertok, Boris. *Rockets and People, Volume I.* NASA SP-2005-4110, 2005.

Chertok, Boris. *Rockets and People: Creating a Rocket Industry, Volume II.* NASA SP-2006-4110, 2006.

Chertok, Boris. *Rockets and People: Hot Days of the Cold War, Volume III.* NASA SP-2009-4110, 2009.

Chertok, Boris. *Rockets and People: The Moon Race, Volume IV.* NASA SP-2011-4110, 2011.

Laufer, Alexander, Todd Post, and Edward Hoffman. *Shared Voyage: Learning and Unlearning from Remarkable Projects.* NASA SP-2005-4111, 2005.

Dawson, Virginia P., and Mark D. Bowles. *Realizing the Dream of Flight: Biographical Essays in Honor of the Centennial of Flight, 1903–2003.* NASA SP-2005-4112, 2005.

Mudgway, Douglas J. *William H. Pickering: America's Deep Space Pioneer.* NASA SP-2008-4113, 2008.

Wright, Rebecca, Sandra Johnson, and Steven J. Dick. *NASA at 50: Interviews with NASA's Senior Leadership.* NASA SP-2012-4114, 2012.

Project Histories, NASA SP-4200

Swenson, Loyd S., Jr., James M. Grimwood, and Charles C. Alexander. *This New Ocean: A History of Project Mercury.* NASA SP-4201, 1966; rep. ed. 1999.

Green, Constance McLaughlin, and Milton Lomask. *Vanguard: A History.* NASA SP-4202, 1970; rep. ed. Smithsonian Institution Press, 1971.

Hacker, Barton C., and James M. Grimwood. *On the Shoulders of Titans: A History of Project Gemini.* NASA SP-4203, 1977; rep. ed. 2002.

Benson, Charles D., and William Barnaby Faherty. *Moonport: A History of Apollo Launch Facilities and Operations.* NASA SP-4204, 1978.

Brooks, Courtney G., James M. Grimwood, and Loyd S. Swenson, Jr. *Chariots for Apollo: A History of Manned Lunar Spacecraft.* NASA SP-4205, 1979.

Bilstein, Roger E. *Stages to Saturn: A Technological History of the Apollo/Saturn Launch Vehicles.* NASA SP-4206, 1980 and 1996.

No SP-4207.

Compton, W. David, and Charles D. Benson. *Living and Working in Space: A History of Skylab.* NASA SP-4208, 1983.

Ezell, Edward Clinton, and Linda Neuman Ezell. *The Partnership: A History of the Apollo-Soyuz Test Project.* NASA SP-4209, 1978.

Hall, R. Cargill. *Lunar Impact: A History of Project Ranger.* NASA SP-4210, 1977.

Newell, Homer E. *Beyond the Atmosphere: Early Years of Space Science.* NASA SP-4211, 1980.

Ezell, Edward Clinton, and Linda Neuman Ezell. *On Mars: Exploration of the Red Planet, 1958–1978.* NASA SP-4212, 1984.

Pitts, John A. *The Human Factor: Biomedicine in the Manned Space Program to 1980.* NASA SP-4213, 1985.

Compton, W. David. *Where No Man Has Gone Before: A History of Apollo Lunar Exploration Missions.* NASA SP-4214, 1989.

Naugle, John E. *First Among Equals: The Selection of NASA Space Science Experiments.* NASA SP-4215, 1991.

Wallace, Lane E. *Airborne Trailblazer: Two Decades with NASA Langley's 737 Flying Laboratory.* NASA SP-4216, 1994.

Butrica, Andrew J., ed. *Beyond the Ionosphere: Fifty Years of Satellite Communications.* NASA SP-4217, 1997.

Butrica, Andrew J. *To See the Unseen: A History of Planetary Radar Astronomy.* NASA SP-4218, 1996.

Mack, Pamela E., ed. *From Engineering Science to Big Science: The NACA and NASA Collier Trophy Research Project Winners.* NASA SP-4219, 1998.

Reed, R. Dale. *Wingless Flight: The Lifting Body Story.* NASA SP-4220, 1998.

Heppenheimer, T. A. *The Space Shuttle Decision: NASA's Search for a Reusable Space Vehicle.* NASA SP-4221, 1999.

Hunley, J. D., ed. *Toward Mach 2: The Douglas D-558 Program.* NASA SP-4222, 1999.

Swanson, Glen E., ed. *"Before This Decade Is Out…" Personal Reflections on the Apollo Program.* NASA SP-4223, 1999.

Tomayko, James E. *Computers Take Flight: A History of NASA's Pioneering Digital Fly-By-Wire Project.* NASA SP-4224, 2000.

Morgan, Clay. *Shuttle-Mir: The United States and Russia Share History's Highest Stage.* NASA SP-2001-4225, 2001.

Leary, William M. *"We Freeze to Please": A History of NASA's Icing Research Tunnel and the Quest for Safety.* NASA SP-2002-4226, 2002.

Mudgway, Douglas J. *Uplink-Downlink: A History of the Deep Space Network, 1957–1997.* NASA SP-2001-4227, 2001.

No SP-4228 or SP-4229.

Dawson, Virginia P., and Mark D. Bowles. *Taming Liquid Hydrogen: The Centaur Upper Stage Rocket, 1958–2002.* NASA SP-2004-4230, 2004.

Meltzer, Michael. *Mission to Jupiter: A History of the Galileo Project.* NASA SP-2007-4231, 2007.

Heppenheimer, T. A. *Facing the Heat Barrier: A History of Hypersonics.* NASA SP-2007-4232, 2007.

Tsiao, Sunny. *"Read You Loud and Clear!" The Story of NASA's Spaceflight Tracking and Data Network.* NASA SP-2007-4233, 2007.

Meltzer, Michael. *When Biospheres Collide: A History of NASA's Planetary Protection Programs.* NASA SP-2011-4234, 2011.

Center Histories, NASA SP-4300

Rosenthal, Alfred. *Venture into Space: Early Years of Goddard Space Flight Center.* NASA SP-4301, 1985.

Hartman, Edwin P. *Adventures in Research: A History of Ames Research Center, 1940–1965.* NASA SP-4302, 1970.

Hallion, Richard P. *On the Frontier: Flight Research at Dryden, 1946–1981.* NASA SP-4303, 1984.

Muenger, Elizabeth A. *Searching the Horizon: A History of Ames Research Center, 1940–1976.* NASA SP-4304, 1985.

Hansen, James R. *Engineer in Charge: A History of the Langley Aeronautical Laboratory, 1917–1958.* NASA SP-4305, 1987.

Dawson, Virginia P. *Engines and Innovation: Lewis Laboratory and American Propulsion Technology.* NASA SP-4306, 1991.

Dethloff, Henry C. *"Suddenly Tomorrow Came…": A History of the Johnson Space Center, 1957–1990.* NASA SP-4307, 1993.

Hansen, James R. *Spaceflight Revolution: NASA Langley Research Center from Sputnik to Apollo.* NASA SP-4308, 1995.

Wallace, Lane E. *Flights of Discovery: An Illustrated History of the Dryden Flight Research Center.* NASA SP-4309, 1996.

Herring, Mack R. *Way Station to Space: A History of the John C. Stennis Space Center.* NASA SP-4310, 1997.

Wallace, Harold D., Jr. *Wallops Station and the Creation of an American Space Program.* NASA SP-4311, 1997.

Wallace, Lane E. *Dreams, Hopes, Realities. NASA's Goddard Space Flight Center: The First Forty Years.* NASA SP-4312, 1999.

Dunar, Andrew J., and Stephen P. Waring. *Power to Explore: A History of Marshall Space Flight Center, 1960–1990.* NASA SP-4313, 1999.

Bugos, Glenn E. *Atmosphere of Freedom: Sixty Years at the NASA Ames Research Center.* NASA SP-2000-4314, 2000.

Bugos, Glenn E. *Atmosphere of Freedom: Seventy Years at the NASA Ames Research Center.* NASA SP-2010-4314, 2010. Revised version of NASA SP-2000-4314.

Bugos, Glenn E. *Atmosphere of Freedom: Seventy Five Years at the NASA Ames Research Center.* NASA SP-2014-4314, 2014. Revised version of NASA SP-2000-4314.

No SP-4315.

Schultz, James. *Crafting Flight: Aircraft Pioneers and the Contributions of the Men and Women of NASA Langley Research Center.* NASA SP-2003-4316, 2003.

Bowles, Mark D. *Science in Flux: NASA's Nuclear Program at Plum Brook Station, 1955–2005.* NASA SP-2006-4317, 2006.

Wallace, Lane E. *Flights of Discovery: An Illustrated History of the Dryden Flight Research Center.* NASA SP-2007-4318, 2007. Revised version of NASA SP-4309.

Arrighi, Robert S. *Revolutionary Atmosphere: The Story of the Altitude Wind Tunnel and the Space Power Chambers.* NASA SP-2010-4319, 2010.

General Histories, NASA SP-4400

Corliss, William R. *NASA Sounding Rockets, 1958–1968: A Historical Summary.* NASA SP-4401, 1971.

Wells, Helen T., Susan H. Whiteley, and Carrie Karegeannes. *Origins of NASA Names.* NASA SP-4402, 1976.

Anderson, Frank W., Jr. *Orders of Magnitude: A History of NACA and NASA, 1915–1980.* NASA SP-4403, 1981.

Sloop, John L. *Liquid Hydrogen as a Propulsion Fuel, 1945–1959.* NASA SP-4404, 1978.

Roland, Alex. *A Spacefaring People: Perspectives on Early Spaceflight.* NASA SP-4405, 1985.

Bilstein, Roger E. *Orders of Magnitude: A History of the NACA and NASA, 1915–1990.* NASA SP-4406, 1989.

Logsdon, John M., ed., with Linda J. Lear, Jannelle Warren Findley, Ray A. Williamson, and Dwayne A. Day. *Exploring the Unknown: Selected Documents in the History of the U.S. Civil Space Program, Volume I: Organizing for Exploration.* NASA SP-4407, 1995.

Logsdon, John M., ed., with Dwayne A. Day and Roger D. Launius. *Exploring the Unknown: Selected Documents in the History of the U.S. Civil Space Program, Volume II: External Relationships.* NASA SP-4407, 1996.

Logsdon, John M., ed., with Roger D. Launius, David H. Onkst, and Stephen J. Garber. *Exploring the Unknown: Selected Documents in the History of the U.S. Civil Space Program, Volume III: Using Space.* NASA SP-4407, 1998.

Logsdon, John M., ed., with Ray A. Williamson, Roger D. Launius, Russell J. Acker, Stephen J. Garber, and Jonathan L. Friedman. *Exploring the Unknown: Selected Documents in the History of the U.S. Civil Space Program, Volume IV: Accessing Space.* NASA SP-4407, 1999.

Logsdon, John M., ed., with Amy Paige Snyder, Roger D. Launius, Stephen J. Garber, and Regan Anne Newport. *Exploring the Unknown: Selected Documents in the History of the U.S. Civil Space Program, Volume V: Exploring the Cosmos.* NASA SP-2001-4407, 2001.

Logsdon, John M., ed., with Stephen J. Garber, Roger D. Launius, and Ray A. Williamson. *Exploring the Unknown: Selected Documents in the History of the U.S. Civil Space Program, Volume VI: Space and Earth Science.* NASA SP-2004-4407, 2004.

Logsdon, John M., ed., with Roger D. Launius. *Exploring the Unknown: Selected Documents in the History of the U.S. Civil Space Program, Volume VII: Human Spaceflight: Projects Mercury, Gemini, and Apollo.* NASA SP-2008-4407, 2008.

Siddiqi, Asif A., *Challenge to Apollo: The Soviet Union and the Space Race, 1945–1974.* NASA SP-2000-4408, 2000.

Hansen, James R., ed. *The Wind and Beyond: Journey into the History of Aerodynamics in America, Volume 1: The Ascent of the Airplane.* NASA SP-2003-4409, 2003.

Hansen, James R., ed. *The Wind and Beyond: Journey into the History of Aerodynamics in America, Volume 2: Reinventing the Airplane.* NASA SP-2007-4409, 2007.

Hogan, Thor. *Mars Wars: The Rise and Fall of the Space Exploration Initiative.* NASA SP-2007-4410, 2007.

Vakoch, Douglas A., ed. *Psychology of Space Exploration: Contemporary Research in Historical Perspective.* NASA SP-2011-4411, 2011.

Ferguson, Robert G., *NASA's First A: Aeronautics from 1958 to 2008.* NASA SP-2012-4412, 2013.

Vakoch, Douglas A., ed. *Archaeology, Anthropology, and Interstellar Communication.* NASA SP-2013-4413, 2014.

Monographs in Aerospace History, NASA SP-4500

Launius, Roger D., and Aaron K. Gillette, comps. *Toward a History of the Space Shuttle: An Annotated Bibliography.* Monographs in Aerospace History, No. 1, 1992.

Launius, Roger D., and J. D. Hunley, comps. *An Annotated Bibliography of the Apollo Program.* Monographs in Aerospace History, No. 2, 1994.

Launius, Roger D. *Apollo: A Retrospective Analysis.* Monographs in Aerospace History, No. 3, 1994.

Hansen, James R. *Enchanted Rendezvous: John C. Houbolt and the Genesis of the Lunar-Orbit Rendezvous Concept.* Monographs in Aerospace History, No. 4, 1995.

Gorn, Michael H. *Hugh L. Dryden's Career in Aviation and Space.* Monographs in Aerospace History, No. 5, 1996.

Powers, Sheryll Goecke. *Women in Flight Research at NASA Dryden Flight Research Center from 1946 to 1995.* Monographs in Aerospace History, No. 6, 1997.

Portree, David S. F., and Robert C. Trevino. *Walking to Olympus: An EVA Chronology.* Monographs in Aerospace History, No. 7, 1997.

Logsdon, John M., moderator. *Legislative Origins of the National Aeronautics and Space Act of 1958: Proceedings of an Oral History Workshop.* Monographs in Aerospace History, No. 8, 1998.

Rumerman, Judy A., comp. *U.S. Human Spaceflight: A Record of Achievement, 1961–1998.* Monographs in Aerospace History, No. 9, 1998.

Portree, David S. F. *NASA's Origins and the Dawn of the Space Age.* Monographs in Aerospace History, No. 10, 1998.

Logsdon, John M. *Together in Orbit: The Origins of International Cooperation in the Space Station.* Monographs in Aerospace History, No. 11, 1998.

Phillips, W. Hewitt. *Journey in Aeronautical Research: A Career at NASA Langley Research Center.* Monographs in Aerospace History, No. 12, 1998.

Braslow, Albert L. *A History of Suction-Type Laminar-Flow Control with Emphasis on Flight Research.* Monographs in Aerospace History, No. 13, 1999.

Logsdon, John M., moderator. *Managing the Moon Program: Lessons Learned from Apollo.* Monographs in Aerospace History, No. 14, 1999.

Perminov, V. G. *The Difficult Road to Mars: A Brief History of Mars Exploration in the Soviet Union.* Monographs in Aerospace History, No. 15, 1999.

Tucker, Tom. *Touchdown: The Development of Propulsion Controlled Aircraft at NASA Dryden.* Monographs in Aerospace History, No. 16, 1999.

Maisel, Martin, Demo J. Giulanetti, and Daniel C. Dugan. *The History of the XV-15 Tilt Rotor Research Aircraft: From Concept to Flight.* Monographs in Aerospace History, No. 17, 2000. NASA SP-2000-4517.

Jenkins, Dennis R. *Hypersonics Before the Shuttle: A Concise History of the X-15 Research Airplane.* Monographs in Aerospace History, No. 18, 2000. NASA SP-2000-4518.

Chambers, Joseph R. *Partners in Freedom: Contributions of the Langley Research Center to U.S. Military Aircraft of the 1990s.* Monographs in Aerospace History, No. 19, 2000. NASA SP-2000-4519.

Waltman, Gene L. *Black Magic and Gremlins: Analog Flight Simulations at NASA's Flight Research Center.* Monographs in Aerospace History, No. 20, 2000. NASA SP-2000-4520.

Portree, David S. F. *Humans to Mars: Fifty Years of Mission Planning, 1950–2000.* Monographs in Aerospace History, No. 21, 2001. NASA SP-2001-4521.

Thompson, Milton O., with J. D. Hunley. *Flight Research: Problems Encountered and What They Should Teach Us.* Monographs in Aerospace History, No. 22, 2001. NASA SP-2001-4522.

Tucker, Tom. *The Eclipse Project.* Monographs in Aerospace History, No. 23, 2001. NASA SP-2001-4523.

Siddiqi, Asif A. *Deep Space Chronicle: A Chronology of Deep Space and Planetary Probes, 1958–2000.* Monographs in Aerospace History, No. 24, 2002. NASA SP-2002-4524.

Merlin, Peter W. *Mach 3+: NASA/USAF YF-12 Flight Research, 1969–1979.* Monographs in Aerospace History, No. 25, 2001. NASA SP-2001-4525.

Anderson, Seth B. *Memoirs of an Aeronautical Engineer: Flight Tests at Ames Research Center: 1940–1970.* Monographs in Aerospace History, No. 26, 2002. NASA SP-2002-4526.

Renstrom, Arthur G. *Wilbur and Orville Wright: A Bibliography Commemorating the One-Hundredth Anniversary of the First Powered Flight on December 17, 1903.* Monographs in Aerospace History, No. 27, 2002. NASA SP-2002-4527.

No monograph 28.

Chambers, Joseph R. *Concept to Reality: Contributions of the NASA Langley Research Center to U.S. Civil Aircraft of the 1990s.* Monographs in Aerospace History, No. 29, 2003. NASA SP-2003-4529.

Peebles, Curtis, ed. *The Spoken Word: Recollections of Dryden History, The Early Years.* Monographs in Aerospace History, No. 30, 2003. NASA SP-2003-4530.

Jenkins, Dennis R., Tony Landis, and Jay Miller. *American X-Vehicles: An Inventory—X-1 to X-50.* Monographs in Aerospace History, No. 31, 2003. NASA SP-2003-4531.

Renstrom, Arthur G. *Wilbur and Orville Wright: A Chronology Commemorating the One-Hundredth Anniversary of the First Powered Flight on December 17, 1903.* Monographs in Aerospace History, No. 32, 2003. NASA SP-2003-4532.

Bowles, Mark D., and Robert S. Arrighi. *NASA's Nuclear Frontier: The Plum Brook Research Reactor.* Monographs in Aerospace History, No. 33, 2004. NASA SP-2004-4533.

Wallace, Lane, and Christian Gelzer. *Nose Up: High Angle-of-Attack and Thrust Vectoring Research at NASA Dryden, 1979–2001.* Monographs in Aerospace History, No. 34, 2009. NASA SP-2009-4534.

Matranga, Gene J., C. Wayne Ottinger, Calvin R. Jarvis, and D. Christian Gelzer. *Unconventional, Contrary, and Ugly: The Lunar Landing Research Vehicle.* Monographs in Aerospace History, No. 35, 2006. NASA SP-2004-4535.

McCurdy, Howard E. *Low-Cost Innovation in Spaceflight: The History of the Near Earth*

Asteroid Rendezvous (NEAR) Mission. Monographs in Aerospace History, No. 36, 2005. NASA SP-2005-4536.

Seamans, Robert C., Jr. *Project Apollo: The Tough Decisions.* Monographs in Aerospace History, No. 37, 2005. NASA SP-2005-4537.

Lambright, W. Henry. *NASA and the Environment: The Case of Ozone Depletion.* Monographs in Aerospace History, No. 38, 2005. NASA SP-2005-4538.

Chambers, Joseph R. *Innovation in Flight: Research of the NASA Langley Research Center on Revolutionary Advanced Concepts for Aeronautics.* Monographs in Aerospace History, No. 39, 2005. NASA SP-2005-4539.

Phillips, W. Hewitt. *Journey into Space Research: Continuation of a Career at NASA Langley Research Center.* Monographs in Aerospace History, No. 40, 2005. NASA SP-2005-4540.

Rumerman, Judy A., Chris Gamble, and Gabriel Okolski, comps. *U.S. Human Spaceflight: A Record of Achievement, 1961–2006.* Monographs in Aerospace History, No. 41, 2007. NASA SP-2007-4541.

Peebles, Curtis. *The Spoken Word: Recollections of Dryden History Beyond the Sky.* Monographs in Aerospace History, No. 42, 2011. NASA SP-2011-4542.

Dick, Steven J., Stephen J. Garber, and Jane H. Odom. *Research in NASA History.* Monographs in Aerospace History, No. 43, 2009. NASA SP-2009-4543.

Merlin, Peter W. *Ikhana: Unmanned Aircraft System Western States Fire Missions.* Monographs in Aerospace History, No. 44, 2009. NASA SP-2009-4544.

Fisher, Steven C., and Shamim A. Rahman. *Remembering the Giants: Apollo Rocket Propulsion Development.* Monographs in Aerospace History, No. 45, 2009. NASA SP-2009-4545.

Gelzer, Christian. *Fairing Well: From Shoebox to Bat Truck and Beyond, Aerodynamic Truck Research at NASA's Dryden Flight Research Center.* Monographs in Aerospace History, No. 46, 2011. NASA SP-2011-4546.

Arrighi, Robert. *Pursuit of Power: NASA's Propulsion Systems Laboratory No. 1 and 2.* Monographs in Aerospace History, No. 48, 2012. NASA SP-2012-4548.

Renee M. Rottner. *Making the Invisible Visible: A History of the Spitzer Infrared Telescope Facility (1971–2003).* Monographs in Aerospace History, No. 47, 2017. NASA SP-2017-4547.

Goodrich, Malinda K., Alice R. Buchalter, and Patrick M. Miller, comps. *Toward a History of the Space Shuttle: An Annotated Bibliography, Part 2 (1992–2011).* Monographs in Aerospace History, No. 49, 2012. NASA SP-2012-4549.

Ta, Julie B., and Robert C. Treviño. *Walking to Olympus: An EVA Chronology, 1997–2011,* Vol. 2. Monographs in Aerospace History, No. 50, 2016. NASA SP-2016-4550.

Gelzer, Christian. *The Spoken Word III: Recollections of Dryden History; The Shuttle Years.* Monographs in Aerospace History, No. 52, 2013. NASA SP-2013-4552.

Ross, James C. *NASA Photo One.* Monographs in Aerospace History, No. 53, 2013. NASA SP-2013-4553.

Launius, Roger D. *Historical Analogs for the Stimulation of Space Commerce.* Monographs in Aerospace History, No 54, 2014. NASA SP-2014-4554.

Buchalter, Alice R., and Patrick M. Miller, comps. *The National Advisory Committee for Aeronautics: An Annotated Bibliography.* Monographs in Aerospace History, No. 55, 2014. NASA SP-2014-4555.

Chambers, Joseph R., and Mark A. Chambers. *Emblems of Exploration: Logos of the NACA and NASA.* Monographs in Aerospace History, No. 56, 2015. NASA SP-2015-4556.

Alexander, Joseph K. *Science Advice to NASA: Conflict, Consensus, Partnership, Leadership.* Monographs in Aerospace History, No. 57, 2017. NASA SP-2017-4557.

Electronic Media, NASA SP-4600

Remembering Apollo 11: The 30th Anniversary Data Archive CD-ROM. NASA SP-4601, 1999.

Remembering Apollo 11: The 35th Anniversary Data Archive CD-ROM. NASA SP-2004-4601, 2004. This is an update of the 1999 edition.

The Mission Transcript Collection: U.S. Human Spaceflight Missions from Mercury Redstone 3 to Apollo 17. NASA SP-2000-4602, 2001.

Shuttle-Mir: The United States and Russia Share History's Highest Stage. NASA SP-2001-4603, 2002.

U.S. Centennial of Flight Commission Presents Born of Dreams—Inspired by Freedom. NASA SP-2004-4604, 2004.

Of Ashes and Atoms: A Documentary on the NASA Plum Brook Reactor Facility. NASA SP-2005-4605, 2005.

Taming Liquid Hydrogen: The Centaur Upper Stage Rocket Interactive CD-ROM. NASA SP-2004-4606, 2004.

Fueling Space Exploration: The History of NASA's Rocket Engine Test Facility DVD. NASA SP-2005-4607, 2005.

Altitude Wind Tunnel at NASA Glenn Research Center: An Interactive History CD-ROM. NASA SP-2008-4608, 2008.

A Tunnel Through Time: The History of NASA's Altitude Wind Tunnel. NASA SP-2010-4609, 2010.

Conference Proceedings, NASA SP-4700

Dick, Steven J., and Keith Cowing, eds. *Risk and Exploration: Earth, Sea and the Stars.* NASA SP-2005-4701, 2005.

Dick, Steven J., and Roger D. Launius. *Critical Issues in the History of Spaceflight.* NASA SP-2006-4702, 2006.

Dick, Steven J., ed. *Remembering the Space Age: Proceedings of the 50th Anniversary Conference.* NASA SP-2008-4703, 2008.

Dick, Steven J., ed. *NASA's First 50 Years: Historical Perspectives.* NASA SP-2010-4704, 2010.

Societal Impact, NASA SP-4800

Dick, Steven J., and Roger D. Launius. *Societal Impact of Spaceflight.* NASA SP-2007-4801, 2007.

Dick, Steven J., and Mark L. Lupisella. *Cosmos and Culture: Cultural Evolution in a Cosmic Context.* NASA SP-2009-4802, 2009.

Dick, Steven J. *Historical Studies in the Societal Impact of Spaceflight.* NASA SP-2015-4803, 2015.

Index